Current Directions and Prospects of Hydrogels for Biomedical Applications

Current Directions and Prospects of Hydrogels for Biomedical Applications

Guest Editors

Zhiyuan Jia
Holger Schönherr
Nowsheen Goonoo

MDPI

Basel • Beijing • Wuhan • Barcelona • Belgrade • Novi Sad • Cluj • Manchester

Guest Editors
Zhiyuan Jia
Biomed Center Innovation gGmbH
Bayreuth
Germany

Holger Schönherr
University of Siegen
Siegen
Germany

Nowsheen Goonoo
Centre for Biomedical and Biomaterials Research
University of Mauritius
Réduit
Mauritius

Editorial Office
MDPI AG
Grosspeteranlage 5
4052 Basel, Switzerland

This is a reprint of the Special Issue, published open access by the journal *Gels* (ISSN 2310-2861), freely accessible at: https://www.mdpi.com/journal/gels/special_issues/H27NA4N5M3.

For citation purposes, cite each article independently as indicated on the article page online and as indicated below:

Lastname, A.A.; Lastname, B.B. Article Title. *Journal Name* **Year**, *Volume Number*, Page Range.

ISBN 978-3-7258-6185-9 (Hbk)
ISBN 978-3-7258-6186-6 (PDF)
https://doi.org/10.3390/books978-3-7258-6186-6

Contents

About the Editors

Zhiyuan Jia

Zhiyuan Jia is a research scientist at Biomed Center Innovation gGmbH in Germany. She began her academic journey with a Bachelor's degree in Materials Science and Engineering from Tianjin Polytechnic University in China in 2012. She then moved to Germany to pursue graduate studies at the University of Siegen, where she received a Master's degree in Chemistry in 2015 and a Doctor degree in natural science in 2020 under the supervision of Prof. Holger Schönherr. During her time at the University of Siegen, her research focused on developing innovative biosensors using hydrogels, polymersomes, anodic oxide alumina, and nanofibers, aiming to improve detection sensitivity and specificity for early-stage wound infection monitoring and bacterial contamination detection in food and water packaging. She collaborated extensively with research groups in Germany, France, the Netherlands, Mauritius, the United Kingdom, and China and continues to maintain these international partnerships in her current work. The research she conducted during her doctoral studies resulted in over six publications in leading international journals, including *Bioactive Materials*, *ACS Applied Materials and Interfaces*, *ACS Applied Bio Materials*, and *Frontiers in Chemistry*. As a senior scientist, she is currently working on microemulsion project for the additive manufacturing of phosphate salt-containing structures and is also participating in an EU project on inkjet printing of regenerative implants. Her ongoing research also explores self-healing hydrogels and porous ceramics for medical applications, aiming to address key challenges in diagnostics and patient care.

Holger Schönherr

Holger Schönherr studied chemistry, polymer chemistry, and polymer physics at the Universities of Mainz and Toronto (1989–1995) and obtained his PhD in 1999 at the University of Twente, the Netherlands. Following a postdoctoral stay at Stanford University (2000–2001), he joined the MESA+ Institute for Nanotechnology in Twente as a tenured assistant professor (later promoted to associate professor) before joining the University of Siegen, Germany, in 2008 as a University Professor in Physical Chemistry. His research interests comprise the modification and characterization of polymeric surfaces and (bio)interfaces, micro- and nanostructured polymeric materials, bacteria detection, antimicrobial strategies, 3D cell culture and surface analysis with atomic force microscopy (AFM), and combined AFM–optical methods. Holger Schönherr has mentored more than 125 M.Sc. and B.Sc. students as well as 38 PhD graduates. He has published more than 310 original peer-reviewed publications, several monographies, and books and currently has an H index of 66 (Google Scholar, 2025). Among other duties, since 2018 he has served as the Dean of the School of Science and Technology at the University of Siegen and since 2024 as the President of POLY-CHAR (World Forum on Advanced Materials). Holger Schönherr has been honored with the Erudite Scholar-in-Residence/Erudite Professorship 2021 (Mahatma Gandhi University, Kottayam, Kerala, India), the 2013 Research Prize of the School of Science and Technology (University of Siegen), the International Polychar Materials Science Prize 2013, the Raphael-Eduard-Liesegang Award of the German Colloid Society in 2011, and an ERC starting (consolidator) grant in 2011. He was also appointed twice as a Guest Professor at the Shanghai Jiao Tong University (2013–2016 and 2020–2022).

Nowsheen Goonoo

Nowsheen Goonoo is a senior researcher with a strong track record in research focused on the intersection of polymers and sustainable biotechnology. She began her academic journey with a PhD degree in Polymer and Biomaterials from the University of Mauritius.

Throughout her career, which included exposure in international laboratories, her research has focused on developing innovative projects that leverage local biomaterials for advanced medical applications, such as wound healing, targeted drug delivery for cancer treatment, and the development of biocompatible implants. The research she has conducted emphasizes sustainability and environmental stewardship, focusing specifically on harnessing local Mauritian resources to create eco-friendly and effective biotechnological solutions addressing local health challenges such as diabetes and cancer. This work has resulted in over 40 co-authored peer-reviewed publications related to biotechnology and includes the successful co-supervision of three PhD projects at the University of Mauritius.

Dr. Goonoo's groundbreaking work has garnered significant international acclaim, including being the first Mauritian awarded the Georg Forster Postdoctoral Fellowship by the prestigious German Alexander Von Humboldt Foundation in 2015. She received further international recognition for her research on fucoidan-based wound dressings through the L'Oreal–UNESCO International Rising Talent Award (2020) and the Inaugural Rising Star Africa Prize (2021). She has collaborated on several joint research projects as a co-investigator, including those funded under FEDER INTERREG OI. As an Affiliate of the African Academy of Sciences (AAS), The World Association of Sciences (TWAS), and a Next Einstein Fellow (NEF) ambassador, she maintains a robust scientific network across the African region, with the goal of translating innovative marine science into tangible, effective biomedical outcomes.

gels

MDPI

Review

Recent Progress in Gels for Neuropathic Pain

Ciprian Pușcașu, Anca Zanfirescu * and Simona Negreș

Faculty of Pharmacy, "Carol Davila" University of Medicine and Pharmacy, Traian Vuia 6, 020956 Bucharest, Romania; ciprian.puscasu@umfcd.ro (C.P.); simona.negres@umfcd.ro (S.N.)
* Correspondence: anca.zanfirescu@umfcd.ro

Abstract: Neuropathic pain is a complex and debilitating condition that affects millions of people worldwide. While several treatment options are available, they often have limited efficacy and are associated with adverse effects. In recent years, gels have emerged as a promising option for the treatment of neuropathic pain. Inclusion of various nanocarriers, such as cubosomes and niosomes, into gels results in pharmaceutical forms with higher drug stability and increased drug penetration into tissues compared to products currently marketed for the treatment of neuropathic pain. Furthermore, these compounds usually provide sustained drug release and are biocompatible and biodegradable, which makes them a safe option for drug delivery. The purpose of this narrative review was to provide a comprehensive analysis of the current state of the field and identify potential directions for future research in the development of effective and safe gels for the treatment of neuropathic pain, ultimately improving the quality of life for patients suffering from neuropathic pain.

Keywords: neuropathic pain; tramadol; capsaicin; cubosomes; niosomes; pregabalin; gabapentin

1. Introduction

Neuropathic pain (NeP) is caused by damage or dysfunction of the nervous system, which can result in abnormal sensory processing and pain signals being sent to the brain [1]. Its etiology is complex, involving nerve damage or injury (owing to trauma, surgery, infection, or diseases, such as diabetes, multiple sclerosis, or HIV), compression or entrapment of nerves (e.g., in carpal tunnel syndrome or herniated discs), and diseases affecting the nervous system (as in stroke, spinal cord injury, or Parkinson's disease) [2].

It may also be the result of exposure to chemotherapy or radiation therapy or of the administration of certain medications (e.g., antivirals or anticonvulsants) [3]. Independent of the etiology, negative (deficits of different somatosensory qualities) and positive (paresthesia and dysesthesia, paroxysmal pain, and ongoing superficial pain) sensory symptoms coexist in NeP [4]. Stimulus-evoked symptoms such as hyperalgesia and allodynia may occur in addition to spontaneous pain and are rarely the only pain manifestation [4].

An increasing trend in the prevalence of NeP is reported: 10% in 2017 vs. 5% in 2008, in the US [5]. Furthermore, a recent study approximated that about 15% of nursing home residents in the US suffer from NeP [6]. A recent research showed that the aged-standardized prevalence of chronic polyneuropathy is 3.3% for the European Union, 3.0% for the United States, and 2.3% for the world population and is expected to increase by ±25% in the next 20 years based on the expected age distributions [7]. Furthermore, in the young population, the incidence of diabetic neuropathy ranges from 2.6% to 11% and cardiac autonomic neuropathy from 4% to 39% of the patients with type 2 diabetes. Thus, diabetic neuropathy and cardiac autonomic neuropathy are the most common forms of neuropathy in adolescents and young adults [8].

The symptoms of NeP can be debilitating and have a significant impact on a person's quality of life [4]. In patients with chronic neuropathic pain, physical and mental health declines more than in those with other types of chronic pain [9,10]. This highlights the impact of neuropathic pain's nature on the quality of life and the need for complex therapy [11]. The costs of NeP are substantial and include medical expenses and productivity loss, adding to the economic burden [12].

Managing Nep can be frustrating and often represents a trial-and-error-based process [12]. Unlike other types of pain, neuropathic pain often does not respond well to traditional pain medications, such as opioids or nonsteroidal anti-inflammatory drugs. Even with the most effective treatment approaches, many individuals with neuropathic pain may experience ongoing symptoms and disability [12,13]. In one investigation, patients with NeP were more willing to take opioids and other pain treatments than those with chronic nonneuropathic pain, yet experienced less pain relief from medications [14]. Besides unsatisfactory effectiveness, even therapeutic agents used as first-line therapy of NeP, such as tricyclic antidepressants, selective inhibitors of serotonin and norepinephrine reuptake, and gabapentinoids, are associated with severe adverse effects. Hence, careful dosage is crucial, particularly in the elderly, as some patients experience adverse effects even with the lowest doses [15].

In the last decade, the use of analgesics-containing gels has gained attention over their oral administration due to several factors. Pregabalin, gabapentin, amitriptyline, and tramadol are among the agents most frequently used today as oral therapy and become the focus of research on developing new gels in the management of NeP [16]. Applied both topically or injectable, gels provide targeted and localized pain relief, allowing for the delivery of medication directly to the affected area—this is particularly beneficial for individuals with neuropathic pain, which is often localized to specific areas [17]. This targeted delivery can result in higher drug concentrations at the site of action, which can lead to improved efficacy and reduced systemic side effects compared to oral administration [18]. The use of new gels can provide sustained drug release over a longer period of time, allowing for more continuous pain relief without the need for frequent dosing [18]. As a result, the use of analgesic-containing gels has emerged as a promising alternative to oral administration for individuals seeking targeted pain relief with fewer systemic side effects [16].

Despite the development of such gels with targeted and sustained drug release and potentially effective in NeP, in the last decade, no recent review summarizing their characteristics exists. Therefore, we aimed to summarize effectively the novelties and improvements made in the field of research of gels destined to treat NeP, focusing on how a certain formulation can influence drug release, bioavailability, and skin permeation, factors which can significantly impact the effectiveness and safety of the medication.

2. Pharmacological Treatment of Neuropathic Pain

Epidemiological surveys indicate that numerous patients with neuropathic pain receive improper treatment [19], owing to inaccurate diagnosis, limited understanding of pain mechanisms, and drugs with poor effectiveness.

A first-line therapy recommended by the guidelines of the European Federation of Neurological Societies (EFNS) [20], the International Association for the Study of Pain (IASP) [21], and the American Academy of Neurology [22] consists in orally administered tricyclic antidepressants, amitriptyline (25–150 mg/day), gabapentin (1200–3600 mg/day), and pregabalin (150–600 mg/day) for various types of painful neuropathy (excluding trigeminal neuralgia) [23].

The mechanism of action of amitriptyline is complex [19]. Besides blocking Na^+, K^+, and Ca^{2+} voltage-gated ion channels, amitriptyline also blocks cholinergic, muscarinic, α2-adrenergic, nicotinic, histaminergic, opioid, adenosine, and N-methyl-D-aspartate receptor channels. The main side effects include dry mouth, weight gain, drowsiness, cognitive impairment, high cardiotoxicity [24], walking disturbances, and falls [25,26].

Gabapentin and pregabalin are used mostly in the management of diabetes-induced peripheral neuropathy and postherpetic neuralgia. The analgesic effect of gabapentin arises from increased GABA synthesis, $\alpha 2\delta$ subunit binding of voltage-gated calcium channels, and non-NMDA receptor antagonism. Gabapentin's adverse reactions occur rather frequently (10–20%) and include somnolence, dizziness, ataxia, and fatigue [26]. Pregabalin reduces the release of neurotransmitters in a manner similar to gabapentin, namely by binding to the α-2-delta subunit of calcium channels [27]. When administered orally, pregabalin has side effects, such as dizziness, drowsiness, dry mouth, blurred vision, poor concentration, decreased levels of thrombocytes, and hypersensitivity [28].

Due to their established efficacy, serotonin and norepinephrine reuptake inhibitors (SNRIs), which block the reabsorption of the neurotransmitters serotonin and norepinephrine in the brain, can also be currently recommended (duloxetine 60–120 mg/day, venlafaxine 150–225 mg/day) as first-line treatment in painful polyneuropathy [22]. SNRIs are mostly recommended in the treatment of peripheral diabetic neuropathy, fibromyalgia, and back pain [23]. Possible adverse effects of SNRI treatment include sickness, dry mouth, vertigo, headache, and enhanced sweating [19]. However, their most severe adverse effects include suicidal thoughts and behavior, hypertension, bleeding, severe hyponatremia, and serotonin syndrome [29]. In addition, IASP guidelines include gabapentin extended-release and gabapentin enacarbil as first-line therapies [21].

Lidocaine is a local anesthetic medication that has been used in the treatment of neuropathic pain. When lidocaine is administered topically or injected into the affected area, it can block the transmission of pain signals from the peripheral nerves to the brain and can provide relief for several hours. It is often used to treat localized neuropathic pain, including postherpetic neuralgia, diabetic neuropathy, and complex regional pain syndrome. It is applied directly to the affected area. Topical lidocaine patches (up to 3 patches/day) are recommended in postherpetic neuralgia, especially in the elderly, as first-line therapy in EFNS guidelines [20] and second-line therapy according to IASP guidelines which consider the quality of evidence supporting the use of lidocaine is weak [21,30]. However, the therapeutic potential of lidocaine patches is modest and side effects are limited to applied areas [31]. Additionally, the American Academy of Neurology included sodium channel blockers for the treatment of neuropathic pain as they demonstrated their efficacy in reducing pain in meta-analyses [22].

Tramadol (200–400 mg/day) and capsaicin cream are used as second-line therapy in all therapy guidelines [22]. Tramadol has a mixt mechanism being an opioid weak agonist and an SNRI and is effective in neuropathic pain [31]. This drug can induce somnolence and confusion in adults, but its advantage compared to potent opioids is the lower risk of abuse [32].

Capsaicin patches (8%) have high-quality evidence and showed prolonged efficacy for approximately 3 months in diabetic or nondiabetic neuropathies [30]. As a natural alkaloid, capsaicin is the active ingredient in the dried fruit or ground powder (paprika) of hot peppers and acts by selectively binding to TRPV1 expressed on Aδ and C fibers [33]. It is systemically absorbed in a low percentage, and the side effects are manifested by transient skin reactions, such as pain, itching, and redness [30]. These side effects together with the potential safety concerns on sensation with long-term use exist, which led to their use as second-line therapy [31].

IASP and EFNS guidelines recommend the use of strong opioids only as a third-line therapy [20]. Despite being extremely effective, the use of these therapeutic agents is limited owing to their high risk of abuse [26] and a recent upward trend in the prescription of opioids causing overdose death and other opioid-related morbidities, particularly in the US, Canada, and the UK [28–30]. Furthermore, the American guide discourages the use of opioids considering them to be nonsuperior to nonopioid medications [31].

Subcutaneous botulinum toxin type A is used as the final therapeutic alternative in refractory cases, because its administration is painful, and long-term effects are limited [28]. Botulinum neurotoxins bind to neuronal cholinergic receptors [32], acting at alpha as well as gamma motor endings and altering spinal as well as cortical motor mechanisms [34]. Botulinum toxin type A, given subcutaneously in doses of 100–300 IU, requires administration in specialized units. Due to painful administration, local anesthetics and 50% nitrous oxide inhalation are recommended before and during treatment to decrease the pain [30].

Drug combinations (gabapentin/opioids or gabapentin/tricyclic antidepressants) are recommended to those patients who are partial responders to analgesic monotherapy [24]. Furthermore, for the first time, EFNS recommends cannabinoids in neuropathic pain in refractory cases [20].

Even though individual agents within the TCA (tricyclic antidepressants), SNRI, and gabapentinoid groups have similar efficacy in neuropathic pain, their adverse effect profile is different [32]. Patients with preexisting constipation, urinary retention, or orthostatic hypotension may show a lower tolerability for TCAs. In addition, cognitive impairment, gait disturbances, and falls may occur, especially in the elderly population [27,34]. On the other hand, the side effects of SNRIs and sodium channel blockers, such as nausea, tiredness, and vertigo, may be intolerable for patients with similar preexisting symptoms [27]. Patients with comorbidities causing peripheral edema should cautiously use gabapentinoids as they can also induce peripheral edema [27].

The average duration for a treatment to show efficacy is approximately 12 weeks. Neuropathic pain treatment should be administered for at least 12 weeks before being evaluated and considered ineffective, appropriate, or intolerable [27]. Regardless of each medication's unique side effect profile, vertigo, somnolence, and tiredness are common in oral therapeutic regimens, while application site reactions occur in topical medication. A summary of the pharmacological therapeutic options is given in Table 1.

Table 1. Pharmacological treatment of NeP.

Pharmacological Treatment of NeP						
First-Line Therapy			Second-Line Therapy		Third-Line Therapy	
TCA Amitriptyline	SNRI Duloxetine Venlafaxine	Gabapentinoids Pregabalin Gabapentin	Topical treatment Capsaicine Lidocaine	Tramadol	Strong opioids Morphine Oxycodone	Botulinum toxin type A
Painful neuropathy Nerve injury pain Postherpetic neuralgia Central postpartum pain Spinal cord injury	Diabetic neuropathy Fibromyalgia Back pain Diabetic pain	Spinal cord injury Herpetic neuralgia Phantom limb syndrome	Localized NeP: Postherpetic neuralgia Diabetic pain Regional pain syndrome Diabetic pain Central Pain	Chemotherapy-induced peripheral neuropathy Phantom limb syndrome	Diabetic pain Postherpetic neuralgia Central pain	Focal peripheral neuropathy and allodynia

Considering that the therapeutic options for treating neuropathic pain have limited effectiveness in relieving pain and numerous side effects, few drugs with novel mechanisms of action are under clinical development for the treatment of this pathology [35], including selective sodium blocking agents such as Nav1.7 antagonists [36], angiotensin type II (EMA) [37], TPRV1, and nerve growth factor antagonists [35]. However, it will take some time before new therapies will be available.

Therefore, developing topical preparations, such as gels, containing substances known to be effective against NeP represents an attractive option as they would provide a faster onset of action, targeted delivery, improved adherence, a lower risk of gastrointestinal side effects, flexibility in dosing, and reduced drug interactions.

3. Development of Analgesic-Containing Gels Useful for NeP

Over the past decade, several types of gels containing analgesics for NeP therapy have been developed, based on the properties of the active substances for drug delivery applications (Table 2). Such gels are based on natural polymers such as chitosan or gelatin, which provide excellent biocompatibility and biodegradability [38]. These gels can be cross-linked with a variety of agents, such as glutaraldehyde, genipin, or sodium tripolyphosphate, to improve their mechanical and swelling properties [39]. Another type of hydrogel is based on synthetic polymers, such as polyethylene glycol or polyvinyl alcohol, which provide tunable properties and controlled drug release capabilities [40]. Furthermore, they can be modified with various functional groups to enhance their stability and biocompatibility. The incorporation of the active substances into these hydrogels can be achieved by different methods, such as physical entrapment, covalent bonding, or electrostatic interactions.

Understanding and optimizing these factors can improve the overall effectiveness of such treatments. Some of these preparations have shown promising analgesic effects after transdermal or subcutaneous delivery in animals, and thus, they might offer an improved therapeutic outcome in NeP.

Table 2. Pharmacokinetic and physicochemical characteristics of analgesic drugs used for neuropathic pain.

Substance	Absorbtion	Metabolism	Elimination	Half-Life	Physicochemical Characteristics	References
Capsaicin	Oral administration 50 to 90% Intravenous/subcutaneous administration in the brain and spinal cord about 5 times higher than in blood; in the liver about 3 times higher than in the blood. Topical administration: rapidly and well absorbed through the skin.	Significant first-pass effect of liver after oral administration	Renal excretion	Following oral ingestion approximately 24.9 ± 5.0 min. Following topical application approximately 24 h	Low water solubility. pKa (Strongest Acidic): 9.93 pKa (Strongest Basic): −1.4 Melting point: 65 °C	[35–37]
Tramadol	Oral administration: rapidly and almost completely absorbed, with a bioavailability of 75%	Extensive first-pass metabolism in the liver	Primarily through metabolism by the liver; metabolites are excreted primarily by the kidneys.	5–6 h	Tramadol hydrochloride salt is highly water-soluble. The pH range of tramadol hydrochloride salt: 4.0–5.5. The melting point of tramadol: 180–184 °C. The solubility of tramadol in ethanol and chloroform makes it possible to incorporate it into hydro-alcoholic or lipophilic gel formulations.	[41,42]
Gabapentin	The oral bioavailability is inversely proportional to the administered dose.	Not appreciably metabolized	In the urine as unchanged drug.	5–7 h	pKa: 3.7 Octanol/water partition coefficient: 1.24, indicating that it is highly hydrophilic.	[43,44]
Pregabalin	Oral administration bioavailability ≥90% regardless of the dose	Less than 2% is metabolized	Almost exclusively eliminated in the urine.	6.3 h	Good water solubility. pKa: 4.2. Octanol/water partition coefficient: −1.35 Low molecular weight −159.23 g/mol	[45–48]

Table 2. *Cont.*

Substance	Absorbtion	Metabolism	Elimination	Half-Life	Physicochemical Characteristics	References
Amitriptyline	Rapidly absorbed following oral administration	Suffers from first-pass metabolism. The main active metabolite is nortriptyline.	Amitriptyline and its metabolites are mainly excreted in the urine.	About 25 h	Highly lipophilic, small-molecular-weight compound. Low water solubility: pKa: 9.4 Octanol/water partition coefficient: 3.3. High molecular weight: 313.87 g/mol.	[49,50]
Ketamine	Absorption is very rapid and the bioavailability is around 93%.	Mainly hepatic metabolism.	Manly urine excretion mainly in the form of metabolites	186 min	pKa: 7.5 Molecular mass: 238 g/mol Ketamine is typically administered intravenously, and its use is associated with several side effects.	[51–53]
Baclofen	Oral bioavailability: 70% to 85%	About 15% of the oral dose is metabolized in the liver.	70–80% is eliminated in an unchanged form by renal excretion.	2–6 h after oral administration.	Water solubility: 4 mg/mL pKa: 9.62 + 0.1 (amino group) 3.87 + 0.1 (carboxyl group) Adverse effects following oral administration affect between 25% and 75% of patients.	[54–56]
Naproxen	Peak plasma concentration after 1 h, peak plasma concentration for naproxen (free acid) is observed after 2 h	Heavily metabolized in the liver.	Approximately 95% of naproxen and its metabolites can be identified in the urine	12–17 h	Water solubility: 15.9 mg/L pKa: 4.15 Melting point: 152 °C	[57–60]
Ibuprofen	Very well absorbed orally.	Rapidly metabolized and biotransformed in the liver to the formation of major metabolites.	Eliminated in the urine.	1.2–2 h	Water solubility: 21 mg/L pKa: 5.3 Melting point: 75–77 °C	[61]
Diclofenac	Totally absorbed from the gastrointestinal tract.	Substantial first-pass metabolism	60–70% urinary elimination and 30% elimination through feces	2 h	Water solubility: 2.37 mg/mL pKa: 4.15 Melting point: 283–285 °C	[62,63]

3.1. Capsaicin

In clinical practice, capsaicin is primarily used as a topical analgesic to treat neuropathic pain. Capsaicin has also been studied for its potential therapeutic effects in other conditions, including osteoarthritis, psoriasis, gastritis, and migraines, although more research is needed to determine its efficacy and safety in these areas [64].

Regarding the currently authorized products containing capsaicin, they include prescription patches of 8%. Usually, transdermal capsaicin implies the use of up to 4 patches for 30–60 min once every 3 months. Overuse of capsaicin patches can lead to severe skin irritation or other side effects. In vitro data demonstrate that the release rate of capsaicin from patches is linear over the duration of application, and approximately 1% of the amount of capsaicin is absorbed into the layers of the epidermis and dermis during an 1-hour application [65]. Several creams containing capsaicin exist worldwide as supplements [66]. Capsaicin cream, available as over-the-counter products with 0.025–0.1% concentrations, with 0.1% considered high potency, can be applied up to four times a day. Although burning, stinging, or itching may occur, these sensations diminish over time [67].

Capsaicin-containing gels have been extensively studied in the last decade. Most of the studies used natural polymers such as chitosan [68], alginate [69], or gelatin [70] to develop capsaicin-containing hydrogels. Several synthetic polymers, including polyethylene glycol (PEG) and polyvinyl alcohol (PVA), were also used. The incorporation of capsaicin into the hydrogels was achieved by different methods, such as physical mixing and more recently, by encapsulation. The release profile of capsaicin from the gels varied depending on the type of polymer and the method of incorporation.

Older studies investigated the formulation of hydrogels containing capsaicin through physical mixing. Wang et al. [71] investigated the in vitro and in vivo skin absorption of capsaicin from chitosan/carboxymethylcellulose-based hydrogels and compared it with that of commercialized creams containing capsaicin. Drug partition between the skin and the gel matrix was critical in the permeation process. As expected, the in vitro permeation of capsaicin from the tested gels is depended on the physicochemical nature and the concentration of the polymer used. Thus, adding the nonionic polymer Pluronic (Plo) F-127 to the gels produced a delayed release of capsaicin. On the other hand, a higher capsaicin permeation rate was obtained in in vitro studies with cationic chitosan and anionic carboxymethyl cellulose hydrogels than with cream bases. In a dose-dependent manner, the cream produced skin erythema in vivo. However, the dose-dependence was not observed in gels which had a lower irritative effect than commercially available cream [71].

Peng et al. [72] designed and investigated the transdermal controlled release of cubic phase gels containing capsaicin. Release studies demonstrated that cubic phase gels imprint a sustained system for capsaicin, the release rate being affected by the initial water content, distribution of capsaicin in the lipid bilayers, and cubic phase gel swelling.

Capsaicin-loaded 1% nanolipoidal carriers (NLCs) were designed to increase permeation and provide analgesic and anti-inflammatory effects with reduced skin irritation [73]. These NLCs and gels with capsaicin-loaded NLCs demonstrated sustained release, noncytotoxic properties, enhanced penetration, and improved pain threshold in a dose-dependent manner, while inhibiting inflammation more effectively than conventional preparations. Reduced skin irritation suggests NLCs as a potential carrier for topical delivery of capsaicin in pain and inflammation therapy [73].

Aylang et al. [74] extracted a natural β-chitin–protein complex film from waste shells of *Ensis* spp. After production and physicochemical characterization of the film, capsaicin was loaded. The loading capacity was 5.79%, and over a period of 120 h capsaicin remained stable with a sustained release rate. Maximum release of capsaicin was recorded as 50.49% (48 h) for pH 4.0, 59.81% (72 h) for pH 5.5, and 59.02% (96 h) for pH 7.4.

Another study developed and characterized a chitosan-based hydrogel containing capsaicinoids-loaded nanocapsules for topical delivery. Several chitosan hydrogels were prepared to determine the optimal composition. The most suitable gel contained the lowest amount of lactic acid (1.5%) and an intermediate amount of chitosan (3.5%), ensuring a pH of 4.34 ± 0.11. After 30 days of storage, the gel exhibited a slight increase in consistency and a decrease in the flow index and pH [68].

Peng et al. used phytantriol- and GMO-based cubosomes as a targeted, sustained transdermal delivery system for capsaicin [75]. Their skin retention of capsaicin (4.32 ± 0.13 μg) was higher than that of capsaicin cream (0.72 ± 0.13 μg). The cubosomes were stable under strong light and high temperatures for up to 10 days and caused minimal irritation. Phytantriol-containing cubosomes exhibited higher permeation and more sustained skin retention of the drug than GMO-based cubosomes and the cream.

To conclude, the newly synthesized gels showed a release rate ranging from 12.96% to 81% of capsaicin and improved skin permeation of the drug. Furthermore, the encapsulation of capsaicin in nanocarriers reduces the irritative effect.

3.2. *Tramadol*

The physicochemical characteristics of tramadol presented in Table 1 make it suitable for incorporation in gels.

3.2.1. Gels for Transdermal Use

Shah et al. [18] investigated the possibility of transdermal delivery of tramadol using a proniosomes-based gel formulation and evaluated its therapeutic potential in vivo. Surfactant-based colloidal drug carriers such as niosomes and their hybrids (proniosomes) effectively transport larger drug quantities to the skin, enabling controlled release for systemic absorption [76]. Dry proniosomes formulations are converted to niosomes after the hydration [76]. For example, the proniosomes embedded in a suitable gel form niosomes by absorbing water [18]. Niosomes increase drug permeation as they overcome the barrier properties of skin and form a drug depot [77]. The formula with the best drug release, stability, and transdermal efficacy contained 100 mg tramadol hydrochloride, 1800 mg Span 80, 1800 mg lecithin, and 200 mg cholesterol [18]

Natori et al. [78] developed a tramadol-containing hydrogel film composed of 20% (w/w) hydroxypropyl methylcellulose (HPMC), which was obtained by irradiation with electron beams. This formula presented similar transparency and elasticity as commercially available dressings. Various electron beam doses imprinted differences in release and permeation rate from hydrogel films containing tramadol; thus, hydrogel films irradiated at 50 kGy showed enhanced release than those irradiated at 30 kGy.

For effective pain management, tramadol-hydrochloride-encapsulated transethosomes were formulated by cold method with different lipoidal carrier systems. A total of 12 formulations were prepared, formulations being designed by using ethanol, edge activator (Span 20 and Cremophor EL-35), and phospholipids (soya lecithin, l-α phosphatidylcholine from egg yolk). The highest encapsulation efficiency was observed for the transethosomes obtained using Cremophor EL-35 0.5% as an edge activator. Their average size ranged between 149.34 and 198.10 nm. As the concentration of edge activator in the formulation increased, the vesicular size decreased. Lecithin-based formulations exhibited a higher particle size and higher viscosity [79].

3.2.2. Gels for Parenteral Use

Hydrogels have emerged as a promising material for parenteral drug delivery due to their ability to absorb large quantities of water while maintaining a three-dimensional network structure. The highly hydrated nature of hydrogels allows them to mimic the extracellular matrix of tissues, promoting biocompatibility and reducing immune reactions [80]. Additionally, hydrogels can be created to respond to various stimuli, such as pH, temperature, and enzymatic activity, enabling controlled drug release [81]. Moreover, the unique physicochemical properties of hydrogels, such as their high water content and tunable mechanical properties, make them ideal candidates for encapsulating and delivering a wide range of therapeutics, including tramadol [82].

Barati et al. [83] developed a chitosan-based thermoresponsive in situ gel-forming formulation with tramadol purposive for subcutaneous injection. Their clear advantage, compared to oral formulations, consists in providing a depot for slow release of the drug over an extended period of time at the site of the injection over 8 h. Furthermore, adding pentasodium triphosphate (TPP) to the formula resulted in the formation of spherical nanocavities in the homogenous containing gel structure, resulting in a higher percentage of the cumulative release than the formula without TPP. The nanostructures were not present in the sol state. They appeared when the sol–gel transition occurred, leading to the emergence of a new concept: pro-nanogels.

Dos Santos et al. studied PL-based binary hydrogels composed of PL 407 and PL 188 [82]. A minimal PL concentration of 35% formed thermo-reversible gels. Drug–micelle interaction studies showed PL 407–PL 188 binary systems with drug partitioning into micelles. The presence of tramadol hydrochloride increased enthalpy variation values during the sol–gel transition phase. Rapid hydrogel dissolution reached 80–90% in 24 h. Tramadol incorporation into the binary system prolonged analgesic effects, extending release for 48–72 h after subcutaneous injection.

3.3. *Gabapentin*

Topical administration of gabapentin is associated with minor adverse effects [84]. Therefore, topical formulations are being developed. However, gabapentin has some physicochemical characteristics that may make it difficult to incorporate into gels, as shown in Table 1 [44].

Despite challenges, successful incorporation of gabapentin into gels has been reported. One clinical trial reported that a 6% gabapentin cream effectively ameliorates vulvodynia [85]. A 10% *w*/*w* topical gabapentin gel applied thrice daily significantly reduced allodynia and hyperalgesia in a rat sciatic nerve constriction model without motor impairment [86]. The same pharmaceutical form also attenuated cisplatin-induced neuropathic allodynia and heat-hypoalgesia [87].

Martin et al. [88] prepared and evaluated several formulations of gabapentin, including preformulated oil-in-water bases and Carbopol-based hydrogels with permeation enhancers. To prevent crystallization, a maximum of 6% (*w*/*w*) gabapentin was incorporated within all Carbopol® hydrogels. The 5% (*w*/*w*) DMSO Carbopol® gels were stable for at least 3 months under ambient conditions.

A new gabapentin formulation was recently developed—chitosan-g-poly(acrylic acid-co-acrylamide) hydrogel composite—containing gabapentin and evaluated at different pH, temperature, and time intervals for drug delivery and controlled release of gabapentin. The research showed a maximum encapsulation of 64% and a drug loading efficiency of the hydrogel of 71% and that the formula imprinted a sustained-release manner of the drug from the hydrogel. In the first 2 h, 90% of the total gabapentin was released according to the dual temperature and pH-responsive hydrogel composite release study [89].

Shakshuki et al. [90] compared three different forms containing gabapentin: Lipoderm cream, Versabase gel, and Emollient cream. All these three forms contained 10% gabapentin.

At 25 °C and 40 °C, the potency of gabapentin in Lipoderm cream highly increased after 28 and 90 days, respectively. In contrast, gabapentin has deteriorated in Emollient cream. At 25 °C, the drug combined with Lipoderm cream did not show changes in organoleptic properties for up to 28 days, but physical changes were observed in other bases. Gabapentin was recrystallized from Versabase gel and Emollient cream within 14 days.

Another study formulated and characterized gabapentin-encapsulated elastic liposomes and compared their efficiency in transdermal delivery of gabapentin with that of the compounded gabapentin-based Plo lecithin organogel. Gabapentin liposomes-containing gel had a significantly slower release rate compared to the Plo lecithin organogel (12 h vs. 4 h). Moreover, after 24 h, liposomes highly increased the percutaneous penetration of gabapentin through the porcine skin leading to greater concomitant drug concentrations compared to the Plo lecithin organogel [91].

3.4. Pregabalin

Currently, there are no authorized topical formulas containing pregabalin. Oral formula has side effects which include dizziness, sleepiness, dry mouth, and blurred vision, among others [92]. Topical formulas could reduce systemic drug exposure and penetration into the brain, minimizing the occurrence of CNS-mediated side effects [93].

Four pregabalin preparations for transdermal application were developed in a recent study [94]: 0.4% aqueous solution, Plo lecithin organogel, hydrophilic cream, and lipophilic cream. The organogel had the highest permeability, followed by the aqueous solution, while creams showed no permeation. Pregabalin was distributed into the dermis 1 h after the application of the organogel. Furthermore, in vivo testing using a mouse model of diabetic neuropathy demonstrated that only the organogel had a significant analgesic effect. This study demonstrated for the first time that pregabalin reached the dermis following topical application of a Plo lecithin-based organogel formulation.

Arafa et al. [92] prepared mucoadhesive topical gels with pregabalin alone or encapsulated in niosomes. Higher cholesterol ratios increased pregabalin entrapment efficiency, as increasing the concentration of cholesterol inhibits the conversion of the gel into liquid and enhances the encapsulation of hydrophilic drugs. Furthermore, the formula with the highest cholesterol content showed the lowest percent of drug release. HPMC and carbopol hydrogel formulations showed a higher release rate (91.2 $\pm$ 0.05%) of PG compared to niosomal PG formulations.

Cevik et al. [95] reported a novel method for synthesizing pH-responsive composite hydrogels using visible light. The pH-responsive layer is formed using poly(methacrylic acid-g-ethylene glycol) [P(MAA-g-EG)] as the macromer, eosin Y as the photoinitiator, and triethanolamine as the co-initiator. The three types of hydrogels, plain, [P(MAA-g-EG)], and P(MAA-g-EG) hydrogels, varied with the composition of the hydrogel prepolymer and the photoinitiation mechanism, with those formed under visible light preserving their integrity better than the ones formed under UV light. Therefore, cross-linked styrene–butadiene–styrene particles enhance the integrity of the hydrogel. Furthermore, in vitro fibroblast viability assay and in vivo implantation experiments indicated the hydrogels were nontoxic and nonirritant.

Another study evaluated different formulas of emulgels-containing pregabalin using Carbopol 940 and other polymers. The best drug release rate was achieved with Carbopol 940 0.4% and HPMC K15M 0.4%, offering rapid analgesic effects while avoiding pregabalin's CNS-mediated side effects [93].

Thus, optimized gel formulations containing pregabalin resulted in a significant release rate of up to 90% and an improved skin permeation for up to 240 h.

3.5. Amitriptyline

Oral amitriptyline can cause serious side effects [50]; therefore, developing topical formulas with amitriptyline for treating NeP is of great interest. However, no registered topical formulas exist currently. This may be due to the inconsistent results seen in clinical studies. Several clinical studies reported no analgesic efficacy of amitriptyline in patients with neuropathy when topically administered as 5% or lower concentration formulas [96]. Furthermore, most of these studies do not report the used formulation. Ho et al. [97] reported the use of a Plo lecithin organogel formulation. Despite reducing pain intensity, topical lidocaine induced minimal clinical improvement, whereas placebo and topical amitriptyline were ineffective. No rationale was given to justify the use of Plo. However, studies using higher concentrations of amitriptyline (10%) reported a more consistent analgesic effect [19].

Conversely, Shakshuki et al. [98] evaluated amitriptyline hydrochloride (1%, 5%, and 10%) compounded with three different bases: Lipoderm base, Emollient Cream, and

Mediflo 30 Plo lecithin organogel. Mean cumulative release after 24 h from the 10% formulation was significantly higher from the Mediflo Plo lecithin organogel than from the Lipoderm base or Emollient Cream: 53.2% vs. 23.9% and 41.8%, respectively. The authors had an extremely important observation: Amitriptyline released from formulas containing less than 1% of active substance did not permeate through the artificial skin membrane. A similar observation was made for the 5% Mediflo Plo lecithin organogel. Furthermore, despite having the highest amitriptyline release rate, 10% Mediflo Plo lecithin had the lowest permeation over 24 h. Amitriptyline 5% in Lipoderm base had the highest flux—this indicates that using a more lipophilic base may be a more suitable choice for creating topical amitriptyline preparations due to its high lipid content and lipophilic properties. By containing a high percentage of lipids, such as isopropyl myristate and caprylic/capric triglyceride, which are highly lipophilic, Lipoderm can improve the solubility and permeability of lipophilic drugs [99]. Another suggestion would be using a pH adjuster. Amitriptyline is a weak base, and its solubility increases at acidic pH. Thus, an appropriate pH adjuster may be used to adjust the pH of the gel to a range where amitriptyline is more soluble.

3.6. Other Substances

Despite not being included in the therapy guidelines for NeP, some reports indicate that the use of topical preparations containing other substances may be useful in the treatment of pain.

Ketamine is primarily used for anesthesia and pain relief. In recent years, it has also gained interest as a treatment for depression, posttraumatic stress disorder, and other psychiatric conditions. The mechanism of action of ketamine is complex and not yet fully understood. It is an N-methyl-D-aspartate receptor antagonist, and it seems it can also act on other receptors in the brain, such as the dopamine and serotonin receptors. Ketamine is typically administered intravenously, and its use is associated with several side effects, including dissociative effects, hallucinations, and changes in blood pressure and heart rate. Therefore, topical formulations may be a great alternative [52]. Wang et al. [52] developed a ketamine-loaded PL F127 stabilized reduced graphene oxide hydrogel for sustained transdermal drug delivery. Adding Plo F127 stabilized reduced graphene oxide sustaining the release of ketamine due to unique π-π stacking interaction. The ex vivo release study showed sustained release of ketamine from hydrogel compared to the control hydrogel, consistent with the results of the in vivo tail-flick evaluation, in which ketamine-loaded reduced graphene oxide had a significantly prolonged analgesic effect (24 h) compared to the ketamine control hydrogel (4 h). Baclofen interacts with two types of receptors, gamma amino butyric acid A and B, inhibiting chemokine-induced chemotaxis [100]. Adverse effects following oral administration include muscle weakness, nausea, somnolence, and paresthesia and affect between 25% and 75% of patients [56].

In a study, a topical gel containing baclofen niosomes was developed by adjusting the ratios between various nonionic surfactants (Span 60, 40), cholesterol, and charge-inducing agents. The entrapment efficiency of the formulations ranged from 4.37% to 80.31%. Encapsulating the drug into niosomes allowed sustained and controlled drug release. The incorporation of free baclofen and two niosomal baclofen formulations into Carbopol 934 led to faster permeation than Pluronic F127. In vivo examination revealed no significant analgesic differences between baclofen niosomal gel and marketed baclofen tablets after 24 h [100].

Baclofen-containing hydrogels with controlled release were prepared using different hydrophilic polymers. A total of 18 formulations were prepared and evaluated. The topical baclofen gels showed good physical properties, and the optimized formulations were stable at room temperature [101].

Another study formulated baclofen-loaded Eudragit® RL100 nanoparticles by nanoprecipitation method. After characterization, particle size decreased with increasing Eudragit® RL100, and the smallest size was observed in formulas containing an organic phase (acetone:

methanol) in a 1:3 ratio. The highest encapsulation efficiency occurred with the highest baclofen-loaded Eudragit® RL100 concentration and the same organic phase ratio [102].

Topically administered nonsteroidal anti-inflammatory drugs reduce proinflammatory PGs by inhibiting cyclooxygenase COX-2, interfering with the nociceptive pathway [103]. Conventional topical preparations are not effective in NeP. However, newer formulations have optimized characteristics. Furthermore, long-term oral administration of NSAIDs can cause gastrointestinal symptoms ranging from vague complaints to duodenal ulcer symptoms.

Naproxen was investigated in a formula using glycofurol as a vehicle-based gel and three various gelling agents (Carbopol 974P, Gantrez AN 119, and polyvinylpyrollidone K30). Skin permeation rates and lag times were evaluated to obtain the best gel formulation. Permeability parameters, steady-state flux, permeability coefficient, and penetration index, showed an increase in optimized formulation containing 2% Transcutol as permeation enhancer. On the other hand, optimized novel glycofurol-based gel formulation showed no topical adverse effects in the skin irritation test. Glycofurol-based gel seems to ensure dermal and transdermal delivery of naproxen and may be used for water-insoluble drugs [104].

As a new approach for delivering therapeutic agents, researchers developed supramolecular hydrogels using small peptides conjugated with NSAIDS. The conjugation did not disrupt the binding of naproxen to COX-2. The presence of D-tyrosine on the D-peptide improved the activity and selectivity of naproxen no matter where the position of naproxen was on the side chain. Furthermore, the conjugation of naproxen greatly reduced its binding to COX-1, which in theory, would diminish the associated adverse gastrointestinal and renal effects [105].

Tryptophan N-capped dipeptides with naproxen were prepared using C-terminal dehydroamino acids. The hydrogels consisted of networks of micro/nanosized fibers formed by peptide self-assembly driven by aromatic group stacking interactions. Hydrophobic peptides (containing C-terminal dehydrophenylalanine) formed more elastic gels at lower critical gelation concentrations and revealed irreversible breakup, while gels with C-terminal dehydroaminobutyric acid and dehydroalanine showed structural recovery and partial healing of the elastic properties. Therefore, these hydrogels are promising drug–nanocarrier candidates [106].

A topical microemulsion-based hydrogel containing ibuprofen was prepared and evaluated. The optimum formulation contained 3% ibuprofen, 6% ethyl oleate, 30% Tween 80/PG (2:1), and water and showed a high permeation rate of 38.06 μg cm^{-2} h^{-1} in vitro using porcine skins. Xanthan gum was used as a gel matrix, for increasing viscosity. The studied microemulsion-based hydrogel presented good stability, being a promising vehicle for topical delivery of ibuprofen [107].

Mauri et al. evaluated the covalent tethering of ibuprofen to a hydrogel matrix via esterification. The COX inhibitory activity of ibuprofen was not affected after the modification of the terminal carboxyl group, ensuring a therapeutic effect that is comparable to that of its salt form. As a result of chemical functionalization, ibuprofen can be given in the form of its free carboxylic acid instead of its sodium salt, which provides a viable alternative. The free carboxylic acid form of ibuprofen is more soluble and diffuses more quickly than the salt form. By incorporating an ibuprofen-diol derivative into the hydrogel formulation, a more sustained release profile was achieved compared to the salt form. These ibuprofen-functionalized hydrogels could be used as injectable tools due to their sol–gel transition, which enables localized drug release and presents promising possibilities for in situ treatments [108].

Using chitosan, lipids, gum arabic, and polyvinyl alcohol, six formulations containing ibuprofen were prepared through ionic interaction, maturation, and freeze–thaw methods. The results showed that the lipid-conjugation-based hydrogel exhibited a higher conjugation efficiency and prolonged drug release than control. Furthermore, ibuprofen effectively reduced LPS-induced PGE2 synthesis [109].

Vivero et al. analyzed poly(hydroxyethyl methacrylate) hydrogels as nonsteroidal anti-inflammatory drugs' delivery systems using 4-vinyl-pyridine and N-(3-aminopropyl) methacrylamide as cross-linkers. Incorporated monomers substantially increased ibuprofen (up to 10-fold) and diclofenac (up to 20-fold) loading, while drug release was limited (less than 10%) due to ionic/hydrophobic interactions.

Water-swollen hydrogels transferred to pH 5.8 or 8.0 phosphate buffer or NaCl solutions showed release driven by competition with environmental ions, sustaining the release process for minimum 24 h for ibuprofen and almost 1 week for diclofenac owing to the remaining hydrophobic interactions and the high polymer density induced by poly(hydroxyethyl methacrylate) [110].

Another study analyzed hydrogel films loaded with ibuprofen using water-soluble polysaccharides such as cellulose sulfate, chitosan sodium, and tripolyphosphate via self-assembly method. The hydrogels formed a dense regularly shaped network due to polyelectrolyte complex formation via electrostatic interaction. Loading and encapsulation efficiency were 43.15 ± 4.88% and 60.65 ± 4.68%, respectively. The hydrogel films showed a sustained release profile during 1440 min test using mice skin [39].

3.7. Associations of Analgesic Substances

The combination of two or more therapeutic agents can be more effective considering the wide variety of pathophysiological mechanisms involved in the development and progression of neuropathic pain. A topical combination of baclofen, amitriptyline, and ketamine was evaluated in a double-blind, placebo-controlled trial, including patients with chemotherapy-induced peripheral neuropathy. The active formulation contained 40 mg of amitriptyline, 20 mg of ketamine, and 10 mg of baclofen in pluronic lecithin organogel and was applied twice a day for 4 weeks (a teaspoonful of gel to each affected area). Although improvements in sensory pain and motor scale were observed, they were not statistically significant, and no adverse effects or systemic toxicity were reported [96].

In an uncontrolled trial, Uzaraga et al. [111] assessed the efficacy of a topical treatment consisting in an amitriptyline 2%, ketamine 1%, and lidocaine 5% (AKL) containing gel in radiation-induced dermatitis and neuropathic pain. The gel was given to 60 individuals, three times per day until 2 weeks post-radiotherapy. A patient assessment was performed every 2–5 days during radiotherapy and at 2 and 6 weeks postradiotherapy. Pain was assessed using the University of Washington neuropathic pain scale. After AKL gel application, a reduction in pain intensity and other symptoms was recorded at 30 min and 2 weeks posttreatment compared to baseline, but fatigue and skin irritation occurred at the application site.

A study case reported the effectiveness of the amitriptyline and ketamine containing gel in a patient with erythromelalgia. Over the next 2 weeks, the patient described "very promising results", with a 60% decrease in pain intensity, primarily burning pain. This patient did not respond previously to oral paracetamol, NSAIDs, gabapentin, and amitriptyline [112]. These clinical trials support the idea of using topical preparations instead of oral therapy.

The method of preparation and the functional excipients used to obtain all these gels are presented in Table 3. As the local conditions within the tissue or cellular environment (pH, temperature) where the gel is applied can influence the efficacy and function of the gel, when available we also included such details in Table 3.

Table 3. Analgesic-containing gels used for neuropathic pain.

Substance	Gel Type	Method of Preparation	Formula with Optimal In Vitro Parameters	Role of Agents	Parameter	Advantages of the New Formulation in the Treatments of Neuropathic Pain	References
Capsaicin	Hydrogel	Dissolution The certain quantity of polymer and drug was added into pH 4 buffer with continuous stirring for 1 h.	0.075% capsaicin, 6% CMC-Na	CMC, synthetic water-soluble cellulose added as matrix for drug delivery systems.	Rate of release 12.69 ± 0.58 g/cm^2 per 1/2 h	Advantages The cumulative quantity of capsaicin 4 h after the application is bigger than that of marketed creams. CMC-Na showed better bio-adhesion to the skin, resulting in a prolonged time of location at the site of application and enhanced permeation efficacy. CMC-Na hydrogels can produce a fast onset and sustained duration of capsaicin release, in contrast to cream bases. Disadvantages All individuals revealed more powerful pungent sensation of CMC-Na hydrogels than cream bases.	[71]
Capsaicin	Biogel	Dissolution Capsaicin was weighed into the monooleate and heated to melt at 45 °C. After the mixture was homogeneously vortexed, propylene glycol and water were added, then homogeneously vortexed.	2.5 mg/g capsaicin, 63% glycerol monooleate, 7% propylene glycol 30%	Monooleate—solvent Propylene glycol, organic solvent decreases the cubic phase gel viscosity and it is a skin penetration enhancer.	Rate of release 33%	Cubic phase gels based on the ternary phase diagram of the monooleate–propylene glycol–water system offer transdermal controlled release of capsaicin. After 108 h, around 30% of capsaicin was released from the gels.	[72]

Table 3. *Cont.*

Substance	Gel Type	Method of Preparation	Formula with Optimal In Vitro Parameters	Role of Agents	Parameter	Advantages of the New Formulation in the Treatments of Neuropathic Pain	References
Capsaicin	Emulsion gel	Hot melt homogenization technique Capsaicin and melted lipids were thoroughly mixed using a magnetic stirrer, forming a clear lipid phase. A separate aqueous phase with Tween 80 was prepared. The hot lipid phase was then gradually added to the hot aqueous phase and homogenized.	Capsaicin, carbopol, ethanol, PEG400	Ethanol—solvent PEG400—solvent Carbopol, a rheology modifier and extended-release polymers that enhance the viscosity of gel and skin permeation and offers sustained release	Rate of release 59.2% Permeation rate (skin of Sprague–Dawley rats) 7.2% (after 24 h)	The capsaicin-loaded nanoparticles containing solution showed a higher analgesic efficacy than the capsaicin-loaded nanoparticles containing gel in the Hot-plate test. The capsaicin-based nanoparticles formulations presented increased analgesic efficacy than capsaicin cream at the same concentration in the Hot-plate test due to the higher permeation ability and improved skin retention. The gel induced slightly less irritation than the solutions.	[73]
Capsaicin	Biogel	Passive loading technique Capsaicin was dissolved in methanol using a sonicator. After adding natural β-chitin–protein complex film, the mix was vortexed at 120 rpm and incubated in a shaker at room temperature for 48 h.	100 mg capsaicin, 8 mL methanol, 200 mg β-chitin–protein complex	Methanol—solvent β-chitin–protein complex—natural polymer β-chitin–protein complex film has a finer structure in contrast to synthetically made chitosan films.	Rate of release pH 4.0: 50.49% (48 h) pH 5.5: 59.81% (72 h) pH 7.4: 59.02% (96 h)	β-chitin–protein complex film offers a prolonged release rate of capsaicin The maximum capsaicin release was at pH 7.4, which is slightly higher than that of the skin, after 96 h.	[74]

Table 3. *Cont.*

Substance	Gel Type	Method of Preparation	Formula with Optimal In Vitro Parameters	Role of Agents	Parameter	Advantages of the New Formulation in the Treatments of Neuropathic Pain	References
Capsaicin	Hydrogel	Interfacial deposition of preformed polymer method An organic phase containing Eudragit RS100®, capsaicinoids mixture, acetone, and capric/caprylic triglyceride was injected in an aqueous phase containing polysorbate 80. After mixing, the solvents were removed with a reduced-pressure evaporator.	5 mg capsaicinoids mixture, 3.5% chitosan, 1.5% lactic acid, 100 mg Eudragit RS100®	Lactic acid-pH modifier Eudragit RS100®-polymer Chitosan, a polysaccharide, that presents enhanced skin bioadhesion, film-forming capability, and wound-healing promotion	Rate of release 81 ± 1% (96 h)	The best formula had the lowest quantity of lactic acid, which provides a tolerable pH value (4.34 ± 0.11) and good biocompatibility.	[68]
Capsaicin	Emulsion gels	High shear mixing Capsaicin was added to a melted mixture of phytantriol and poloxamer 407 at 60 °C. This mixture was stirred with water, and after equilibrating for 48 h at room temperature, a cubic phase gel formed.	10.88 mg capsaicin, 0.3047 g PL F127, 3.0096 g phytantriol.	Phytantriol—solvent	Rate of release F1: 41% F2: 33% Permeation rate (skin of Sprague–Dawley rats) F1: 0.32 ± 0.05 $\mu g \cdot cm^{-1} \cdot h^{-1}$ F2: 0.18 ± 0.02 $\mu g \cdot cm^{-2} \cdot h^{-1}$	Phytantriol forms cubosomes determining a sustained release of capsaicine (longer than that of on-market products). Cubosomes formulations induced no obvious irritation to the skin.	[75]

Table 3. *Cont.*

Substance	Gel Type	Method of Preparation	Formula with Optimal In Vitro Parameters	Role of Agents	Parameter	Advantages of the New Formulation in the Treatments of Neuropathic Pain	References
Tramadol	Organogel	Coacervation phase separation method Proniosome components and tramadol were mixed with absolute ethyl alcohol and heated in a thermostatic water bath. Further, phosphate buffer (pH 7.4) was added on the water bath. After cooling down to room temperature, the solutions was mixed with HPMC.	100 mg tramadol, 1800 mg Span 80, 1800 mg lecithin, 100 mg cholesterol	Lecithin—nonionic surfactant Cholesterol—nonionic surfactant Span 80—hydrophobic surfactant	Rate of release 60% (6 h) Permeation rate (skin of New Zealand Wistar albino rats) 2300 $\mu g/cm^{-1}$ (in 24 h)	The tramadol gel demonstrated higher analgesic efficacy than oral tablets in rat tests and showed significant anti-inflammatory effects. This formula had improved entrapment efficiency and transdermal flux compared to tramadol cream. The low transition temperature of Span80 contributed to a fluid state, facilitating drug transport into the skin.	[18]
Tramadol	Hydrogel	Irradiation with electron beams Tramadol solution was created by diluting tramadol injection in purified water. HPMC hydrogel films were purified and dried before being turned into xerogels. The xerogels were then submerged in tramadol solution for 24 h to obtain hydrogel films containing tramadol (30 or 50 kGy dose).	20% tramadol, 20% hydroxypropyl methylcellulose	HPMC—cellulosic polymer HPMC interaction with the drug is minimum and do not inhibit the sustained release of the drug HPMC produced a transparent film	Rate of release 100% (after 240 min) Permeation rate (skin of mice) 1.36 mg/cm^2 (in 240 min)	The tramadol-containing gel showed a higher analgesic efficacy than oral tramadol tablets in acetic-acid-induced abdominal writhing test in rats. The new formula presented an enhanced release and skin permeation compared to standard formulations. The amount of tramadol released from the hydrogel film and the amount of skin permeation increased by changing the electron dose.	[78]

Table 3. *Cont.*

Substance	Gel Type	Method of Preparation	Formula with Optimal In Vitro Parameters	Role of Agents	Parameter	Advantages of the New Formulation in the Treatments of Neuropathic Pain	References
Tramadol	Organogel	Cold method The aqueous phase was heated at 30 °C and then delivered to the organic phase dropwise with constant mixing. The stir was continued for 45 min to bring the transethosomal dispersions that were supplemented for size reduction by probe sonication.	100 mg tramadol, 3% α phosphatidylcholine from egg yolk, 0.5% cremophor EL-35	α phosphatidylcholine—lipid carrier; showed low viscosity Cremophor EL-35—edge activator	Rate of release 79.98% (8 h)	The new formula exhibited a controlled rate of release compared to standard formulations. The use of edge activator in formulation increased the skin permeability.	[79]
Tramadol	Hydrogel	Dispersion Chitosan was dispersed in an acetic acid solution, and tramadol and poloxamer F-127 were dissolved in it. The mixture was placed in an ice bath for 30 min, then glycerophosphate disodium salt hydrate and pentasodium triphosphate were added. Finally, the mixture was moved to a 37 °C water bath, where gel formation occurred after about 1–1.5 min.	20% tramadol, 1% chitosan, 20% PL F-127, 14.5% glycerophosphate disodium salt hydrate, 0.5% pentasodium triphosphate	Chitosan, a natural polysaccharide that can form gels by interacting with other polymers or through physical and chemical cross-linking. It can form hydrogels by swelling in water.	Rate of release 80% (8 h)	Chitosan extends the release of drug compared to tramadol-containing gel.	[83]

Table 3. *Cont.*

Substance	Gel Type	Method of Preparation	Formula with Optimal In Vitro Parameters	Role of Agents	Parameter	Advantages of the New Formulation in the Treatments of Neuropathic Pain	References
Tramadol	Hydrogel	Cold method Tramadol was dispersed in various solutions containing PL 407 alone or in binary systems with PL 188 and were retained at 4 °C under magnetic stirring. The PL concentrations were selected in order to obtain a thermoreversible gel at minimum possible final concentration with a maximum final PL concentration of 35% (weight per weight [*w*/*w*]).	20 mg × mL $^{-1}$ tramadol, 25% PL 407, 10% PL 188	PL 407 and PL 188—nonionic surfactants	Rate of release 100% (after 4 h)	PL 407 and PL 188-based binary hydrogels showed controlled release of tramadol. Subsequently, they had extended duration of analgesia (72 h) compared to tramadol solution in tail-flick test. This leads to the possibility of reapplying every 48–72 h at lower doses. The gel showed an enhanced release rate at pH 7.4 and 37 °C.	[82]
Gabapentin	Hydrogel	Dissolution Gabapentin was dissolved into de-ionized water, and methyl and propyl hydroxybenzoate were dissolved in a permeation enhancer solvent. The enhancer mixture was added to the aqueous mixture and mixed for 5 min.	6% gabapentin, 0.75% carbopol	Carbopol—synthetic polymer carbomer	Mean flux value 2661.62 ± 50.39 Permeation rate ("nonhy-drated" human epidermal membrane) 7.56 ± 5.50 mcg/cm^2/h	Carbopol can incorporate low-molecular-weight compounds, such as gabapentin. Prehydrated and non-prehydrated membranes had similar gabapentin permeability, but prehydrated ones showed less variability.	[88]

Table 3. *Cont.*

Substance	Gel Type	Method of Preparation	Formula with Optimal In Vitro Parameters	Role of Agents	Parameter	Advantages of the New Formulation in the Treatments of Neuropathic Pain	References
Gabapentin	Emulsion gel	-	10% gabapentin, xanthan gum hydrocolloid, polyacrylamide	-	-	A 10% *w*/*w* topical gabapentin gel applied thrice daily demonstrated strong antiallodynic and antihyperalgic effects in Hot-plate and von Frey test. Topical use of the gel potentially avoids dose titration in neuropathic pain patients, reduces pain similarly to systemic gabapentin, and avoids related side effects.	[86]
Gabapentin	Emulsion gel	-	10% gabapentin, xanthan gum hydrocolloid, polyacrylamide	-	-	Topical application of gabapentin highly reduced cisplatin-associated neuropathic allodynia and heat-hypoalgesia. It might offer an alternative for neuropathic pain relief in patients treated with chemotherapy or those intolerant to systemic medications' side effects.	[87]

Table 3. *Cont.*

Substance	Gel Type	Method of Preparation	Formula with Optimal In Vitro Parameters	Role of Agents	Parameter	Advantages of the New Formulation in the Treatments of Neuropathic Pain	References
Gabapentin	Hydrogel	Copolymerization The chitosan-g-poly(acrylic acid-co-acrylamide) hydrogel was obtained following the reaction between a radical initiator (ammonium persulfate) and a cross-linking agent (N,N′-methylene bisacrylamide).	Gabapentin, acrylic acid, acrylamide monomers, chitosan	Chitosan—natural polymer Acrylamide monomers—cross-linkers	Rate of release 90% in the first 2 h	Acrylic acid and acrylamide monomers overcome the disadvantages of chitosan by cross-linking. In vitro simulation of gabapentin release from hydrogel in stomach-like (pH 1.2) and physiological buffer (pH 7.4) conditions showed direct drug diffusion from loaded gel samples in both pHs, with faster diffusion in pH 1.2 due to swelled samples.	[89]
Pregabalin	Organogel	Mixing A 1% aqueous solution of pregabalin was mixed with propylene glycol. This solution was then mixed with the oil phase solution of a PLO gel kit.	0.4% pregabalin, lecithin, isopropyl palmitate, PL 407	PL 407—block copolymer Lecithin—absorption enhancer Isopropyl palmitate—solvent for lecithin	Permeation rate (skin of mice) 0.7 μg/mL (after 120 h)	A significant analgesic effect was observed 1.5 h after application using von Frey test in mice streptozotocin-induced diabetic neuropathy.	[94]

Table 3. *Cont.*

Substance	Gel Type	Method of Preparation	Formula with Optimal In Vitro Parameters	Role of Agents	Parameter	Advantages of the New Formulation in the Treatments of Neuropathic Pain	References
Pregabalin	Hydrogel	High shear mixing A transparent gel was obtained by spraying with continuous stirring on the water surface 2% *w*/*w* gel bases (HPMC and carbopol 934). Then a uniform and clear solution was formed under continuous stirring by dissolving the pregabalin in the polymer dispersion.	5 g niosomes (1.5 g pregabalin F1 Span 60: cholesterol, 4:1 F2 Span 60: cholesterol, 4:4 F3 Span 60: cholesterol, 4:7) in HPMC or carbopol	Cholesterol enhanced the entrapment efficiency. Span 60, nonionic surfactant, gave the highest entrapment efficiency and stability of niosomes.	Release rate 32.2 ± 0.02% Permeation rate (skin of rats) 28.34% (in a period of 8 h)	-	[92]
Pregabalin	Hydrogel	Photopolymerization under both UV lights The pH-responsive layer was prepared using poly(methacrylic acid-g-ethylene glycol) as the macromer, eosin Y as the photoinitiator, and triethanolamine as the co-initiator. Hydrophobic domains were added by incorporating cross-linked styrene-butadiene-styrene (SBS) 30 copolymer in the pH-sensitive prepolymer.	3.6 g pregabalin, 2.0 g Poly(ethylene glycol) monomethyl ether monomethacrylate	-	Rate of release 86.4% at neutral pH Permeation rate (human fibroblast cell line): 28.34% over a period of 8 h	The gel was swollen and transparent at pH 7.0, while opaque at pH 2.2. Hydrogels formed with UV and visible light showed reversibility of swelling, responding to repeated pH variations in both low (pH 2.2) and high (pH 7.0) buffers.	[95]

Table 3. *Cont.*

Substance	Gel Type	Method of Preparation	Formula with Optimal In Vitro Parameters	Role of Agents	Parameter	Advantages of the New Formulation in the Treatments of Neuropathic Pain	References
Pregabalin	Emulsion gel	High shear mixing Span 80 was dissolved in castor oil (oil phase). Pregabalin and tween 80 were dissolved in distilled water, mixed with methyl and propyl parabens dissolved in propylene glycol (aqueous phase). Both phases were heated separately to 70–80 °C, before combining the oil phase with the aqueous phase.	2 g pregabalin, 10 g propylene glycol, 4 g Tween 80, 1 g Span 80, 10 g castor oil, 0.4 g carbopol 940.	Carbopol—polymer	Rate of release 30% in 20 min and 93% in 360 min	Carbopol in the specified amount resulted in the best drug release rate.	[93]
Amitriptyline	Organogel	High shear mixing Amitriptyline was mixed with Poloxamer 30% after being dissolved in sterile water. Isopropyl lecithin was mixed with the poloxamer component.	5% amitryptiline, 30% PL, lecithin-isopropyl myristate	-	-	No significant change in pain intensity vs. topical lidocaine in patients with postherpetic neuralgia, chronic postsurgical pain, and painful peripheral neuropathy	[97]
Amitriptyline	Organogel	High shear mixing Amitriptyline was finely powdered and mixed with ethoxy diglycol to obtain a smooth paste. Mediflo PLO gel was added.	10% amitryptiline, ethoxy diglycol, Mediflo PLO gel	-	Rate of release 53.2% Permeation rate (Strat-M membrane) 9.3% (over 24 h)	Mediflo PLO gel resulted in the highest release rate at 32 ± 0.5 °C and pH 7.4. However, Lipoderm base and Emollient Cream resulted in a lower cumulative permeation relative to Mediflo PLO gel.	[98]

Table 3. *Cont.*

Substance	Gel Type	Method of Preparation	Formula with Optimal In Vitro Parameters	Role of Agents	Parameter	Advantages of the New Formulation in the Treatments of Neuropathic Pain	References
Ketamine	Hydrogel	High shear mixing Ketamine was added in the Pluronic® F127 stabilized reduced graphene oxide dispersion and kept under mild stirring for 24 h. The viscosity of the ketamine loaded Pluronic® F127 stabilized reduced graphene oxide dispersion was increased by adding 2% Carbopol 94 for topical application.	5% ketamine, 0.1 µg/mL pluronic F127 stabilized reduced graphene oxide, 2% Carbopol 940	Pluronic F127 stabilized reduced graphene oxide offered a prolonged release of ketamine due to the unique π-π stacking interaction between ketamine and reduced graphene oxide.	Permeation rate 120.0 $\mu g/cm^2/h$	Ketamine-loaded reduced graphene oxide demonstrated sustained analgesic effect (24 h) compared to control hydrogel (4 h) in the tail-flick study. This alternative for neuropathic pain treatment using Pluronic® F127 graphene oxide hydrogel avoids side effects and skin irritation associated with other administration routes.	[52]
Baclofen	Organogel	Thin-film hydration method	5% baclofen niosomes (baclofen/Span 60/40/cholesterol) in 1% Carbopol 934	Span60/40—nonionic surfactants Cholesterol, enhancer of niosomal membrane rigidity.	Rate of release 62.75% (in 24 h)–for niosomes Permeation rate (cellulose membrane) 99.51% (in 24 h) at a maintained temperature of 37 ± 0.5 °C and pH 5.5	Reduced carrageenan-induced paw edema similar to oral marketed tablets. Topically applied niosomes improve the residence time of drugs in the stratum corneum and epidermis, while decreasing the systemic absorption of drug Carbopol prints faster permeation to the skin. Niosomes offer a prolonged and controlled release of the drug.	[100]

Table 3. *Cont.*

Substance	Gel Type	Method of Preparation	Formula with Optimal In Vitro Parameters	Role of Agents	Parameter	Advantages of the New Formulation in the Treatments of Neuropathic Pain	References
Baclofen	Hydrogel	High shear mixing	F13: baclofen in Carbopol 934 (1:4) F18: baclofen in xanthan (1:6)	Carbopol/Xanthan-thickening agent, stabilizer, controlled release	Rate of release F13: 98.98% (after 24 h) Permeation rate (skin of Albino rats) F13: 4.27 μg/cm^2/h	Carbopol and xanthan as gelling agents showed enhanced permeation through the skin and release rate.	[101]
Naproxen	Inorganic gel	Dissolution Gantrez AN 119 and glycofurol were homogenized to form a clear dispersion, which was degassed under vacuum and stored at room temperature for 1 day before use. Naproxen was dissolved before the addition of the polymer.	5% naproxen, 2.5% Gantrez, glycofurol	Gantrez—copolymer, gelling and bioadhesive agent. Glycofurol—good adhesiveness and spreadability.	Rate of release 584.78 ± 32.8 μg/cm^2 Permeation rate (skin of Sprague–Dawley rats) 161.168 ± 29.2 μg/cm^2/h	No skin irritation Glycofurol use results in high drug permeation.	[104]
Naproxen	Hydrogel	Synthetic procedures	Naproxen, diphenylalanine, N-hydroxysuccinimide	Diphenylalanine-D-amino acid	Rate of release 35.8% after 24 h	Minimal adverse effects	[105]

Table 3. *Cont.*

Substance	Gel Type	Method of Preparation	Formula with Optimal In Vitro Parameters	Role of Agents	Parameter	Advantages of the New Formulation in the Treatments of Neuropathic Pain	References
Ibuprofen	Hydrogel	High shear mixing The clear microemulsion-based hydrogel was prepared by completely dissolving the xanthan gum in the microemulsion under stirring.	3% ibuprofen, 3% ethyl oleate, 20% Tween 80, 10% propylene glycol, 1.5% xanthan gum	Xanthan gum increases viscosity. Ethyl oleate enhances the solubilizing capacity of microemulsion systems. Tween 80 acts as a surfactant, and propylene glycol as a cosurfactant.	Rate of release 38.06 ± 1.04 ($\mu g \times cm^{-2}\, h^{-1}$) Permeation rate (porcine ear skin) $7.61 \pm 0.21 \times 10^{-3}$ $cm \times h^{-1}$	Ethyl oleate determines excellent skin permeation rate of ibuprofen.	[107]
Ibuprofen	Hydrogel	Dissolution Carbomer 974P was dissolved in a phosphate-buffered saline solution and pH was adjusted to 7.8 with 1 M NaOH. Agarose was added and the mixture was microwave-irradiated (500 W) for 30 s at 80 °C to initiate the condensation reaction. After cooling to 55 °C, the mixture was poured into steel cylindrical molds for gelation.	2.8% ibuprofen in carbomer 974P	-	Rate of release 80% after 48 h	Rate of release was pH-sensitive: at acidic or basic pH levels, increased hydrolytic rate was reported, with the alkaline conditions showing the fastest release (+10% cumulative release in acidic medium and +20% cumulative release in alkaline medium at 24 h, compared to neutral pH)-	[108]

Table 3. *Cont.*

Substance	Gel Type	Method of Preparation	Formula with Optimal In Vitro Parameters	Role of Agents	Parameter	Advantages of the New Formulation in the Treatments of Neuropathic Pain	References
Ibuprofen	Hydrogel	Freeze–thaw cycle Ibuprofen was dissolved in sodium hydroxide and polyethylene glycol. Chitosan was dissolved in 2% acetic acid and brought to 20 mL with distilled water to give a 2% *w*/*v* concentration. The drug–polymer suspension was obtained by adding the Ibuprofen solution to the chitosan solution. A nanoconjugate with a gum arabic gel matrix was formed by dispersing gum arabic propylene glycol solution and polyvinylpyrrolidone under continuous stirring and then incorporating the ibuprofen nanoconjugate into the gum arabic matrix solution.	50 mg ibuprofen, 4% phospholipon 90G, 2% chitosan, 10% polyvinyl alcohol, 2.5% gum arabic	Gum arabic—reduced the crystallinity of ibuprofen	Rate of release 90% (after approximately 13 h)	Nanoconjugate hydrogel ibuprofen-loaded chitosan–PC90G showed sustained and controlled release, surpassing the disadvantages associated with the oral dosage form.	[109]

Table 3. *Cont.*

Substance	Gel Type	Method of Preparation	Formula with Optimal In Vitro Parameters	Role of Agents	Parameter	Advantages of the New Formulation in the Treatments of Neuropathic Pain	References
Ibuprofen/Diclofenac	Inorganic gel	Dissolution Polymerization Ethyleneglycol dimethacrylate, N-(3-aminopropyl) methacrylamide, and 4-vinyl-pyridine were dissolved in 2-Hydroxyethyl methacrylate. The initiator 2, 20-azobis(isobutyronitrile) was added, and the monomer solution was injected into a mold. Polymerization occurred for 12 h at 50 °C followed by 24 h at 70 °C. The gels were then submerged in boiling water for 15 min to remove unreacted monomers. The resulting 10.5 mm diameter discs were washed in water, 0.9% NaCl, and 0.1 M HCl, and finally dried at 40 °C.	5 mg/g diclofenac, 18 mg/g ibuprofen, 4-vinyl-pyridine, N-(3-aminopropyl) methacrylamide, ethyleneglycol dimethacrylate	4-vinyl-pyridine and N-(3-aminopropyl) methacrylamide remarkably increased the amount of ibuprofen and diclofenac loaded.	Rate of release 60% for both ibuprofen and diclofenac Sustained release for 24 h for ibuprofen Sustained release for 1 week for diclofenac.	No difference was observed in the release rate at both pH 5.8 and 8.0.	[110]

Table 3. *Cont.*

Substance	Gel Type	Method of Preparation	Formula with Optimal In Vitro Parameters	Role of Agents	Parameter	Advantages of the New Formulation in the Treatments of Neuropathic Pain	References
Ibuprofen	Biogel	Layer-by-layer self-assembly method The chitosan hydrochloride and sodium cellulose sulfate were dissolved in distilled water and ultrasonized. This mixture was used as the intermediate layer solution. Aqueous solutions of 2% (*w*/*v*) sodium cellulose sulfate and 2% (*w*/*v*) sodium tripolyphosphate were used as the upper and lower layer solutions, respectively.	5% ibuprofen 2% sodium tripolyphosphate in chitosan hydrochloride/sodium cellulose sulfate (4:1)	Sodium cellulose sulfate—polyanionic polymer chitosan—natural polymer	Rate of release 27.2 ± 1.0% (in the first 60 min) 35.3 ± 1.88% (after 1440 min) Permeation rate (skin of mice) 3140.44 ± 159.89 mg/cm (24 h)	Sodium cellulose sulfate, chitosan hydrochloride, and sodium tripolyphosphate determine a favorable sustained release profile, no cytotoxicity, and good biocompatibility.	[39]
Baclofen, amitriptyline, ketamine	Organogel	-	10 mg baclofen, 40 mg amitriptyline, 20 mg ketamine in PL lecithin	-	-	Double-blind, placebo-controlled trial (chemotherapy-induced peripheral neuropathy) Slight improvement in sensory pain and motor scale vs. placebo Topical gel was well tolerated, without evident systemic toxicity	[96]

Table 3. *Cont.*

Substance	Gel Type	Method of Preparation	Formula with Optimal In Vitro Parameters	Role of Agents	Parameter	Advantages of the New Formulation in the Treatments of Neuropathic Pain	References
Baclofen, amitriptyline, ketamine	Organogel	-	10 mg baclofen, 10 mg amitriptyline, and 20 mg ketamine in PL lecithin	-	-	Prospective, single-arm, cohort pilot study (neuropathic pain caused by radiation skin reaction): reduction in pain intensity, sharpness, burning, sensitivity, itchiness, and unpleasantness, at 30 min posttreatment and at 2 weeks posttreatment. The gel may prove effective in relieving pain in subjects who do not respond to standard treatment, such as opioids. The efficacy of gel in reducing burning pain may be explained by blocking the release of chemical mediators or stimulation of ion channels by drugs, such as amitriptyline, ketamine	[111]

Table 3. *Cont.*

Substance	Gel Type	Method of Preparation	Formula with Optimal In Vitro Parameters	Role of Agents	Parameter	Advantages of the New Formulation in the Treatments of Neuropathic Pain	References
Amitriptyline, ketamine	-	-	2% amitriptyline and 0.5% ketamine in methylcellulose	-	-	Case report (patient with erythromelalgia): topical gel decreased the intensity of pain, and enhanced the functional status and quality of life of a patient (which had not responded satisfactory to oral paracetamol, NSAIDs, gabapentin, and amitriptyline). The gel made it possible to stop all other pain medications of patient. Due to biochemical effects of amitriptyline and ketamine, the gel could avert vasodilation and decrease the redness and high skin temperature characteristic of this disorder.	[112]

Legend: CMC-Na, sodium carboxymethyl cellulose; COX, cytochrome c oxidase; DMSO, dimethyl sulfoxide; EDTA, ethylenediaminetetraacetic acid; HEPES, 4-1-piperazineethanesulfonic acid; HPMC, hydroxypropyl methylcellulose; NSAID, nonsteroidal anti-inflammatory drugs; PEG400, polyethylene glycol 400; PL188, poloxamer 188; PL407, poloxamer 407; PLO, pluronic lecithin organogel; P(MAA625 g-EG), poly(methacrylic acid-g-ethylene glycol; SBS, styrene–butadiene–styrene block copolymer; UV light, ultraviolet light.

4. Conclusions and Perspectives

Management of NeP is often challenging and may require a multimodal approach involving early diagnosis, psychological therapy, as well as treatment to relieve symptomatic pain [113]. According to the latest guidelines, oral therapies are used as first-line therapies for the treatment of NeP [114], with topical therapies being the second line owing mostly to low efficacy. Furthermore, oral analgesics are often associated with high toxicity.

Gels represent a promising platform for transdermal drug delivery due to their unique physicochemical properties. Gels can provide a sustained drug release, improved bioavailability, and controlled delivery of drugs through the skin. The high water content of gels allows for better hydration of the stratum corneum, which can increase the permeability of the skin and enhance drug absorption [115]. In addition, gels can be designed to adhere to the skin, minimizing drug loss due to rubbing or sweat [116]. The gel matrix can also protect the drug from degradation and improve its stability during storage and transport [66]. Transdermal delivery systems prevent the occurrence of adverse effects by preventing peak and trough plasma concentrations and decreasing drug accumulation [67].

Over the years, several studies investigated the development of gels with capsaicine, amitryptiline, gapabentin, pregabalin, tramadol, and ketamine to be used in the treatment of NeP [117].

In recent years, different new forms of gels with therapeutic agents useful for NeP have been discovered and tested. Nanoemulsion containing capsaicin, proniosomes gel formulation containing tramadol, tramadol hydrochloride encapsulated transethosomes, chitosan-g-poly(acrylic acid-co-acrylamide) hydrogel composite containing gabapentin, pluronic lecithis organogel formulation containing pregabalin, Mediflo pluronic lecithin organogel containing amitriptyline, and ketamine-loaded Pluronic® F127 stabilized reduced graphene oxide hydrogel for sustain drug delivery via transdermal route are among the pharmaceutical gels characterized, tested for the analgesic effect in vivo in preclinical or clinical studies and demonstrating a real potential to replace oral therapy.

The recent developments made in the formulation of gels containing analgesics include:

- incorporation of vesicular carriers, such as niosomes, transethosomes, and cubosomes, with enhanced skin penetration and drug delivery compared to conventional liposomes [118]. These carriers protected the encapsulated drug from degradation and acted as local reservoirs of the drug. All the included studies revealed that the utilization of vesicular carriers as drug delivery systems enables a profound interaction with the skin layers, leading to sustained and efficient drug release for a wide range of drugs. However, a better understanding of this interaction would allow further optimization of the formulas.
- use of in situ gel-forming formulations which also ensured a prolonged and constant drug release. Furthermore, thermoresponsive polymers that can undergo a reversible sol–gel transition in response to temperature changes have been used in the preparation of injectable gels for drug delivery. In situ cross-linking was used as a trigger to stimulate this transformation.
- use of multicomponent gels that contain more than one type of polymer. They offer better mechanical stability and tunable properties.
- co-delivery of multiple drugs in a single gel formulation can improve treatment efficacy and reduce side effects.

Our investigation encountered several inherent limitations due to the heterogeneous nature of the study designs evaluated. For instance, disparate methodologies were employed to quantify drug release rates and skin permeation profiles, potentially leading to inconsistent results. Furthermore, although several gels were developed, they were not tested in preclinical or clinical studies. Thus, it is difficult to evaluate their real analgesic utility. We consider that creating uniform protocols of evaluation of drug release for gels and performing an appropriate preclinical and clinical assessment of these products would provide the very much needed scientific base for identifying new gels useful in the therapy of NeP.

In conclusion, recent advances in the formulation of analgesic gels have brought about significant improvements in drug delivery systems for pain management. The development of new gelling agents, additives, and drug delivery technologies has led to enhanced solubility, bioavailability, and controlled release of analgesics, resulting in better pain relief and improved patient compliance. Additionally, the development of personalized medicine and the exploration of nonopioid analgesics in gel formulations offer new directions for future research. While there are still challenges to be overcome, such as ensuring safety and addressing the need for targeted drug delivery, the progress made in recent years provides a strong foundation for the continued development of advanced analgesic gel formulations.

Author Contributions: Conceptualization, S.N.; data curation, C.P.; writing—original draft preparation, C.P.; writing—review and editing, A.Z. All authors have read and agreed to the published version of the manuscript.

Funding: This work was co-financed by the European Social Fund, through the Operational Program Human Capital, project number POCU/993/6/13/154722.

Institutional Review Board Statement: Not applicable.

Informed Consent Statement: Not applicable.

Data Availability Statement: All data generated or analyzed during this study are included in this published article.

Conflicts of Interest: The authors declare no conflict of interest.

References

1. Terminology | International Association for the Study of Pain. Available online: https://www.iasp-pain.org/resources/terminology/ (accessed on 17 October 2022).
2. RTreede, R.-D.; Jensen, T.S.; Campbell, J.N.; Cruccu, G.; Dostrovsky, J.O.; Griffin, J.W.; Hansson, P.; Hughes, R.; Nurmikko, T.; Serra, J. Neuropathic pain: Redefinition and a grading system for clinical and research purposes. *Neurology* **2007**, *70*, 1630–1635. [CrossRef] [PubMed]
3. Colloca, L.; Ludman, T.; Bouhassira, D.; Baron, R.; Dickenson, A.H.; Yarnitsky, D.; Freeman, R.; Truini, A.; Attal, N.; Finnerup, N.; et al. Neuropathic pain. *Nat. Rev. Dis. Prim.* **2017**, *16*, 17002. [CrossRef] [PubMed]
4. Nickel, F.T.; Seifert, F.; Lanz, S.; Maihöfner, C. Mechanisms of neuropathic pain. *Eur. Neuropsychopharmacol.* **2012**, *22*, 81–91. [CrossRef] [PubMed]
5. DiBonaventura, M.D.; Sadosky, A.; Concialdi, K.; Hopps, M.; Kudel, I.; Parsons, B.; Cappelleri, J.C.; Hlavacek, P.; Alexander, A.H.; Stacey, B.R.; et al. The prevalence of probable neuropathic pain in the US: Results from a multimodal general-population health survey. *J. Pain Res.* **2017**, *10*, 2525–2538. [CrossRef] [PubMed]
6. Mbrah, A.K.; Nunes, A.P.; Hume, A.L.; Zhao, D.; Jesdale, B.M.; Bova, C.; Lapane, K.L. Prevalence and treatment of neuropathic pain diagnoses among U.S. nursing home residents. *Pain* **2021**, *163*, 1370–1377. [CrossRef] [PubMed]
7. Taams, E.N.; Drenthen, J.; Hanewinckel, R.; Ikram, M.A.; van Doorn, A.P. Prevalence and Risk Factor Profiles for Chronic Axonal Polyneuropathy in the General Population. *Neurology* **2022**, *20*, e2234–e2240. [CrossRef] [PubMed]
8. Franceschi, R.; Mozzillo, E.; Di Candia, F.; Rosanio, F.M.; Leonardi, L.; Liguori, A.; Micheli, F.; Cauvin, V.; Franzese, A.; Piona, C.A.; et al. A systematic review of the prevalence, risk factors and screening tools for autonomic and diabetic peripheral neuropathy in children, adolescents and young adults with type 1 diabetes. *Acta Diabetol.* **2022**, *59*, 293–308. [CrossRef] [PubMed]
9. Langley, P.C.; Van Litsenburg, C.; Cappelleri, J.C.; Carroll, D. The burden associated with neuropathic pain in Western Europe. *J. Med. Econ.* **2012**, *16*, 85–95. [CrossRef]
10. Freynhagen, R.; Baron, R.; Gockel, U.; Tölle, T.R. pain*DETECT*: A new screening questionnaire to identify neuropathic components in patients with back pain. *Curr. Med. Res. Opin.* **2006**, *22*, 1911–1920. [CrossRef]
11. Freynhagen, R.; Bennett, I.M. Diagnosis and management of neuropathic pain. *BMJ* **2009**, *339*, b3002. [CrossRef]
12. O'connor, A.B. Neuropathic Pain. *Pharmacoeconomics* **2009**, *27*, 95–112. [CrossRef]
13. Hans, G.; Masquelier, E.; De Cock, P. The diagnosis and management of neuropathic pain in daily practice in Belgium: An observational study. *BMC Public Health* **2007**, *7*, 170. [CrossRef] [PubMed]
14. Torrance, N.; Smith, B.H.; Watson, M.C.; Bennett, I.M. Medication and treatment use in primary care patients with chronic pain of predominantly neuropathic origin. *Fam. Pract.* **2007**, *24*, 481–485. [CrossRef] [PubMed]
15. Bates, D.; Schultheis, B.C.; Hanes, M.C.; Jolly, S.M.; Chakravarthy, K.V.; Deer, T.R.; Levy, R.M.; Hunter, C.W. A Comprehensive Algorithm for Management of Neuropathic Pain. *Pain Med.* **2019**, *20*, S2–S12. [CrossRef] [PubMed]

16. Yang, X.; Fang, P.; Xiang, D.; Yang, Y. Topical treatments for diabetic neuropathic pain (Review). *Exp. Ther. Med.* **2019**, *17*, 1963–1976. [CrossRef] [PubMed]
17. Jorge, L.L.; Feres, C.C.; Telles-Dias, P. Topical preparations for pain relief: Efficacy and patient adherence. *J. Pain Res.* **2010**, *4*, 11–24. [CrossRef] [PubMed]
18. Shah, J.; Nair, A.B.; Shah, H.; Jacob, S.; Shehata, T.; Morsy, M.A. Enhancement in antinociceptive and anti-inflammatory effects of tramadol by transdermal proniosome gel. *Asian J. Pharm. Sci.* **2019**, *15*, 786–796. [CrossRef]
19. Kopsky, D.J.; Hesselink, J.M.K. High Doses of Topical Amitriptyline in Neuropathic Pain: Two Cases and Literature Review. *Pain Pract.* **2011**, *12*, 148–153. [CrossRef]
20. Cruccu, G.; Sommer, C.; Anand, P.; Attal, N.; Baron, R.; Garcia-Larrea, L.; Haanpaa, M.; Jensen, T.S.; Serra, J.; Treede, R.-D. EFNS guidelines on neuropathic pain assessment: Revised 2009. *Eur. J. Neurol.* **2010**, *17*, 1010–1018. [CrossRef]
21. Finnerup, N.B.; Attal, N.; Haroutounian, S.; McNicol, E.; Baron, R.; Dworkin, R.H.; Gilron, I.; Haanpää, M.; Hansson, P.; Jensen, T.S.; et al. Pharmacotherapy for neuropathic pain in adults: A systematic review and meta-analysis. *Lancet Neurol.* **2015**, *14*, 162–173. [CrossRef]
22. Price, R.; Smith, D.; Franklin, G.; Gronseth, G.; Pignone, M.; David, W.S.; Armon, C.; Perkins, B.A.; Bril, V.; Rae-Grant, A.; et al. Oral and Topical Treatment of Painful Diabetic Polyneuropathy: Practice Guideline Update Summary. *Neurology* **2022**, *98*, 31–43. [CrossRef] [PubMed]
23. Shinu, P.; Morsy, M.A.; Nair, A.B.; Al Mouslem, A.K.; Venugopala, K.N.; Goyal, M.; Bansal, M.; Jacob, S.; Deb, P.K. Novel Therapies for the Treatment of Neuropathic Pain: Potential and Pitfalls. *J. Clin. Med.* **2022**, *11*, 3002. [CrossRef] [PubMed]
24. Moore, R.A.; Derry, S.; Aldington, D.; Cole, P.; Wiffen, P.J. Amitriptyline for neuropathic pain and fibromyalgia in adults. *Cochrane Database Syst. Rev.* **2012**, *12*, 8242. [CrossRef]
25. Macone, A.; Otis, J.A.D. Neuropathic Pain. *Semin. Neurol.* **2018**, *38*, 644–653. [CrossRef]
26. Maloney, J.; Pew, S.; Wie, C.; Gupta, R.; Freeman, J.; Strand, N. Comprehensive Review of Topical Analgesics for Chronic Pain. *Curr. Pain Headache Rep.* **2021**, *25*, 7. [CrossRef]
27. Cavalli, E.; Mammana, S.; Nicoletti, F.; Bramanti, P.; Mazzon, E. The neuropathic pain: An overview of the current treatment and future therapeutic approaches. *Int. J. Immunopathol. Pharmacol.* **2019**, *33*, 1–10. [CrossRef]
28. Attal, N.; de Andrade, D.C.; Adam, F.; Ranoux, D.; Teixeira, M.J.; Galhardoni, R.; Raicher, I.; Üçeyler, N.; Sommer, C.; Bouhassira, D. Safety and efficacy of repeated injections of botulinum toxin A in peripheral neuropathic pain (BOTNEP): A randomised, double-blind, placebo-controlled trial. *Lancet Neurol.* **2016**, *15*, 555–565. [CrossRef]
29. Gabriel, M.; Sharma, V. Antidepressant discontinuation syndrome. *Can. Med. Assoc. J.* **2017**, *189*, E747. [CrossRef]
30. Attal, N. Pharmacological treatments of neuropathic pain: The latest recommendations. *Rev. Neurol.* **2018**, *175*, 46–50. [CrossRef]
31. Nolano, M.; Simone, D.A.; Wendelschafer-Crabb, G.; Johnson, T.; Hazen, E.; Kennedy, W.R. Topical capsaicin in humans: Parallel loss of epidermal nerve fibers and pain sensation. *Pain* **1999**, *81*, 135–145. [CrossRef]
32. Aoki, K.R.; Guyer, B. Botulinum toxin type A and other botulinum toxin serotypes: A comparative review of biochemical and pharmacological actions. *Eur. J. Neurol.* **2001**, *8*, 21–29. [CrossRef]
33. Mitchell, K.; Bates, B.D.; Keller, J.M.; Lopez, M.; Scholl, L.; Navarro, J.; Madian, N.; Haspel, G.; Nemenov, I.M.; Iadarola, M.J. Ablation of Rat TRPV1-Expressing Adelta/C-Fibers with Resiniferatoxin: Analysis of Withdrawal Behaviors, Recovery of Function and Molecular Correlates. *Mol. Pain* **2010**, *6*, 94. [CrossRef] [PubMed]
34. Trompetto, C.; Abbruzzese, G.; Berardelli, A. Central effects of botulinum toxin type A: Evidence and supposition. *Mov. Disord.* **2004**, *19*, S60–S64. [CrossRef]
35. O'Neill, J.; Brock, C.; Olesen, A.E.; Andresen, T.; Nilsson, M.; Dickenson, A.H. Unravelling the Mystery of Capsaicin: A Tool to Understand and Treat Pain. *Pharmacol. Rev.* **2012**, *64*, 939–971. [CrossRef] [PubMed]
36. Zhao, Y.; Sun, C.; Shi, F.; Firempong, C.K.; Yu, J.; Xu, X.; Zhang, W. Preparation, characterization, and pharmacokinetics study of capsaicin via hydroxypropyl-beta-cyclodextrin encapsulation. *Pharm. Biol.* **2015**, *54*, 130–138. [CrossRef] [PubMed]
37. Capsaicin: Uses, Interactions, Mechanism of Action | DrugBank Online. Available online: https://go.drugbank.com/drugs/DB06774 (accessed on 21 April 2023).
38. Zhuang, H.; Zheng, J.P.; Gao, H.; De Yao, K. In vitro biodegradation and biocompatibility of gelatin/montmorillonite-chitosan intercalated nanocomposite. *J. Mater. Sci. Mater. Med.* **2007**, *18*, 951–957. [CrossRef]
39. Wang, X.-H.; Su, T.; Zhao, J.; Wu, Z.; Wang, D.; Zhang, W.-N.; Wu, Q.-X.; Chen, Y. Fabrication of polysaccharides-based hydrogel films for transdermal sustained delivery of Ibuprofen. *Cellulose* **2020**, *27*, 10277–10292. [CrossRef]
40. MAbu Ghalia, M.; Dahman, Y. Radiation crosslinking polymerization of poly (vinyl alcohol) and poly (ethylene glycol) with controlled drug release. *J. Polym. Res.* **2015**, *22*, 3319–3326. [CrossRef]
41. Grond, S.; Sablotzki, A. Clinical pharmacology of tramadol. *Clin. Pharmacokinet.* **2004**, *43*, 879–923. [CrossRef]
42. Smyj, R.; Wang, X.-P.; Han, F. Tramadol Hydrochloride. *Profiles Drug Subst. Excip. Relat. Methodol.* **2013**, *38*, 463–494. [CrossRef]
43. FDA. MedWatch: The FDA Safety Information and Adverse Event Reporting Program. Available online: www.fda.gov/medwatch (accessed on 20 April 2023).
44. Bockbrader, H.N.; Wesche, D.; Miller, R.; Chapel, S.; Janiczek, N.; Burger, P. A Comparison of the Pharmacokinetics and Pharmacodynamics of Pregabalin and Gabapentin. *Clin. Pharmacokinet.* **2010**, *49*, 661–669. [CrossRef]
45. Ben-Menachem, E. Pregabalin Pharmacology and Its Relevance to Clinical Practice. *Epilepsia* **2004**, *45*, 13–18. [CrossRef] [PubMed]

46. Verma, V.; Singh, N.; Jaggi, A. Pregabalin in Neuropathic Pain: Evidences and Possible Mechanisms. *Curr. Neuropharmacol.* **2014**, *12*, 44–56. [CrossRef] [PubMed]
47. Lee, N.-W.; Lee, H.J.; Kim, H.-J.; Chang, S.-H.; Park, D.J. Two cases of pregabalin neurotoxicity in chronic kidney disease patients. *Clin. Kidney J.* **2011**, *4*, 138. [CrossRef] [PubMed]
48. FDA; CDER. High Blood Pressure. Available online: https://www.fda.gov/ (accessed on 20 April 2023).
49. Amitriptyline Active Not Recruiting Phase 4 Trials for Depressive Disorder Treatment | DrugBank Online. Available online: https://go.drugbank.com/drugs/DB00321/clinical_trials?conditions=DBCOND0024243&phase=4&purpose=treatment&status=active_not_recruiting (accessed on 20 April 2023).
50. Graff-Radford, S.B.; Shaw, L.R.; Naliboff, B.N. Amitriptyline and Fluphenazine in the Treatment of Postherpetic Neuralgia. *Clin. J. Pain* **2000**, *16*, 188–192. [CrossRef] [PubMed]
51. Clements, J.; Nimmo, W.; Grant, I. Bioavailability, Pharmacokinetics, and Analgesic Activity of Ketamine in Humans. *J. Pharm. Sci.* **1982**, *71*, 539–542. [CrossRef]
52. Wang, R.; Gan, J.; Li, R.; Duan, J.; Zhou, J.; Lv, M.; Qi, R. Controlled delivery of ketamine from reduced graphene oxide hydrogel for neuropathic pain: In vitro and in vivo studies. *J. Drug Deliv. Sci. Technol.* **2020**, *60*, 101964. [CrossRef]
53. Mion, G.; Villevieille, T. Ketamine Pharmacology: An Update (Pharmacodynamics and Molecular Aspects, Recent Findings). *CNS Neurosci. Ther.* **2013**, *19*, 370–380. [CrossRef]
54. Baclofen | DrugBank Online. Available online: https://go.drugbank.com/articles/A245323 (accessed on 20 April 2023).
55. Romito, J.W.; Turner, E.R.; Rosener, A.J.; Coldiron, L.; Udipi, A.; Nohrn, L.; Tausiani, J.; Romito, B.T. Baclofen therapeutics, toxicity, and withdrawal: A narrative review. *SAGE Open Med.* **2021**, *9*, 2197. [CrossRef]
56. Ertzgaard, P.; Campo, C.; Calabrese, A. Efficacy and safety of oral baclofen in the management of spasticity: A rationale for intrathecal baclofen. *J. Rehabil. Med.* **2017**, *49*, 193–203. [CrossRef]
57. Davies, N.M.; Anderson, K.E. Clinical Pharmacokinetics of Naproxen. *Clin. Pharmacokinet.* **1997**, *32*, 268–293. [CrossRef]
58. Falany, C.N.; Ström, P.; Swedmark, S. Sulphation of o-desmethylnaproxen and related compounds by human cytosolic sulfotransferases. *Br. J. Clin. Pharmacol.* **2005**, *60*, 632–640. [CrossRef] [PubMed]
59. Hall, D.; Bolinske, T.; Sinatra, E.; Akhtar, S.; Albert, G.; Allison, S.; Anwar, M.; Arita, H.; Barker, A.; Bekhit, M.H.; et al. Naproxen. In *The Essence of Analgesia and Analgesics*; Cambridge University Press: Cambridge, UK, 2011; pp. 221–225. [CrossRef]
60. Naproxen: Uses, Interactions, Mechanism of Action | DrugBank Online. Available online: https://go.drugbank.com/drugs/DB00788 (accessed on 21 April 2023).
61. Bushra, R.; Aslam, N. An Overview of Clinical Pharmacology of Ibuprofen. *Oman Med. J.* **2010**, *25*, 155–161. [CrossRef] [PubMed]
62. Davies, N.M.; Anderson, K.E. Clinical Pharmacokinetics of Diclofenac. *Clin. Pharmacokinet.* **1997**, *33*, 184–213. [CrossRef]
63. Todd, P.A.; Sorkin, E.M. Diclofenac Sodium. A reappraisal of its pharmacodynamic and pharmacokinetic properties, and therapeutic efficacy. *Drugs* **1988**, *35*, 244–285. [CrossRef]
64. Campbell, J.N.; Stevens, R.; Hanson, P.; Connolly, J.; Meske, D.S.; Chung, M.-K.; Lascelles, B.D.X. Injectable Capsaicin for the Management of Pain Due to Osteoarthritis. *Molecules* **2021**, *26*, 778. [CrossRef]
65. CHMP. Qutenza; INN-Capsaicin. Available online: www.ema.europa.eu/contact (accessed on 21 March 2023).
66. The 6 Best Capsaicin Creams: Uses, Options, and Risks. Available online: https://www.medicalnewstoday.com/articles/best-capsaicin-creams (accessed on 21 March 2023).
67. Groninger, H.; Schisler, R.E. Topical Capsaicin for Neuropathic Pain #255. *J. Palliat. Med.* **2012**, *15*, 946–947. [CrossRef]
68. Contri, R.V.; Katzer, T.; Pohlmann, A.R.; Guterres, S.S. Chitosan Hydrogel Containing Capsaicinoids-Loaded Nanocapsules: An Innovative Formulation for Topical Delivery. *Soft Mater.* **2010**, *8*, 370–385. [CrossRef]
69. Choi, A.-J.; Kim, C.-J.; Cho, Y.-J.; Hwang, J.-K.; Kim, C.-T. Characterization of Capsaicin-Loaded Nanoemulsions Stabilized with Alginate and Chitosan by Self-assembly. *Food Bioprocess Technol.* **2011**, *4*, 1119–1126. [CrossRef]
70. Wang, J.; Chen, S.; Xu, Z.C. Synthesis and Properties Research on the Nanocapsulated Capsaicin by Simple Coacervation Method. *J. Dispers. Sci. Technol.* **2008**, *29*, 687–695. [CrossRef]
71. Wang, Y.-Y.; Hong, C.-T.; Chiu, W.-T.; Fang, J.-Y. In vitro and in vivo evaluations of topically applied capsaicin and nonivamide from hydrogels. *Int. J. Pharm.* **2001**, *224*, 89–104. [CrossRef] [PubMed]
72. Peng, X.; Wen, X.; Pan, X.; Wang, R.; Chen, B.; Wu, C. Design and In Vitro Evaluation of Capsaicin Transdermal Controlled Release Cubic Phase Gels. *AAPS PharmSciTech* **2010**, *11*, 1405–1410. [CrossRef]
73. Wang, X.-R.; Gao, S.-Q.; Niu, X.-Q.; Li, L.-J.; Ying, X.-Y.; Hu, Z.-J.; Gao, J.-Q. Capsaicin-loaded nanolipoidal carriers for topical application: Design, characterization, and in vitro/in vivo evaluation. *Int. J. Nanomed.* **2017**, *12*, 3881–3898. [CrossRef] [PubMed]
74. Aylanc, V.; Ertosun, S.; Akyuz, L.; Bilican, B.K.; Gokdag, S.; Bilican, I.; Cakmak, Y.S.; Yilmaz, B.A.; Kaya, M. Natural β-chitin-protein complex film obtained from waste razor shells for transdermal capsaicin carrier. *Int. J. Biol. Macromol.* **2020**, *155*, 508–515. [CrossRef] [PubMed]
75. Peng, X.; Zhou, Y.; Han, K.; Qin, L.; Dian, L.; Li, G.; Pan, X.; Wu, C. Characterization of cubosomes as a targeted and sustained transdermal delivery system for capsaicin. *Drug Des. Dev. Ther.* **2015**, *9*, 4209–4218. [CrossRef] [PubMed]
76. Ahmad, M.Z.; Mohammed, A.A.; Ibrahim, M.M. Technology overview and drug delivery application of proniosome. *Pharm. Dev. Technol.* **2016**, *22*, 302–311. [CrossRef]
77. Jacob, S.; Nair, A.B.; Aldhubiab, B.E. Preparation and evaluation of niosome gel containing acyclovir for enhanced dermal deposition. *J. Liposome Res.* **2017**, *27*, 283–292. [CrossRef]

78. Natori, N.; Shibano, Y.; Hiroki, A.; Taguchi, M.; Miyajima, A.; Yoshizawa, K.; Kawano, Y.; Hanawa, T. Preparation and Evaluation of Hydrogel Film Containing Tramadol for Reduction of Peripheral Neuropathic Pain. *J. Pharm. Sci.* **2022**, *112*, 132–137. [CrossRef]
79. Sundar, V.D.; Divya, P.; Dhanaraju, M.D. Design Development and Characterisation of Tramadol Hydrochloride Loaded Transethosomal Gel Formulation for Effective Pain Management. *Indian J. Pharm. Educ. Res.* **2020**, *54*, s88–s97. [CrossRef]
80. Tavakoli, J.; Wang, J.; Chuah, C.; Tang, Y. Natural-based Hydrogels: A Journey from Simple to Smart Networks for Medical Examination. *Curr. Med. Chem.* **2020**, *27*, 2704–2733 [CrossRef] [PubMed]
81. Jacob, S.; Nair, A.B.; Shah, J.; Sreeharsha, N.; Gupta, S.; Shinu, P. Emerging Role of Hydrogels in Drug Delivery Systems, Tissue Engineering and Wound Management. *Pharmaceutics* **2021**, *13*, 357. [CrossRef]
82. De Araújo, D.; dos Santos, A.C.M.; Akkari, A.C.S.; Ferreira, I.R.S.; Páscoli, M.; Guilherme, V.A.; de Paula, E.; Fraceto, L.; de Lima, R.; Melo, P.D.S.; et al. Poloxamer-based binary hydrogels for delivering tramadol hydrochloride: Sol-gel transition studies, dissolution-release kinetics, in vitro toxicity, and pharmacological evaluation. *Int. J. Nanomed.* **2015**, *10*, 2391–2401. [CrossRef] [PubMed]
83. Barati, M.; Samani, S.M.; Jahromi, L.P.; Ashrafi, H.; Azadi, A. Controlled-release in-situ gel forming formulation of tramadol containing chitosan-based pro-nanogels. *Int. J. Biol. Macromol.* **2018**, *118*, 1449–1454. [CrossRef] [PubMed]
84. Cline, A.E.; Turrentine, J.E. Compounded Topical Analgesics for Chronic Pain. *Dermatitis* **2016**, *27*, 263–271. [CrossRef]
85. Boardman, L.A.; Cooper, A.S.; Blais, L.R.; Raker, C.A. Topical Gabapentin in the Treatment of Localized and Generalized Vulvodynia. *Obstet. Gynecol.* **2008**, *112*, 579–585. [CrossRef]
86. Shahid, M.; Subhan, F.; Ahmad, N.; Ali, G.; Akbar, S.; Fawad, K.; Sewell, R. Topical gabapentin gel alleviates allodynia and hyperalgesia in the chronic sciatic nerve constriction injury neuropathic pain model. *Eur. J. Pain* **2016**, *21*, 668–680. [CrossRef] [PubMed]
87. Shahid, M.; Subhan, F.; Ahmad, N.; Sewell, R.D.E. Efficacy of a topical gabapentin gel in a cisplatin paradigm of chemotherapy-induced peripheral neuropathy. *BMC Pharmacol. Toxicol.* **2019**, *20*, 51. [CrossRef]
88. Martin, C.J.; Alcock, N.; Hiom, S.; Birchall, J.C. Development and Evaluation of Topical Gabapentin Formulations. *Pharmaceutics* **2017**, *9*, 31. [CrossRef]
89. Khosravi, N.; Youseftabar-Miri, L.; Divsar, F.; Hallajian, S.; Hafezi, K. Development and evaluation of chitosan-g-poly(acrylic acid-co-acrylamide) hydrogel composite containing gabapentin for in vitro controlled release. *J. Mol. Struct.* **2022**, *1270*, 133934. [CrossRef]
90. Shakshuki, A.; Yeung, P.; Agu, R.U. Compounded gabapentin for neuropathic pain: Stability and beyond-use date (BUD) in some commonly used bases. *J. Am. Pharm. Assoc.* **2019**, *59*, 514–520. [CrossRef]
91. Le, U.M.; Baltzley, S.; AlGhananeem, A. Gabapentin in Elastic Liposomes: Preparation, Characterization, Drug Release, and Penetration Through Porcine Skin. *Int. J. Pharm. Compd.* **2018**, *22*, 498–503. [PubMed]
92. Arafa, M.G.; Ayoub, B.M. DOE Optimization of Nano-based Carrier of Pregabalin as Hydrogel: New Therapeutic & Chemometric Approaches for Controlled Drug Delivery Systems. *Sci. Rep.* **2017**, *7*, 41503. [CrossRef]
93. Haddad, N.; Hasian, J. Effect of combination of some Polymers with Carbopol 940 on Pregabalin Release Rate from Emulgels. *Res. J. Pharm. Technol.* **2022**, *15*, 2003–2009. [CrossRef]
94. Nagao, M.; Tajima, M.; Sugiyama, E.; Shinouchi, R.; Shibata, K.; Yoshikawa, M.; Yamamoto, T.; Sato, V.H.; Nobe, K.; Sato, H. Evaluation of in vitro transdermal permeation, mass spectrometric imaging, and in vivo analgesic effects of pregabalin using a pluronic lecithin organogel formulation in mice. *Pharmacol. Res. Perspect.* **2022**, *10*, e00919. [CrossRef] [PubMed]
95. Cevik, O.; Gidon, D.; Kizilel, S. Visible-light-induced synthesis of pH-responsive composite hydrogels for controlled delivery of the anticonvulsant drug pregabalin. *Acta Biomater.* **2015**, *11*, 151–161. [CrossRef]
96. Barton, D.L.; Wos, E.J.; Qin, R.; Mattar, B.I.; Green, N.B.; Lanier, K.S.; Bearden, J.D.; Kugler, J.W.; Hoff, K.L.; Reddy, P.S.; et al. A double-blind, placebo-controlled trial of a topical treatment for chemotherapy-induced peripheral neuropathy: NCCTG trial N06CA. *Support. Care Cancer* **2010**, *19*, 833–841. [CrossRef]
97. Ho, K.-Y.; Huh, B.K.; White, W.D.; Yeh, C.-C.; Miller, E.J. Topical Amitriptyline Versus Lidocaine in the Treatment of Neuropathic Pain. *Clin. J. Pain* **2008**, *24*, 51–55. [CrossRef] [PubMed]
98. Shakshuki, A.; Yeung, P.; Agu, R.U. Compounded Topical Amitriptyline for Neuropathic Pain: In Vitro Release from Compounding Bases and Potential Correlation with Clinical Efficacy. *Can. J. Hosp. Pharm.* **2020**, *73*, 133. [CrossRef]
99. Feturi, F.G.; Weinstock, M.; Zhao, W.; Zhang, W.; Schnider, J.T.; Erbas, V.E.; Oksuz, S.; Plock, J.A.; Rohan, L.; Spiess, A.M.; et al. Mycophenolic Acid for Topical Immunosuppression in Vascularized Composite Allotransplantation: Optimizing Formulation and Preliminary Evaluation of Bioavailability and Pharmacokinetics. *Front. Surg.* **2018**, *5*, 20. [CrossRef]
100. Nasr-Alla, S.M.; El-Nabarawi, M.; Bendas, E.; El-Ridy, M.; Abdel-Jaleel, G.; Nasr-Alla, S.M. Formulation and evaluation of topical niosomal gel of baclofen. *J. Chem. Pharm. Res.* **2015**, *7*, 277–288.
101. Syed, M.; Rao, D.; Rao, V.; Nagarani, G. Formu Lation and E Va Lu Ation of Contr O Lle D Re Le Ase of Top Ical ge l bac lofe N. *EMCare Cover. J. Pharmanest Int. J. Adv. Pharm. Sci.* **2023**, *6*, 5.
102. Yussef, A.; Fayez, S.; Sakran, W. Formulation and Evaluation of Baclofen Polymeric Nanoparticles for Transdermal Delivery In-vitro and Ex-vivo Optimization. *J. Adv. Pharm. Res.* **2021**, *5*, 285–296. [CrossRef]
103. Kocot-Kępska, M.; Zajączkowska, R.; Mika, J.; Kopsky, D.J.; Wordliczek, J.; Dobrogowski, J.; Przeklasa-Muszyńska, A. Topical Treatments and Their Molecular/Cellular Mechanisms in Patients with Peripheral Neuropathic Pain—Narrative Review. *Pharmaceutics* **2021**, *13*, 450. [CrossRef]

104. Barakat, N.S. Evaluation of Glycofurol-Based Gel as a New Vehicle for Topical Application of Naproxen. *AAPS PharmSciTech* **2010**, *11*, 1138–1146. [CrossRef] [PubMed]
105. Li, J.; Kuang, Y.; Gao, Y.; Du, X.; Shi, J.; Xu, B. d-Amino Acids Boost the Selectivity and Confer Supramolecular Hydrogels of a Nonsteroidal Anti-Inflammatory Drug (NSAID). *J. Am. Chem. Soc.* **2012**, *135*, 542–545. [CrossRef] [PubMed]
106. Vilaça, H.; Hortelão, A.C.; Castanheira, E.M.; Queiroz, M.J.R.; Hilliou, L.; Hamley, I.W.; Martins, J.A.; Ferreira, P.M. Dehydrodipeptide Hydrogelators Containing Naproxen N-Capped Tryptophan: Self-Assembly, Hydrogel Characterization, and Evaluation as Potential Drug Nanocarriers. *Biomacromolecules* **2015**, *16*, 3562–3573. [CrossRef]
107. Chen, H.; Chang, X.; Du, D.; Li, J.; Xu, H.; Yang, X. Microemulsion-based hydrogel formulation of ibuprofen for topical delivery. *Int. J. Pharm.* **2006**, *315*, 52–58. [CrossRef] [PubMed]
108. Mauri, E.; Rossetti, A.; Mozetic, P.; Schiavon, C.; Sacchetti, A.; Rainer, A.; Rossi, F. Ester coupling of ibuprofen in hydrogel matrix: A facile one-step strategy for controlled anti-inflammatory drug release. *Eur. J. Pharm. Biopharm.* **2019**, *146*, 143–149. [CrossRef]
109. Mahmood, S.; Almurisi, S.H.; Al-Japairai, K.; Hilles, A.R.; Alelwani, W.; Bannunah, A.M.; Alshammari, F.; Alheibshy, F. Ibuprofen-Loaded Chitosan–Lipid Nanoconjugate Hydrogel with Gum Arabic: Green Synthesis, Characterisation, In Vitro Kinetics Mechanistic Release Study and PGE2 Production Test. *Gels* **2021**, *7*, 254. [CrossRef]
110. Andrade-Vivero, P.; Fernandez-Gabriel, E.; Alvarez-Lorenzo, C.; Concheiro, A. Improving the Loading and Release of NSAIDs from pHEMA Hydrogels by Copolymerization with Functionalized Monomers. *J. Pharm. Sci.* **2007**, *96*, 802–813. [CrossRef] [PubMed]
111. Uzaraga, I.; Gerbis, B.; Holwerda, E.; Gillis, D.; Wai, E. Topical amitriptyline, ketamine, and lidocaine in neuropathic pain caused by radiation skin reaction: A pilot study. *Support. Care Cancer* **2011**, *20*, 1515–1524. [CrossRef] [PubMed]
112. Portugal, F.; Araújo, A.; Silva, C.; Campos, M.; Valentim, A. Combination gel of 2% amitriptyline and 0.5% ketamine to treat refractory erythromelalgia pain—A case report of pain control success. *Rev. Esp. Anestesiol. Reanim.* **2021**, *68*, 293–296. [CrossRef]
113. Kaku, M.; Vinik, A.; Simpson, D.M. Pathways in the Diagnosis and Management of Diabetic Polyneuropathy. *Curr. Diabetes Rep.* **2015**, *15*, 35–609. [CrossRef]
114. Sawynok, J.; Zinger, C. Topical amitriptyline and ketamine for post-herpetic neuralgia and other forms of neuropathic pain. *Expert Opin. Pharmacother.* **2015**, *17*, 601–609. [CrossRef]
115. Silva, C.; Topgaard, D.; Kocherbitov, V.; Sousa, J.; Pais, A.; Sparr, E. Stratum corneum hydration: Phase transformations and mobility in stratum corneum, extracted lipids and isolated corneocytes. *Biochim. Biophys. Acta Biomembr.* **2007**, *1768*, 2647–2659. [CrossRef]
116. Bhowmik, D. Topical Drug Delivery System. Available online: https://www.researchgate.net/publication/304716203 (accessed on 20 March 2023).
117. Van Nooten, F.; Treur, M.; Pantiri, K.; Stoker, M.; Charokopou, M. Capsaicin 8% Patch Versus Oral Neuropathic Pain Medications for the Treatment of Painful Diabetic Peripheral Neuropathy: A Systematic Literature Review and Network Meta-analysis. *Clin. Ther.* **2017**, *39*, 787–803.e18. [CrossRef]
118. Das, B.; Nayak, A.K.; Mallick, S. Nanovesicles for delivery of antifungal drugs. *Appl. Nanovesicular Drug Deliv.* **2022**, *2022*, 383–397. [CrossRef]

gels

MDPI

Review

Natural Polymeric Hydrogels Encapsulating Small Molecules for Diabetic Wound Healing

Elena Iulia Oprita *, Andreea Iosageanu and Oana Craciunescu

National Institute of R&D for Biological Sciences, 296, Splaiul Independentei, 060031 Bucharest, Romania; andreea.iosageanu@incdsb.ro (A.I.); oana.craciunescu@incdsb.ro (O.C.)
* Correspondence: iulia.oprita@incdsb.ro

Abstract: Diabetes is a condition correlated with a high number of diagnosed chronic wounds as a result of a complex pathophysiological mechanism. Diabetic chronic wounds are characterized by disorganized and longer stages, compared to normal wound healing. Natural polymer hydrogels can act as good wound dressings due to their versatile physicochemical properties, represented mainly by high water content and good biocompatibility. Natural bioactive hydrogels are polymers loaded with bioactive compounds providing antibacterial and antioxidant properties, modulation of inflammation and adherence to wounded tissue, compared to traditional dressings, which enables promising future applications for diabetic wound healing. Natural bioactive compounds, such as polyphenols, polysaccharides and proteins have great advantages in promoting chronic wound healing in diabetes due to their antioxidant, anti-inflammatory, antimicrobial, anti-allergic and wound healing properties. The present paper aims to review the wound healing mechanisms underlining the main issues of chronic wounds and those specifically occurring in diabetes. Also, the review highlights the recent state of the art related to the effect of hydrogels enriched with natural bioactive compounds developed as biocompatible functional materials for improving diabetic-related chronic wound healing and providing novel therapeutic strategies that could prevent limb amputation and increase the quality of life in diabetic patients.

Keywords: hydrogel; flavonoids; collagen; alginate; cellulose; wound healing; diabetes

1. Introduction

Diabetes mellitus (DM) represents a group of chronic metabolic diseases characterized by defective insulin secretion and insulin resistance, which determines increased blood glucose, a condition called hyperglycemia. In addition, hyperglycemia is frequently found associated with various complications, such as diabetic neuropathy, retinopathy, cardiomyopathy and nephropathy, leading to loss of sensation in the lower extremities, blindness, renal failure, chronic wounds and lower extremity amputations [1].

One in ten adults is diabetic and just above half a billion people are living with this condition worldwide, according to the 10th edition IDF Diabetes Atlas [2]. It is estimated that by 2045 the number of individuals suffering from diabetes will increase to 783.2 million people [3]. With this increasing trend, the global costs of diabetes are also expected to rise. Research shows that the total economic burden of diabetes reached about $966 billion in 2021 and is expected to reach $1054 billion by 2045 [4]. Regarding regional distribution, the countries with the highest prevalence of diabetes are China, India, the United States, Indonesia and Mexico, of which India is the country with the highest mortality rate from this condition [5].

Chronic wounds associated with diabetes can be classified as foot, venous and pressure ulcers, with diabetic foot ulcer (DFU) being among the most common complications of patients with diabetes [6]. Due to loss of sensation in the lower extremities caused by diabetic neuropathy, patients may not feel heat or abrasions and, thus, they can get hurt. The wound becomes infected, which, in conjunction with the hyperglycemic microenvironment, disrupts the healing process [7]. DFUs affect 19–34% of patients with diabetes and up to 25% of cases require amputation [8]. After amputation, the wound becomes larger and more difficult to heal, which is the reason for diabetic foot ulcer having just about 50% mortality within 5 years of amputation [9].

Conventional treatment uses the "TIME" principle to treat chronic wounds. It includes tissue debridement, infection control, moisture balance and edges of the wound [3]. First of all, wound bed preparation is very important to facilitate chronic wound healing. Tissue debridement involves the removal of non-viable tissue from the wound bed, in order to allow the tissue to heal. If there are any signs of infection, they must be treated promptly, as pathogens cause tissue damage and increase inflammatory response. Moisture balance refers to ensuring a warm and moist environment that is required for healing, while wound edges assessment can indicate the progress of healing and confirms if the treatment is effective [10].

However, the current strategy in the management of diabetic wounds is not entirely effective, with almost a quarter of patients undergoing low limb amputations [8]. Therefore, tissue engineering strategies are urgently needed, including the development of hydrogels and other bioactive platforms for the restoration of skin integrity and the improvement of patients' quality of life.

In contrast to sponges and conventional bandages that only cover the wound and absorb its exudate, natural polymeric hydrogels are considered highly suitable candidates for diabetic wound dressings due to their good biocompatibility and antibacterial properties [1]. They can provide a moist environment, a microporous structure ensuring a good milieu for gas exchange and nutrient transport and a suitable environment mimicking the composition and/or structure of the extracellular matrix (ECM) for cell proliferation and migration [11]. Moreover, hydrogels, encapsulating small molecules or drugs, such as polyphenols, growth factors, or antibiotics, can further augment the stimulation of wound healing [12]. Accordingly, they can facilitate the healing process of chronic wounds and reduce the likelihood of amputations.

This paper aims to review the potential of novel natural polymeric hydrogels encapsulating small molecules and the involved mechanisms to effectively manage the healing of diabetic wounds. First, the differences between the biological and biochemical processes in normal vs. diabetic wound healing were summarized. Recent studies on in vitro and in vivo effects of small molecules and natural gelling polymers were reviewed. The primary focus of this review lay in exploring the natural polymeric hydrogels transformed into delivery platforms for the administration of incorporated beneficial small molecules that can facilitate the complex process of diabetic wound healing. The ultimate objective of this comprehensive review was to offer scientific insights and guidance for further management of diabetic wounds with a special emphasis on advanced natural hydrogels for reducing the incidence of amputations in individuals suffering from diabetes.

2. Biological and Biochemical Processes during Normal vs. Diabetic Wound Healing

Skin lesions are classified as acute wounds occurring after injury or trauma and going through normal wound healing processes, and complex chronic wounds constantly affected by abnormal factors that prevent the unfolding of normal stages of the wound healing process. Thus, chronic wounds may stall in one particular stage of the healing process (e.g., inflammatory stage) for a longer time, which desynchronizes the cascade events, causing prolongation of the healing process, in particular in diabetic wounds [12].

2.1. Stages of Normal Wound Healing

Normal wound healing is a continuous and dynamic process that involves a multitude of biochemical interactions at the cellular and molecular levels. According to Singer and Clark [13], the process can be divided into four overlapping stages: hemostasis, inflammation, proliferation and remodeling (maturation) [14] (Figure 1). The involvement of particular cell types is required and varies during different stages of repair. There is also the need for growth factors, cytokines, regulatory molecules, enzymes and components of the ECM for repairing and restoring the damaged tissue, according to the injury degree [15].

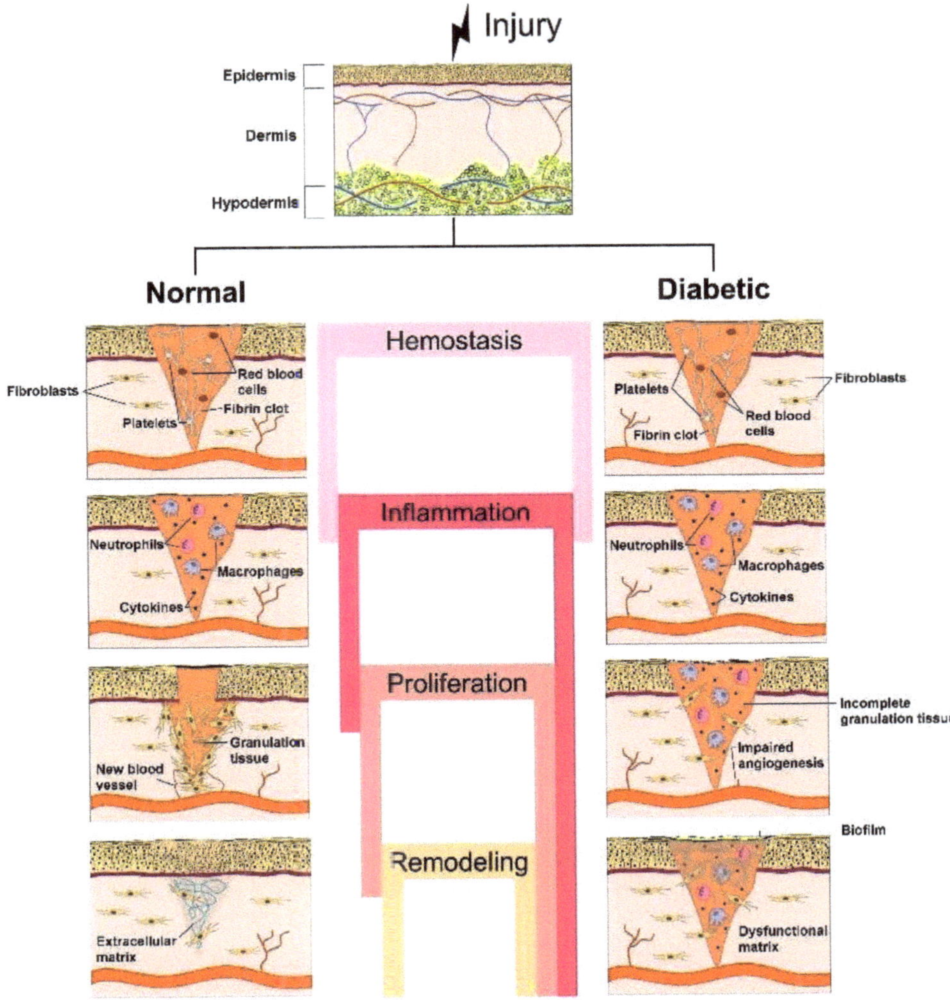

Figure 1. Schematic representation of the stages of normal wound healing, i.e., hemostasis, inflammation, proliferation and remodeling, vs. the pathogenesis of diabetic wounds, considering the chronic inflammatory phase that leads to incomplete granulation of tissue, impaired angiogenesis, formation of dysfunctional matrix and, finally, delaying the healing process.

The hemostasis stage (few hours) begins immediately after an injury to stop bleeding and is characterized by:

- Vascular constriction that decreases blood circulation at the wound site;
- Platelet aggregation at the wound site owing to the interaction with proteins (collagen (COL) and fibronectin);
- Degranulation;
- Conversion of soluble fibrinogen into insoluble fibrin to arrest bleeding;
- Mediation of hemostasis through key agents, such as fibrin, fibronectin and vitronectin;

- Production of growth factors, such as transforming growth factor β (TGF-β), platelet-derived growth factor (PDGF), fibroblast growth factor (FGF), epidermal growth factor (EGF) and chemokines, by the clot's surrounding area to efficiently aid the wound healing [15,16]. PDGF acts also in vascularization by attracting fibroblasts, which stimulate tissue repair through COL deposition [17]. In the first phase of the homeostasis stage, prostaglandin H2 is converted into thromboxane A2 (TXA2) by the action of thromboxane synthase. Then, TXA2 acts as a powerful platelet activator and vasoconstrictor, in addition to participating in the release of macrophages, neutrophils and endothelial cells, playing an important role in the following stages of the wound healing process [18,19].

The inflammation stage (1–6 days), initiated when the bleeding has stopped, has the goal of avoiding the entrance of pathogens, preventing infections and more severe complications. This stage is characterized by:

- Leucocytes' (especially neutrophils) migration to the injured site to eliminate debris and bacteria;
- Proinflammatory cytokines secretion by neutrophils to promote the expression of adhesion molecules;
- Monocytes' migration into the wound site and differentiation into macrophages.

Neutrophils, monocytes and macrophages play an important role due to the release of proinflammatory cytokines, such as interleukin-1 (IL-1), IL-6, IL-8, tumor necrosis factor α (TNF-α) and growth factors (PDGF, TGF-α, TGF-β, FGF and insulin-like growth factor-1 (IGF-1)) involved in the activation of fibroblasts and epithelial cells needed in the proliferation stage [19,20].

The proliferation stage (epidermal regeneration) (4–21 days) diminishes the wound area by epidermal regeneration, granulation tissue formation, COL synthesis to provide tissue strength and angiogenesis [21]. The activated fibroblasts migrate to the wound bed and synthesize ECM constituents (hyaluronan, fibronectin and proteoglycans), leading to the replacement of the fibrin clot. The granulation tissue is a connective tissue in which the predominant protein is COL type III, which provides a scaffold for novel ECM and blood vessel synthesis. The process of re-epithelialization taking place in this stage creates a new epidermal barrier after keratinocytes' proliferation and migration to the wound tissue mediated by cytokines secreted by innate immune cells [19]. Endothelial cell growth and chemotaxis are regulated by low oxygen tension and the production of lactic acid and biogenic amines. Furthermore, fibroblasts are differentiated into myofibroblasts, a process that results in lower cell proliferation together with an increase in COL synthesis [12]. In consequence, in this stage, wound contraction takes place to reduce it to a smaller size [19].

The remodeling (scar maturation) stage (3 weeks–2 years) is characterized by:

- Slow transformation of ECM into a mature scar;
- COL production reorganization in the ECM by replacing COL type III with COL type I and closure of the wound;
- Decrease of blood supply and formation of new blood vessels;
- Formation of a cellular environment and mature avascular tissue [19].

After all four stages of normal wound healing, it was reported that skin could only recover a maximum tensile strength of 80% [16].

2.2. Pathology of Diabetic Wound Healing

The primary factors hindering wound healing in diabetic patients do not deviate from those involved in the development of other chronic wounds. However, the underlying causes for the emergence of these factors differ. Diabetic wounds can evolve into chronic wounds due to a hyperglycemic environment that is ideal for bacterial growth and increases the risk of microbial infection. Untreated infections promote inflammation at the wound bed, as well as oxidative stress, both altering the wound healing process [9]. In addition, diabetic wounds become complicated to heal due to the following disturbing conditions:

hypoxia, chronic inflammation leading to excessive expression of matrix metalloproteinases (MMPs), and impaired angiogenesis [22].

The main factor of diabetic wound damage is severe hypoxia occurring after tissue injury. Oxygen is very important in wound healing to promote fibroblast proliferation, enhance immune function and stimulate angiogenesis. Diabetic patients have an inadequate oxygen supply due to vascular dysfunction and neuropathy, and a high oxygen consumption rate by recruitment of inflammatory cells in the wounded area [23,24]. Thus, long-term oxygen deprivation impairs the wound healing process [24,25]. In a milieu with low blood oxygen levels, the expression of the hypoxia-inducible factor 1α (HIF-1α) gene and the synthesis of HIF-1α are lowered, the cell response is hindered, and the angiogenesis is affected, thus causing the wound healing process to be slowed down [16]. Also, hypoxia stimulates the inflammatory response and increases the level of free oxygen radicals, leading to an extension of the duration of effective wound healing [23,24].

Compared with normal wound healing, diabetic wound healing is characterized by a chronic inflammatory phase due to uncontrolled inflammation and the persistence of inflammatory cells at the wounded area (Figure 1). In chronic inflammation, a significant decrease of fibroblast proliferation, function and differentiation into myofibroblasts takes place, together with lowering of TGF-β type II receptor expression and COL synthesis, thus also hindering the final phase of skin remodeling [24,26]. In addition, macrophages and their phenotype play an important role in the transition between wound healing stages, in particular in the inflammatory stage. M1 macrophages have been defined as having an active role in the secretion of pro-inflammatory cytokines, such as TNF-α, IL-1, IL-6 and IL-8 during inflammation, while M2 macrophages were involved in anti-inflammatory activities during remodeling [18]. Macrophages' transition from the M1 to M2 phenotype is modulated by activation of p38/MKP-1, microRNA-21 and protein kinase B (AKT) signaling pathways. In cases of this transition being blocked and the failure of M2 macrophages to be made available, the healing process is prevented, the remodeling stage is not reached and the entire wound healing process is endangered [18,20]. Besides, the inflammatory phase in diabetic wounds with high glucose levels results in increased expression of MMPs, liable for excessive ECM degradation and regulation of keratinocyte migration. The increase of GFs and GF receptors' levels along with that of integrins and integrin receptors results in the slowing down of diabetic wound healing [24].

Diabetic wounds are, also, defined by a slow formation of new blood vessels at the wound situs that disturbs the blood supply. The mechanisms of this pathological process are correlated to bacterial contamination, imbalanced angiogenic factors, e.g., TGF-α, TGF-β, FGF-2, vascular endothelial growth factor (VEGF), EGF, HIF-1α, and vascular inhibitory factors, such as platelet-activating protein and vasopressors. Also, diabetics develop macroangiopathy, which leads to reduced blood flow. Thus, the production of NO decreases because of reduced levels of nitric oxide synthase (NOS), abnormal capillary regulation and impaired diastolic function of the vasculature, leading to microvascular malfunction [24]. Diabetic patients with sustained hyperglycemia states suffer alterations in the endothelial cells' metabolism and function, which results in various defective micro- and macro-circulatory issues triggering a hindered angiogenesis process. A damaged vasculature obstructs oxygen delivery to the tissue, creating a hypoxic microenvironment and causing persistent inflammation at the wound situs [27]. Moreover, this microenvironment is able to impair endothelial cell function, thereby disrupting the balance between pro-angiogenic and anti-angiogenic factors, leading to reduced angiogenesis and creating a loop [11].

The healing time of diabetic wounds is longer due to an excessive and abnormal accumulation of several compounds like advanced glycation end products (AGEs) and reactive oxygen species (ROS), pro-/anti-inflammatory cytokines balance, interleukins, leukotrienes and complement factors at the wound site, but also to lack of ECM proteins [28]. The accumulation of AGEs can alter keratinocyte and fibroblast function. The formation of free oxygen radicals can produce harmful effects on blood supply and the structure and metabolism of peripheral nerves. Thus, in diabetic patients, hand–foot neuropathy is the most common form and one of the main reasons for impaired healing of foot wounds [16]. In diabetic foot ulcers and diabetic nephropathy, activators of nuclear factor erythroid 2-related factor 2 (Nrf2) reduce oxidative stress and improve the process of wound healing. The transcription factors Nrf2 and nuclear factor kappa B (NF-kB) promote wound healing through anti-inflammatory and antioxidant effects or via the immune system. They have a key and reciprocal role [29]. Nrf2, the primary regulator of intracellular redox homeostasis, is a redox-sensitive transcription factor, which regulates the expression of cytoprotective genes, leading to keratinocyte apoptosis, repair-related inflammation and protection against excessive accumulation of ROS. In turn, NF-kB activates the immune response, cell proliferation and migration, and regulates the expression of MMPs, secretion and the stability of cytokines and growth factors for wound healing [23,26].

3. Effect of Small Molecules and Natural Polymers in Diabetic Wound Healing

Plant-derived bioactive compounds or small molecules are mainly classified as polyphenols, triterpenes, carotenoids and tocopherols, alkaloids, glucosinolates and capsaicinoids. They manifest antioxidant, anti-inflammatory, antimicrobial, anti-allergic, antiproliferative and antimutagenic properties of interest for the human health. The structure–activity correlation allows their various applications as main ingredients of products for the food and pharmaceutical industry [30,31].

3.1. *Small Molecules*

Polyphenols, a large family of plant secondary metabolites, are characterized by the presence of a huge number of hydroxyl groups attached to aromatic rings and are mainly divided into flavonoid and non-flavonoid compounds [32]. It was demonstrated that phenolic compounds could provide useful properties in the wound healing process. Polyphenols modulate the inflammatory agents, such as cytokines, increase angiogenesis and vascular genesis, can promote epithelization and ameliorate wound contraction rates. Polyphenols have a role in the hemostasis phase due to some polyphenol moieties (catechol or pyrogallol groups), which can interact with serum proteins in the blood, forming complexes that can stop bleeding by creating a physical barrier [33]. These hemostatic effects can also be due to the antibacterial properties of polyphenols.

3.1.1. Flavonoids

Flavonoids, important bioactive phenolic compounds abundant in plants, have chemical structures based on the fifteen-carbon skeleton, a common basic structure of diphenylpropanes (C6-C3-C6), with two benzene rings (A and B) linked by a heterocyclic pyran ring (C) (Figure 2). They are subdivided into 12 major classes based on the level of oxidation and pattern substitution of the pyran ring, of which the more significant are flavones (apigenin, luteolin), flavonols (quercetin, kaempferol), flavanones (hesperidin, naringenin), anthocyanidins, isoflavones (genistein), flavan-3-ols (flavanol) (catechin) [30,34,35].

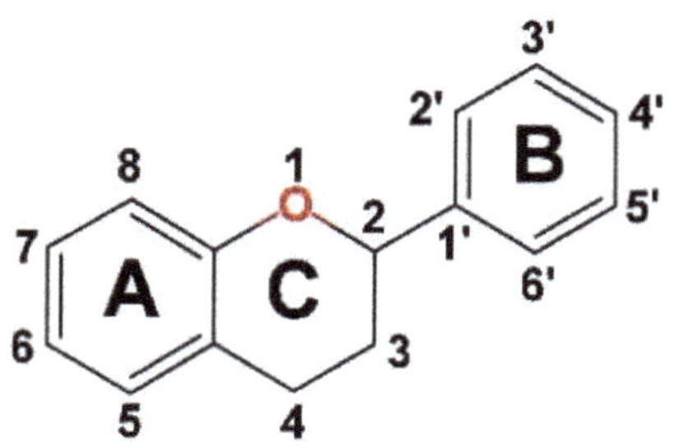

Figure 2. The basic chemical structure of flavonoids with two benzene rings (labeled A and B) and a heterocyclic pyran ring (labeled C).

Tannins, an important group of flavonoid compounds, comprise four main categories according to their chemical structure: condensed tannins (proanthocyanidins), hydrolysable tannins (gallotannins and ellagitannins), phlorotannins and complex tannins (catechin glycosidic linked to gallotannins and ellagitannins). Condensed tannins are oligomers and polymers of flavan-3-ol monomers and their most common basic units are represented by (+)-catechin, (−)-epicatechin, (+)-gallocatechin, (−)-epigallocatechin and (−)-epigallocatechin gallate [30,36].

Flavonoids possess an extensive spectrum of pharmacological activity, including antioxidative, anti-inflammatory, antimicrobial, antidiabetic, antimutagenic, anticarcinogenic properties. They have also demonstrated photoprotective, chemoprotective and wound healing properties, which led to several studies related to potential applications in skin diseases, such as aging and skin cancer. Much research has shown that flavonoids, such as curcumin, hesperidin, luteolin, quercetin, kaempferol, apigenin, rutin, naringin, morin, mangiferin have potential for the management of diabetic wounds through stimulation of ECM synthesis and angiogenesis, and modulation of growth factors (Table 1). Thus, curcumin accelerated the healing process by reducing inflammation and had an apoptotic effect in the early stages of wound healing [28]. Luteolin significantly reduced the expression of inflammatory factors, such as IL-1, IL-6 and TNF-α and reduced the levels of prostaglandin E2, interferon and leukotriene B4 [28]. Also, luteolin down-regulated the expression of NF-kB, while inducing an increase in antioxidative enzyme activity, such as superoxide dismutase 1 (SOD1) and glutathione peroxidase (GSH-Px). An in vivo treatment with kaempferol of excisional wounds in diabetic rats showed remarkable results after 14 days, with a higher healing rate of 92.12% and re-epithelization scores, compared to the control [37]. Flavonoids also regulated MMP-2, MMP-8, MMP-9, MMP-13 through the Ras/Raf/ MEK/ERK, PI3K/Akt signaling pathways in diabetic wounds and activated Nrf2 expression, stimulating diabetic wound healing through cellular stress decrease, acceleration of cell proliferation and neovascularization [28].

3.1.2. Non-Flavonoid Compounds

The group of non-flavonoid compounds includes phenolic acids and stilbenes. Phenolic acids are characterized by a phenol ring containing at least one carboxylic acid group and derived from two main phenolic compounds, benzoic acid and cinnamic acid, respectively [30]. Hydroxybenzoic acids containing seven carbon atoms (C6-C1) are the simplest phenolic acids found in nature and occur in their free or conjugated forms. Some representative examples: caffeic, ferulic, p-coumaric and sinapic acids. Hydroxycinnamic acids are characterized by a C6-C3 chain, rarely occur in their free form in plants and are represented by gallic, syringic and vanillic acids [30]. In vivo studies showed several roles of phenolic acids in speeding up the wound healing by blocking leukocytes infiltration, ROS scavenging, improving cell attachment and promoting tissue regeneration even in diabetic wounds [38]. Thus, it was found that gallic acid embedded in glucomannan composite hydrogel diminished inflammatory markers and favored COL aggregation and re-epithelization of diabetic wounds [39], while sinapic acid-loaded hydrogels showed a similar effect on full-thickness excisional wounds in streptozotocin (STZ)-induced diabetic rats to the ointment used as positive control [40].

Stilbenes, derived from the phenylpropanoid pathway, make up a polyphenol group characterized by two phenyl rings connected through a two-carbon methylene bridge (C6-C2-C6). The most studied compound is resveratrol, recognized for its potent antioxidant and anti-inflammatory activities [41]. Both in vitro and in vivo studies in human umbilical vein endothelial cell culture and diabetic rat models showed that a topical administration of pterostilbene exhibited a stronger effect than resveratrol in the suppression of HIF-1α and the normalizing of oxidative stress and, thus, accelerated diabetic wound healing [42].

3.1.3. Other Plant Bioactive Compounds

Triterpenes and triterpenoids (a functionalized form of triterpenes), an important group of terpenoids, contain at least 18 subclasses of compounds, of which the most known are saponins, squalene derivatives and phytosterols [28,43]. Carotenoids and β-carotene as a predominant compound are direct precursors of vitamin A, while tocopherols consisting of α- and γ-tocopherols are the precursors of vitamin E. A broad compound group is represented by alkaloids with caffeine as a major component, having a heterocyclic ring structure with at least one nitrogen atom and exhibiting alkali-like properties. Glucosinolates are negatively charged compounds consisting of an S-linked thioglucose unit, an O-sulfated thiohydroximate group and a highly variable side chain, responsible for different biological properties, such as chemopreventive and anti-inflammatory activity [30].

The medicinal plant-derived naphthoquinone plumbagin (5-hydroxy-2-methyl-naphthalene-1,4-dione) isolated from the roots of *Plumbago zeylanica* is a promising compound for diabetic wound healing [33]. Shao et al. (2019) showed that plumbagin promoted the wound closure and contraction of diabetic rat wounds by epithelialization acceleration, COL type I deposition, promotion of insulin secretion and improvement of the antioxidant enzymes' activity [44]. At the gene level, plumbagin up-regulated the expression of Nrf-2, TGF-β and α-SMA and down-regulated the expression of Keap1 in diabetic rats [44]. Also, plumbagin increased EGF, VEGF and FGF, decreased MMP-2, COX-2, iNOS, CD8, CD163 and NF-kB p65 and suppressed IL-6 and IL-1β [33].

Nerolidol or peruviol (3,7,11-trimethyl-1,6,10-dodecatrien-3-ol) is a naturally occurring sesquiterpene alcohol found in the essential oils of different types of aromatic medicinal plants. Several biological and pharmacological effects have been reported, such as anti-inflammatory, antimicrobial, antibiofilm and repellent properties [45].

Ginsenosides isolated from *Panax ginseng* have a four-ring structure with a steroidal body and sugar moieties. It has been demonstrated that ginsenoside Rb1 displayed a promising effect in diabetic chronic wound healing due to an in vitro significant increase in cell proliferation, COL synthesis and up-regulation of VEGF, TGF-β1 and TIMP-1 in fibroblast culture obtained from patients with DFUs [46]. In addition, in vivo treatment with ginsenoside Rb1 confirmed the role of secreted VEGF in the formation of thick granulation tissue with more new blood vessels, while TGF-β1 and TIMP-1 increased COL synthesis [46]. Similarly, notoginsenoside R1 from *Panax notoginseng* facilitated wound healing in diabetic rats by elevating COL synthesis, increasing ECM secretion, accelerating the wound closure rate, up-regulating the expression of CD31 and down-regulating the expression of caspase-3. An administration of notoginsenoside R1 promoted down-regulation of MMP-3, MMP-9, IL-1β and IL-6 and up-regulation of TIMP1 and TGF-β1 [46].

Table 1. Action of polyphenols in diabetic wound healing.

Polyphenols Class	Compound	Activity	Reference
Flavonoids	*Luteolin*	In vivo decrease in blood glucose levels, accelerated skin wounds' re-epithelization in diabetic rats by inhibiting the inflammatory cell infiltration, decreasing IL1-β, IL-6, TNF-α expression, and reducing oxidative stress, down-regulated NF-kB and up-regulated SOD1 and glutathione peroxidase (GSH-Px) expression mediated by p-Nrf2; In vivo intraperitoneal administration treated diabetes-associated wounds by targeting NF-kB/MMP-9 axis and Nrf2-mediated antioxidant system.	[46]
	Quercetin	In vivo modulated fibroblast activity, up-regulated VEGF and TGF-β1 in diabetic scars; In vivo oral application increased COL synthesis, deposition and orientation and decreased inflammatory cytokines (IL1-β and TNF-α) in rat diabetic wounds; In vivo topical administration modulated cytokines and GFs, and inhibited inflammatory reactions in rat diabetic wounds; promoted macrophages' M1-M2 phenotype switch during wound healing in diabetic mice.	[33,35,46]

Table 1. *Cont.*

Polyphenols Class	Compound	Activity	Reference
	Rutin (quercetin-3-O-rutoside	In vivo prevented oxidative stress and inflammatory response, improving wound healing in hyperglycemic rats; In vivo decreased the number of inflammatory cells; stimulated Nrf-2 activity and antioxidant enzymes (SOD1 and GSH-Px) expression; In vivo down-regulated IL-1β, IL-6, TNF-α, NF-kB, MMP-2, MMP-9, TGF-β1, VEGF expression levels; In vivo intraperitoneal administration elevated neurogenic-related protein expression.	[46]
	Myricetin	In vitro prevented cellular oxidative stress by regulating antioxidant enzymes; In vitro enhanced pro-COL I and III levels, inhibited MMP-1, MMP-2 and MMP-9 synthesis, increased TIMP1/MMPs ratio by enhancing TIMP-1 mRNA expression, suppressed catalase (CAT) and SOD1 in diabetic fibroblasts.	[46,47]
	Icariin	In vivo anti-inflammatory and pro-angiogenic activities in diabetic rats by down-regulating NF-kB, TNF-α, MMP-2, MMP-9 levels and increasing IL-10 and CD31 levels; In vivo topical administration stimulated normal ECM formation in the healing tissue by increasing the relative COL deposition.	[46,48]
	Vicenin-2 (VCN-2)	In vivo inhibited oxidative and inflammatory stress in a dose-dependent manner, stimulating wound healing in STZ-induced DM rats; In vitro increased cell proliferation, reduced inflammatory cells, down-regulated proinflammatory cytokines (IL-1β, IL-6, TNF-α), mediators (iNOS, COX2) and nitric oxide (NO) expression via NF-kB pathway; improved epithelialization and remodeling; stimulated fibroblast proliferation and migration, neoangiogenesis and wound contraction, down-regulated MMP-9, VEGF and TGF-1β levels via HIF-1α pathway; In vivo topical administration reduced food and fluid intakes, decreased blood glucose level and increased insulin level, body weight and percentage of wound closure.	[46,49]
	Mangiferin	In vivo topical administration inhibited oxidative stress, decreased the wound area and increased skin thickness; enhanced EGF, FGF, Nrf-2, TGF-β, VEGF and PI3K expression and decreased MMP-2, TNF-α and NF-kB p65 expression in diabetic wound; reduced the inflammatory phase in hyperglycemic conditions.	[46,50]
	Curcumin	In vivo topical administration accelerated re-epithelialization rate, accelerated wound closure through down-regulation of TNF-α, IL-1β and MMP-9 levels, up-regulation of IL-10 level and elevation of SOD, CAT and GSH-Px activity, improved thick granulation tissue formation, COL synthesis, deposition and orientation in rat models of diabetic ulcer and wound.	[46,51]
	Kaempferol	In vivo topical agent with 92.12% wound healing rate in diabetic rats.	[37]
	Epigallocatechin-3 gallate	In vivo enhanced wound healing through acceleration of re-epithelization and angiogenesis, reduced cytokines level and inhibited macrophage accumulation, inflammation response and Notch signaling in diabetic mouse wounds.	[52]
	Hesperidin	In vivo accelerated angiogenesis and vasculogenesis via up-regulation of VEGF-C, TGF-β, Ang-1/Tie-2 and Smad-2/3 mRNA expression, increased COL deposition and suppressed IL-6 and TNF-α inflammatory mediators, enhancing wound healing of chronic DFUs in STZ-induced Sprague Dawley rats.	[53]

Table 1. *Cont.*

Polyphenols Class	Compound	Activity	Reference
	Genistein	In vivo subcutaneous administration modulated oxidative stress, improved angiogenesis by FoxO1 and iNOS suppression; oral administration supported wound healing, lowered oxidative stress and inflammation in diabetic mice wounds.	[35]
	Puerarin	In vitro down-regulated inflammatory cytokines expression by inhibition of MAPK and NF-κB inflammatory signaling pathways in high-glucose cell culture; improved polarization to M2 macrophages at cellular level.	[54]
Stilbenes	*Resveratrol*	In vitro exerted antioxidant effect and inhibited TNF-α and NF-kB.	[55]
Phenolic acids	*Ferulic acid*	In vivo inhibited lipid peroxidation, increased CAT, SOD and glutathione expression, elevated NO and serum zinc and copper, improving the healing process in diabetic ulcers.	[56]
	Syringic acid	In vivo topical administration improved wound healing by promoting cell migration and proliferation in STZ-induced diabetic rats; significantly reduced MMP-2, MMP-8 and MMP-9 levels, up-regulated TIMP-1 and TIMP-2 levels; elevated COL I, CD31, CD68, α-SMA, TGF-β1 and VEGF content in diabetic wounds.	[46,56,57]
	Chlorogenic acid	In vivo stimulated COL production, reduced the level of oxidative and inflammation markers (MDA/NO), increased GSH level, maintained SOD/CAT level, accelerating wound healing in STZ-induced diabetic rats.	[58]
	Gallic acid	In vivo ROS scavenger, exerted antioxidant activity, promoting wound healing in a diabetic mouse model.	[59]
Tannins	*Tannic acid*	In vivo antioxidant, hemostatic, anti-inflammatory, antimicrobial activity useful in skin wounds and ulcers.	[58]
Terpenes	*Kirenol*	In vivo exerted anti-inflammatory, antioxidant and wound healing activity by regulation of MMP-2 and MMP-9 expression, inhibition of NF-kB, COX-2 and iNOS expression and MDA content, elevated antioxidant enzymes activity, favoring angiogenesis and formation of granulation tissue in STZ-induced diabetic rats.	[46]
Alkaloids	*Berberine*	In vivo topical application accelerated novel ECM synthesis and wound healing process through modulation of TrxR1 and its downstream JNK signaling, expression of MMP9 and TIMP1, up-regulation of TGF-β1, resulting in promotion of fibroblast proliferation and inhibition of oxidative stress and apoptosis in HFD- and STZ-induced diabetic rats.	[46,60]

3.2. Natural Polymers

Polysaccharides, also named glycans, comprise monosaccharides and their derivatives linked by glycosidic bonds in long polymeric chains [30]. They are categorized into monopolysaccharides (one type of monosaccharide molecule) and heteropolysaccharides (two or more monosaccharide molecules). Also, polysaccharides differ in their molecular chain length, the degree of chain branching and their positive or negative charge [17]. Polysaccharides elicit promising medical applications due to their extensive obtainability, biodegradability and innocuousness. Moreover, proteins derived from animal ECM are natural polymers of great interest for the development of biocompatible and biodegradable hydrogels.

3.2.1. Cellulose

Cellulose, the most abundant biological polymer, is a polysaccharide made up of D-glucose units connected by β-(1 → 4)-glycosidic linkages. It has a high absorbent capacity, which makes it very effective for intermediate to severe exudate wounds.

Carboxymethyl cellulose (CMC) is a polysaccharide synthesized in alkali conditions for hydroxyl groups' activation by etherification with monochloroacetic acid, acting as a hydrophilic, biocompatible and biodegradable material that stimulates fibroblast proliferation and migration, being recommended in diabetic wound healing.

Bacterial cellulose (BC), a type of nano-polymer produced both by Gram-negative bacteria (*Komagataeibacter xylinus*, *Komagataeibacter intermedius*, *Agrobacterium*, *Azotobacter*, *Achromobacter* spp.) and Gram-positive bacteria (*Komagataeibacter hansenii*) comprises glucose as monomers bonded by β-1,4-glucopyranosyl links. BC has excellent processing properties and is suitable for use as a wound dressing due to its water content (close to 99%), high tensile strength, good thermal stability and good elasticity [61]. In addition, BC maintains a moist wound environment and controls wound exudate and has good biocompatibility and biodegradability. Also, BC possesses unique physical and chemical properties, such as an ultrafine nanofiber network and high crystallinity and water retention capacity [62]. It was reported that the BC-based dressings require specific treatments (e.g., fusidic acid) to ensure their efficacy in wound healing or antimicrobial capacities, but this treatment may cause allergic reactions. The *K. xylinus* strain is commonly used in the production of BC-based dressings for diabetic wound healing [63]. The properties of BC produced by *K. intermedius* showed dramatic differences from that produced by *K. xylinus* [61]. There is a lack of knowledge regarding BC produced from *K. intermedius* for treating diabetic wounds, especially without modification or addition of antimicrobial substances [61].

3.2.2. Alginate

Structurally, alginate, a natural anionic biopolymer, is a linear hydrophilic polysaccharide consisting of D-mannuronic acid (M) and L-guluronic acid (G) units. The units are present in the form of blocks with a variable M/G ratio [63], either as continuous G chain segments (GGGG), M chain segments (MMMM), or alternating MGMG chain segments [1,64]. It is considered that only the G unit is involved in multivalent ion-induced cross-linking of alginate, and a high G content improves the mechanical strength of alginate-based hydrogels [1]. Alginate has valuable biological properties, such as stopping bleeding, moisturizing and allowing gas exchange [14].

3.2.3. Chitosan (CS)

CS, a natural linear copolymer and the only alkaline polysaccharide in nature, is a product of chitin N-deacetylation, which consists of glucosamine and N-acetylglucosamine units. The great versatility of CS has been reported due to its derivatives resulted through the functionalization of its hydroxyl and amino groups by carboxymethyl, carboxyethyl, carboxybutyl, succinyl, acyl, or 5-methyl pyrrolidone moieties [65]. CS has many important biomedical characteristics, such as excellent biocompatibility, biodegradation (could be completely degraded within 14 days after being implanted in animals), non-toxicity, adhesion, inertness, non-antigenicity, antioxidative and good hemostatic properties and wound healing characteristics [1,65]. The cationic nature of chitosan ensures its antimicrobial activity against a wide range of microorganisms (*Escherichia coli*, *Staphylococcus aureus*, *Salmonella typhimurium* and *Listeria monocytogenes*) due to which CS is widely used as wound dressings for diabetics [1,66]. CS mimics the normal ECM environment, induces COL and elastin secretion by promoting fibroblast proliferation and stimulates the growth of granulation tissue and tissue repair [1].

3.2.4. Hyaluronic Acid (HA)

The natural polysaccharide HA is the simplest non-sulfated glycosaminoglycan (GAG) (a class of negatively charged polysaccharides) and the main component of the ECM. HA is biodegradable, non-immunogenic, biocompatible with minimal toxicity and has a high capacity for water absorption and water retention. Also, HA actively participates in all stages of wound healing: inflammation, re-epithelization, granulation tissue formation, proliferation and remodeling [12,67]. HA's action on intracellular signaling pathways mediated by cellular receptors, such as receptor differentiation cluster 44 (CD44), HA-mediated motor receptor, HA endocytosis receptor and lymphatic vessel endothelial cell receptor 1, has been reported, and it decreased the inflammatory response, favoring the tissue repair. This was supported by observations made upon stimulation of anti-inflammatory cytokines IL-2, IL-10 and TGF-β through Foxp3 expression in regulatory T cells [64]. HA was also found to promote angiogenesis and endothelial cell migration by modulation of the MAPK/ERK signaling pathway after binding to CD44 [67]. Recently, HA's involvement in macrophage activation and polarization into M1 pro-inflammatory or M2 anti-inflammatory phenotypes was demonstrated [1]. This observation allowed for a novel therapeutic strategy using HA to modulate local immunity by favoring polarization to the M2 phenotype in diabetic wound healing.

3.2.5. Other Polysaccharides

Hsian-tsao polysaccharides, the major functional component in *Mesona procumbens* Hemsl., have strong antioxidant and anti-inflammatory effects. It was reported that Hsian-tsao polysaccharides displayed a promising effect on wound healing in diabetic wounds by decreasing crust and improving the formation of thick granulation tissue with more new blood vessels, and re-epithelialization [46]. In addition, they up-regulated IL-8, TIMP-1, VEGF, MIP-2 and MCP-1 and down-regulated MMP-2 and MMP-9, and suppressed MG-triggered protein glycation and ROS accumulation [68]. Also, they enhanced the methylglyoxal-impeded phagocytosis of *Staphylococcus aureus* and *Pseudomonas aeruginosa* driven by macrophages, which improved the impaired wound healing. It has been shown that polysaccharides were more effective than crude extracts in regulating the factors associated with diabetic wound repair [46].

3.2.6. Collagen

COL is the main protein found in the connective tissue and the most important polymeric component of ECM. It has many functions, including providing tissue strength and structural stability, possesses flexibility and thermal and enzymatic stability, directs cell adhesion and migration, promotes protein secretion and regulates cellular growth and metabolism during development and repair [69]. Also, COL is biocompatible, biodegradable and hemostatic, all important features for wound healing. Among the 28 types of identified and described COL, in the skin, there are mainly COL types I (70%), III (15%) and V. COL plays a significant role in each of the four stages of wound healing. Due to injury, COL exposure activates the clotting cascade during the hemostatic and inflammatory stages of wound healing, resulting in a fibrin clot that stops the initial bleeding [69]. Also, COL activates microRNA signaling pathways to promote the development of a macrophage phenotype that is anti-inflammatory and pro-angiogenic. During wound healing, COL type I replaces COL type III, which is first generated in considerable amounts, creating the scar.

3.2.7. Gelatin

Gelatin (Gel) is the product of partial denaturation and hydrolysis of COL and is widely used in wound healing applications due to its hemostatic properties, good biodegradability and histocompatibility. Gel, generated by the thermal denaturation of COL, has lower immunogenicity than COL and is less likely to cause an immune response in the body [1]. Due to the presence of the arginine-glycine-aspartate (RGD) cell binding sequence and MMP reactive peptide sequence retained in the Gel fabrication process, it promotes cell adhesion mediated by integrins [70].

4. Recent Development of Hydrogels Encapsulating Small Molecules for Healing of Diabetic Wounds

Natural polymer hydrogels are polymeric networks loaded with natural bioactive compounds that can act as good wound dressings due to their versatile physicochemical properties, represented mainly by a high water absorption capacity. Moreover, they provide a temporary structure facilitating cell infiltration and proliferation, and due to their biodegradability, they are gradually replaced by new ECM [71]. The stretching properties confer to hydrogels good adherence to tissue wounds of different shapes. Natural polymer hydrogels loaded with various bioactive compounds provide not only good biocompatibility, but also antibacterial and antioxidant properties and modulation of inflammation at the wound situs to a higher extent compared to traditional dressings, affording promising future applications in diabetic wound healing [17,72] (Figure 3).

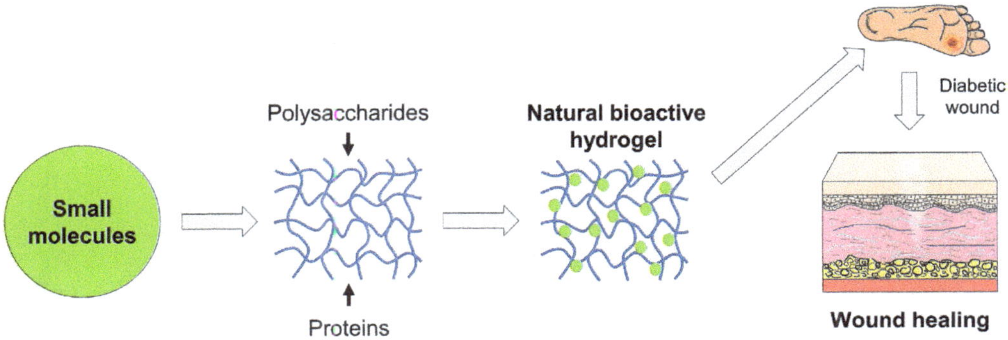

Figure 3. Schematic representation of fabrication of natural bioactive hydrogels, consisting of natural polymeric networks of polysaccharides and proteins loaded with small molecules for their use in diabetic wound healing.

Functional hydrogels have been developed and evaluated to overcome the wound infections of diabetic ulcers:

- Antibacterial hydrogels loaded with antibiotics or drug-like components to control bacterial growth;
- Antifouling hydrogels by creating superhydrophilic or more rarely superhydrophobic surfaces, which can reduce bacterial attachment and biofilm formation [22];
- Antioxidant hydrogels loaded with antioxidant scavengers or drugs, such as curcumin and gallic and tannic acid, to improve the inherent antioxidant properties of the constituent macromolecules against ROS formation [73].

4.1. Polysaccharide-Based Hydrogels

Various outstanding features of polysaccharides can be used to fabricate biomimetic and multifunctional hydrogels as efficient wound dressings. These hydrogels mimic the natural ECM, are biodegradable, biocompatible and non-toxic, stimulate cell proliferation and have renewability and water-retention abilities. Polysaccharide-based hydrogels have exceptional physicochemical properties and unique therapeutic interventions owing to distinctive architectures and an abundance of functional groups. Multifunctional polysaccharide-based hydrogels are promising for infected and diabetic wound healing. Hydrogels designed using polysaccharides can effectively protect wounds from bacterial attack [17].

4.1.1. Cellulose-Based Hydrogels

Due to their low mechanical strength, cellulose-based hydrogels are usually blended with other polymeric materials, and chemically modified for enhanced wound healing action. Even if cellulose-based hydrogels have revolutionized the design of antibacterial dressings, intensive research is still needed to produce more advanced hydrogels composed of cellulose that can combat the rise of antibiotic-resistant strains and aid in the development of biofilms in wound regeneration [17]. Thus, CMC plant fibers were used to develop hydrofiber dressings that produce a soft gel whenever they contact wound exudates [65]. A combination of CMC and carbomer 940 containing a polysaccharidic extract of *Periplaneta americana* was very efficient in M2 macrophage polarization, COL deposition, angiogenesis and accelerating wound healing in diabetic rat wounds [74]. CMC/sericin-based hydrogels with intrinsic antioxidant, anti-inflammatory and antibacterial properties promoted re-epithelization of diabetic rat wounds [75]. Cellulose-derived cotton gauze dressings have been traditionally used in clinical settings and have proved to have the capability to accelerate the wound healing process by a controlled release of several growth factors at the injury site [65]. Thus, the release of basic fibroblast growth factor (FGF), EGF and phosphodiesterase growth factor promoted the migration and proliferation of fibroblasts and inhibited bacteria proliferation in the wound. A hydroxymethyl cellulose-based hydrogel also containing uniformly dispersed MnO_2 nanosheets was proved to act as an ROS scavenger, converting them to O_2, needed in different stages of wound healing to lower in situ inflammation, while an addition of silk fibroin inhibited MMPs gene expression for a synergistic promotion of matrix remodeling [76]. To reinforce the structural and mechanical properties of hydrogels mimicking the microenvironment of natural ECM, a BC matrix was impregnated with carbon nanotubes and showed higher antibacterial activity compared to the ointment used as positive control and faster healing of diabetic wounds in an STZ-induced model in BALBc mice, compared to the untreated control [77].

4.1.2. Alginate-Based Hydrogels

Alginate hydrogels are recommended as dressings for diabetic wound healing due to their unique advantages, compared to other hydrogels, i.e., they are highly hydrophilic and very useful for excessive exuding wounds, and they can change their fine porosity in contact with exudate, increasing their volume, and do not present tissue adhesion, allowing a safe repair process [1].

An injectable hydrogel with therapeutic efficiency was made from sodium alginate and chondroitin sulfate using a solvent casting method and enriched in curcumin to exert ROS scavenging ability, biocompatibility and antibacterial activity in an excisional wound model developed in diabetic rats [78].

In a clinical study, sodium alginate hydrogel enriched with fatty acids and vitamins A and E proved its efficacy in foot wound healing of diabetic patients with a 22.2% reduction of the lesion area, the pressure ulcer scale for healing (PUSH) score dropping from 9.8 to 6.6, or even complete wound healing [79]. The great potential of alginate dressings combined with EGF was found in a randomized controlled trial with 18 patients with refractory wounds as increased proliferation and differentiation of epidermal stem cells, compared to an EGF-alone treatment [80].

Currently, Purilon^R Gel and Nu-Gel^{TM} are the commercialized alginate saline gels for facilitating natural autolytic debridement and faster healing of leg ulcers and non-infected DFU [81].

4.1.3. CS-Based Hydrogels

CS hydrogels have excellent antibacterial properties and good biocompatibility and biodegradability, and several combinations with natural bioactive compounds were designed to demonstrate improved functions during diabetic wound treatment (Table 2). Thus, CS-based hydrogel containing a flavonoidic extract of *Passiflora edulis* Sims leaves showed stimulation of the antioxidant defense system due to the controlled release of bioactive compounds, enhancing wound healing in diabetic rats. However, a film covering the wounded site for 14 days can lead to hypoxia and high oxidative stress with increased lipid peroxidation [82]. CS-based hydrogel impregnated with nerolidol promoted significant wound healing [46]. It has been reported to have a diabetic ulcer healing effect within 14 days of treatment with nerolidol-functionalized gold nanoparticles ointment in albino rats, compared with the untreated control group [83]. An injectable hydrogel was developed as CS-CMC-g-PF127, loaded or not with Cur, presenting viscoelastic behavior, good swelling properties, a controlled release profile, thermal stability and crystalline characteristics [84]. The Cur-loaded CS-CMC-g-PF127 injectable hydrogel exhibited fast wound repair potential by the stimulation of the cell migration and proliferation at the damage site and by providing a sustained drug delivery platform for hydrophobic moieties [84]. CS hydrogels encapsulating quercetin or tannin as bioactive compounds significantly promoted the wound healing process in diabetic wounds [85].

A clinical trial of Velazco et al. (2012) reported the application of completely biodegradable CS-based dressings on patients with diabetic foot complications [86]. After 8 days of treatment, the granulation process began and the size of the lesion decreased after 22 days, while total closure of the ulcer was observed after 45 days.

There is already a market product ChitoHeal based on CS, which proved in clinical trials several advantages in burns, cuts, scratches and DFU treatment due to its biocompatibility, being effective in scar and healing rate reduction [87].

Table 2. Activity of chitosan-based hydrogels in diabetic wound healing.

Hydrogel Type	Experiment Type	Activity	Reference
Quaternized CS/tannic acid hydrogels	STZ-induced diabetic rat model	Good injectability and self-healing, cytocompatibility, hemostatic capability and biodegradability, radical scavenging activity, COL deposition, no scar formation, skin regeneration.	[88]
CS-puerarin hydrogel	STZ-induced diabetic rat model	Promoted diabetic wound healing and accelerated angiogenesis, inhibition of the miR-29 mediated inflammation response.	[1,89]
Sulfated chitosan (SCS)-doped COL type I (Col I/SCS) hydrogel	STZ-induced diabetic rat model	Reduced inflammation through minimizing macrophages' polarization into M1 phenotype, decreased production of pro-inflammatory IL-6 and increased production of anti-inflammatory cytokines IL-4, TGF-β1 in chronic diabetic wounds; stimulated COL synthesis, angiogenesis and cell migration for wound closure in diabetic wounds.	[1,90]
Apigenin loaded gellan gum-CS (GGCH) hydrogel	STZ-induced diabetic rat model	Increased level of SOD, GSH, CAT, protein content in granuloma tissue; biocompatibility, biodegradability, moist nature, antioxidant effectiveness; increased hydroxyproline level and collagen turnover; decreased epithelialization period; higher wound healing in diabetics.	[91]
CS/ HA-based hydrogel with MOF-loaded lipoic acid	In vitro and in vivo analysis in diabetic Sprague Dawley rats	Antibacterial activity and antioxidant performance, promoted cell proliferation and migration, wound healing process, better granulation tissue formation and more COL deposition.	[92]
Bio-multifunctional benzaldehyde-terminated 4-arm PEG (4-arm-PEG-CHO)/carboxymethyl CS (CMCS)/basic fibroblast growth factor (bFGF) hydrogels (BP/CS-bFGF)	STZ-induced diabetic rat model	Strong wet-tissue adhesion, self-mending fast hemostasis capacity, excellent biocompatibility, antibacterial property, increased production of Ki67, promoted the generation of epithelialization and COL, induced formation of hair follicles, enhanced neovascularization by up-regulating the production of CD31 and CD34.	[93]
CS hydrogels functionalized with either unfractionated heparin or bemiparin (a low molecular weight heparin, LMWH)	STZ-induced diabetic rat model	Accelerated inflammation, improved the epithelization process, formation of high-quality cicatricial tissue, improved diabetes-associated impaired wound healing.	[94]

4.1.4. HA-Based Hydrogels

HA-based hydrogels are promising tools in wound healing applications due to their composition, structure and size. Due to HA functional groups, including carboxylic acids (–COOH), primary and secondary hydroxyl groups (–OH) and *N*-acetyl groups, useful cross-linking can increase the mechanical strength of the hydrogel. Also, the combination of HA hydrogel with small molecules can improve its physical and chemical characteristics, like biocompatibility, biodegradability, mass transferability and antibacterial activity [95] (Table 3).

Table 3. Activity of hyaluronic acid-based hydrogels in diabetic wound healing.

Hydrogel Type	Experiment Type	Activity	Reference
Paeniflorin-loaded HA-based hydrogel	Ex vivo and in vivo experimental approaches in diabetic mice model	Stimulated transition of macrophages from M1 pro-inflammatory phenotype to M2 anti-inflammatory/pro-healing phenotype, lowered inflammation and promoted COL synthesis, new blood vessel formation, re-epithelialization of cutaneous wounds.	[96]
Hyaluronan/COL-based hydrogels containing high-sulfated hyaluronan	In vitro and in vivo studies in diabetic db/db mice	Reduced inflammation, augmented pro-regenerative macrophage activation, increased vascularization, accelerated new tissue formation and wound closure.	[97]
Nanotechnologically-modified curcumin and EGF encapsulated into HA and CS-based hydrogel	In vitro and in vivo studies in STZ-induced diabetic mice	High antioxidant, anti-inflammatory and migration-promoting effects, improved wound healing by granulation tissue formation, re-epithelialization and skin regeneration.	[98]
Glucose-responsive HA derivate (HAMA-PBA)/catechin (HMPC) hydrogel	In vitro and in vivo studies in diabetic wound model	High antioxidant capability, increased expression of VEGF and CD31, stimulated angiogenesis, decreased inflammatory responses by low IL-6 level and high IL-10 level, fast wound repair in three weeks.	[99]

4.2. Protein Hydrogels

4.2.1. Collagen-Based Hydrogels

COL-based hydrogels are characterized by a fibrillar network forming a porous structure, biomimetic to initial ECM, acting to stimulate cell migration and colonization and new ECM constituent synthesis for deep and surface wound healing. Despite these advantages, COL-based hydrogels do not present antibacterial and antioxidant activity for more efficient skin wound repair [69]. However, Apligraf® (Organogenesis Inc., Canton, MA, USA) and Dermagraft® (Organogenesis Inc., Canton, MA, USA) are two commercialized COL-based hydrogels approved by the Food and Drug Administration (FDA). The use of Apligraf in a randomized controlled trial on neuropathic DFU patients promoted cell proliferation and 51.5% of cases reached complete wound closure in 12 weeks [100]. There is currently research being conducted on novel compositions of COL-based hydrogels loaded with bioactive compounds that can improve the biological properties required for skin wound treatment, in particular for diabetic patients with chronic wounds [72,101]. Clinical studies indicated that pre-vascularized COL/fibrin hydrogels increased hypodermis thickness and accelerated the wound healing process in DFU patients [102]. Ulcers presented higher healing rates compared to standard treatment with saline gauze and healed completely after 12 weeks of treatment with COL/CS hydrogels in neuropathic DFU patients [102].

4.2.2. Gelatin-Based Hydrogels

Narisepalli et al. (2023) developed a Gel-based biodegradable hydrogel embedding pentacyclic triterpene asiaticoside nanoparticles for diabetic wound healing. In vitro cell culture studies demonstrated significant proliferation and migration of fibroblasts, while application to the wounds of diabetic rats improved wound healing in terms of improving COL biosynthesis, up-regulating the COL type I protein level and enhancing α-SMA gene expression, compared to control untreated groups [103]. Recently, Gel-based thermoresponsive injectable hydrogels with antioxidative, antiinfection, antibacterial, biocompatibility and tissue regeneration characteristics have become of great interest as potential candidates for effective regeneration of diabetic wounds. Liu et al. (2018) developed a thermosensitive hydrogel containing Cur nanoparticles loaded in Gel microspheres, a system for efficient, safe and fast healing of standardized skin wounds created in STZ-induced diabetic mice [104]. Using an emerging technology of 3D bioprinting, Xia et al. (2022) designed GelMA hydrogel loaded with Cur, an antioxidant and anti-apoptosis compound, to regulate the microenvironment of diabetic wounds [105]. This hydrogel encapsulating adipose-derived stem cells (ADSCs) could repair full-thickness skin defects of diabetic nude mice by significantly down-regulating AGEs and ROS production, reducing ADSC apoptosis, promoting angiogenesis and wound re-epithelialization, and targeting the AGEs/AGER/p65 pathway.

4.2.3. Other Protein Hydrogels

Recently, Bai et al. (2023) developed a sericin-based hydrogel containing injectable platelet-rich fibrin (i-PRF), a biocompatible material with anti-inflammatory properties. i-PRF is rich in growth factors, platelets, leukocytes and fibrin, which play important roles in different wound healing stages, being useful in diabetic wound repair [103]. The controlled release of i-PRF-derived bioactive growth factors from the hydrogel reduced inflammation levels, increased COL synthesis, promoted angiogenesis and led to wound healing in diabetic nude mice [106].

5. Conclusions and Future Remarks

In conclusion, there are remarkable effects of the hydrogels enriched with natural bioactive compounds for improving diabetic-related chronic wound healing. They could provide novel therapeutic strategies to prevent limb amputation and increase the quality of life in diabetic patients. In vitro and in vivo studies in rat models of diabetic wounds demonstrated the positive effect of polyphenols as natural bioactive compounds in all stages of healing, alone, complementary or even synergistic to the functional properties of natural polysaccharidic or protein hydrogels. Limitations of clotting, oxidative stress and inflammation, cell proliferation and tissue remodeling found in natural polymeric hydrogels were improved after small molecules encapsulation, as found by specific markers analysis at the gene expression and protein secretion levels. Future studies will contribute to further improving the composition and structure of biocompatible functional hydrogels and help us face the challenges of revealing the complex mechanisms of diabetic wound healing.

Author Contributions: Conceptualization, E.I.O. and O.C.; resources, E.I.O.; writing—original draft preparation, E.I.O. and A.I.; writing—review and editing, O.C. All authors have read and agreed to the published version of the manuscript.

Funding: This work was funded by Program Nucleu, project no. 23020101/2023 and Program 1, Development of the National R&D System, Subprogram 1.2, Institutional Performance, Projects for Excellence Financing in RDI, contract no. 2PFE/2021.

Institutional Review Board Statement: Not applicable.

Informed Consent Statement: Not applicable.

Data Availability Statement: Not applicable.

Conflicts of Interest: The authors declare no conflict of interest.

References

1. Xu, Y.; Hu, Q.; Wei, Z.; Ou, Y.; Cao, Y.; Zhou, H.; Wang, M.; Yu, K.; Liang, B. Advanced polymer hydrogels that promote diabetic ulcer healing: Mechanisms, classifications, and medical applications. *Biomater. Res.* **2023**, *2023*, 27. [CrossRef]

2. Sun, H.; Saeedi, P.; Karuranga, S.; Pinkepank, M.; Ogurtsova, K.; Duncan, B.B.; Stein, C.; Basit, A.; Chan, J.C.; Mbanya, J.C.; et al. IDF Diabetes Atlas: Global, regional and country-level diabetes prevalence estimates for 2021 and projections for 2045. *Diabetes Res. Clin. Pract.* **2022**, *183*, 109119. [CrossRef]
3. Oyebode, O.A.; Jere, S.W.; Houreld, N.N. Current Therapeutic Modalities for the Management of Chronic Diabetic Wounds of the Foot. *J. Diabetes Res.* **2023**, *2023*, 1359537. [CrossRef]
4. Liu, Y.; Liu, Y.; He, W.; Mu, X.; Wu, X.; Deng, J.; Nie, X. Fibroblasts: Immunomodulatory factors in refractory diabetic wound healing. *Front. Immunol.* **2022**, *13*, 918223. [CrossRef]
5. Lin, X.; Xu, Y.; Pan, X.; Xu, J.; Ding, Y.; Sun, X.; Song, X.; Ren, Y.; Shan, P.F. Global, regional, and national burden and trend of diabetes in 195 countries and territories: An analysis from 1990 to 2025. *Sci. Rep.* **2020**, *10*, 14790. [CrossRef] [PubMed]
6. Gianino, E.; Miller, C.; Gilmore, J. Smart Wound Dressings for Diabetic Chronic Wounds. *Bioengineering* **2018**, *5*, 51. [CrossRef] [PubMed]
7. International Diabetes Federation. *IDF Diabetes Atlas 2021*, 10th ed.; IDF Diabetes Atlas: Brusseles, Belgium, 2021. Available online: http://www.diabetesatlas.org/ (accessed on 6 August 2023).
8. Zhang, J.; Liu, H.; Che, T.; Zheng, Y.; Nan, X.; Wu, Z. Nanomaterials for diabetic wound healing: Visualization and bibliometric analysis from 2011 to 2021. *Front. Endocrinol.* **2023**, *14*, 1124027. [CrossRef]
9. McDermott, K.; Fang, M.; Boulton, A.J.; Selvin, E.; Hicks, C.W. Etiology, Epidemiology, and Disparities in the Burden of Diabetic Foot Ulcers. *Diabetes Care* **2023**, *46*, 209–221. [CrossRef]
10. Harries, R.L.; Bosanquet, D.C.; Harding, K.G. Wound bed preparation: TIME for an update. *Int. Wound J.* **2016**, *13* (Suppl. S3), 8–14. [CrossRef] [PubMed]
11. Zhang, S.; Ge, G.; Qin, Y.; Li, W.; Dong, J.; Mei, J.; Ma, R.; Zhang, X.; Bai, J.; Zhu, C.; et al. Recent advances in responsive hydrogels for diabetic wound healing. *Mater. Today Bio* **2023**, *18*, 100508. [CrossRef]
12. Shah, S.A.; Sohail, M.; Khan, S.; Minhas, M.U.; de Matas, M.; Sikstone, V.; Hussain, Z.; Abbasi, M.; Kousar, M. Biopolymer-based biomaterials for accelerated diabetic wound healing: A critical review. *Int. J. Biol. Macromol.* **2019**, *139*, 975–993. [CrossRef] [PubMed]
13. Singer, A.J.; Clark, R.A. Cutaneous wound healing. *N. Engl. J. Med.* **1999**, *341*, 738–746. [CrossRef] [PubMed]
14. Wang, F.; Zhang, W.; Li, H.; Chen, X.; Feng, S.; Mei, Z. How Effective are nano-based dressings in diabetic wound healing? A comprehensive review of literature. *Int. J. Nanomed.* **2022**, *17*, 2097–2119.
15. Guo, S.; DiPietro, L. Factors affecting wound healing. *J. Dent. Res.* **2010**, *89*, 219–229. [CrossRef]
16. Huang, C.; Yuan, W.; Chen, J.; Wu, L.-P.; You, T. Construction of Smart Biomaterials for Promoting Diabetic Wound Healing. *Molecules* **2023**, *28*, 1110. [CrossRef]
17. Raina, N.; Pahwa, R.; Thakur, V.K.; Gupta, M. Polysaccharide-based hydrogels: New insights and futuristic prospects in wound healing. *Int. J. Biol. Macromol.* **2022**, *223*, 1586–1603. [CrossRef]
18. Holl, J.; Kowalewski, C.; Zimek, Z.; Fiedor, P.; Kaminski, A.; Oldak, T.; Moniuszko, M.; Eljaszewicz, A. Chronic Diabetic Wounds and Their Treatment with Skin Substitutes. *Cells* **2021**, *10*, 655. [CrossRef]
19. Criollo-Mendoza, M.S.; Contreras-Angulo, L.A.; Leyva-López, N.; Gutiérrez-Grijalva, E.P.; Jiménez-Ortega, L.A.; Heredia, J.B. Wound Healing Properties of Natural Products: Mechanisms of Action. *Molecules* **2023**, *28*, 598. [CrossRef]
20. Fernandes, A.; Rodrigues, P.; Pintado, M.; Tavaria, F. A systematic review of natural products for skin applications: Targeting inflammation, wound healing, and photo-aging. *Phytomedicine* **2023**, *115*, 154824. [CrossRef]
21. Gonzalez, A.C.D.O.; Costa, T.F.; de Araújo Andrade, Z.; Medrado, A.R.A.P. Wound healing—A literature review. *An. Bras. Dermatol.* **2016**, *91*, 614–620. [CrossRef]
22. Gao, D.; Zhang, Y.; Bowers, D.T.; Liu, W.; Ma, M. Functional hydrogels for diabetic wound management. *APL Bioeng.* **2021**, *5*, 031503. [CrossRef] [PubMed]
23. Desmet, C.M.; Préat, V.; Gallez, B. Nanomedicines and gene therapy for the delivery of growth factors to improve perfusion and oxygenation in wound healing. *Adv. Drug Deliv. Rev.* **2018**, *129*, 262–284. [CrossRef] [PubMed]
24. Guan, Y.; Niu, H.; Liu, Z.; Dang, Y.; Shen, J.; Zayed, M.; Ma, L.; Guan, J. Sustained oxygenation accelerates diabetic wound healing by promoting epithelialization and angiogenesis and decreasing inflammation. *Sci. Adv.* **2021**, *7*, eabj0153. [CrossRef] [PubMed]
25. Qin, W.; Wu, Y.; Liu, J.; Yuan, X.; Gao, J.A. Comprehensive Review of the Application of Nanoparticles in Diabetic Wound Healing: Therapeutic Potential and Future Perspectives. *Int. J. Nanomed.* **2022**, *17*, 6007–6029. [CrossRef]
26. Bai, Q.; Han, K.; Dong, K.; Zheng, C.; Zhang, Y.; Long, Q.; Lu, T. Potential Applications of Nanomaterials and Technology for Diabetic Wound Healing. *Int. J. Nanomed.* **2020**, *15*, 9717–9743. [CrossRef]
27. A Elafros, M.; Andersen, H.; Bennett, D.L.; Savelieff, M.G.; Viswanathan, V.; Callaghan, B.C.; Feldman, E.L. Towards prevention of diabetic peripheral neuropathy: Clinical presentation, pathogenesis, and new treatments. *Lancet Neurol.* **2022**, *21*, 922–936. [CrossRef]
28. Zulkefli, N.; Zahari, C.N.M.C.; Sayuti, N.H.; Kamarudin, A.A.; Saad, N.; Hamezah, H.S.; Bunawan, H.; Baharum, S.N.; Mediani, A.; Ahmed, Q.U.; et al. Flavonoids as Potential Wound-Healing Molecules: Emphasis on Pathways Perspective. *Int. J. Mol. Sci.* **2023**, *24*, 4607. [CrossRef]
29. Su, L.; Li, X.; Wu, X.; Hui, B.; Han, S.; Gao, J.; Li, Y.; Shi, J.; Zhu, H.; Zhao, B.; et al. Simultaneous deactivation of FAK and Src improves the pathology of hypertrophic scar. *Sci. Rep.* **2016**, *6*, 26023. [CrossRef]

30. Câmara, J.S.; Albuquerque, B.R.; Aguiar, J.; Corrêa, R.C.G.; Gonçalves, J.L.; Granato, D.; Pereira, J.A.M.; Barros, L.; Ferreira, I.C.F.R. Food Bioactive Compounds and Emerging Techniques for Their Extraction: Polyphenols as a Case Study. *Foods* **2020**, *10*, 37. [CrossRef]
31. Pai, S.; Hebbar, A.; Selvaraj, S. A critical look at challenges and future scopes of bioactive compounds and their incorporations in the food, energy, and pharmaceutical sector. *Environ. Sci. Pollut. Res. Int.* **2022**, *29*, 35518–35541. [CrossRef]
32. Mssillou, I.; Bakour, M.; Slighoua, M.; Laaroussi, H.; Saghrouchni, H.; Amrati, F.E.-Z.; Lyoussi, B.; Derwich, E. Investigation on wound healing effect of Mediterranean medicinal plants and some related phenolic compounds: A review. *J. Ethnopharmacol.* **2022**, *298*, 115663. [CrossRef] [PubMed]
33. Falbo, F.; Spizzirri, U.G.; Restuccia, D.; Aiello, F. Natural Compounds and Biopolymers-Based Hydrogels Join Forces to Promote Wound Healing. *Pharmaceutics* **2023**, *15*, 271. [CrossRef] [PubMed]
34. Singla, R.K.; Dubey, A.K.; Garg, A.; Sharma, R.K.; Fiorino, M.; Ameen, S.M.; A Haddad, M.; Al-Hiary, M. Natural Polyphenols: Chemical Classification, Definition of Classes, Subcategories, and Structures. *J. AOAC Int.* **2019**, *102*, 1397–1400. [CrossRef] [PubMed]
35. Ma, E.Z.; Khachemoune, A. Flavonoids and their therapeutic applications in skin diseases. *Arch. Dermatol. Res.* **2023**, *315*, 321–331. [CrossRef]
36. Koleckar, V.; Kubikova, K.; Rehakova, Z.; Kuca, K.; Jun, D.; Jahodar, L.; Opletal, L. Condensed and hydrolysable tannins as antioxidants influencing the health. *Mini Rev. Med. Chem.* **2008**, *8*, 436–447. [CrossRef]
37. Özay, Y.; Güzel, S.; Yumrutaş, Ö.; Pehlivanoğlu, B.; Erdoğdu, I.H.; Yildirim, Z.; Türk, B.A.; Darcan, S. Wound healing effect of kaempferol in diabetic and nondiabetic Rats. *J. Surg. Res.* **2019**, *233*, 284–296. [CrossRef]
38. Rakotondrabe, T.F.; Fan, M.-X.; Muema, F.W.; Guo, M.-Q. Modulating inflammation-mediated diseases via natural phenolic compounds loaded in nanocarrier systems. *Pharmaceutics* **2023**, *15*, 699. [CrossRef]
39. Hou, Y.; Huang, H.; Gong, W.; Wang, R.; He, W.; Wang, X.; Hu, J. Co-Assembling of Natural Drug-Food Homologous Molecule into Composite Hydrogel for Accelerating Diabetic Wound Healing. *Biomat. Adv.* **2022**, *140*, 213034. [CrossRef]
40. Kaltalioglu, K. Sinapic acid-loaded gel accelerates diabetic wound healing process by promoting re-epithelialization and attenuating oxidative stress in rats. *Biomed. Pharmacother.* **2023**, *163*, 114788. [CrossRef]
41. Sirerol, J.A.; Rodríguez, M.L.; Mena, S.; Asensi, M.A.; Estrela, J.M.; Ortega, A.L. Role of Natural Stilbenes in the Prevention of Cancer. *Oxidative Med. Cell. Longev.* **2016**, *2016*, 3128951. [CrossRef]
42. Hu, W.; Yu, H.; Zhou, X.; Li, M.; Xiao, L.; Ruan, Q.; Huang, X.; Li, L.; Xie, W.; Guo, X.; et al. Topical administration of pterostilbene accelerates burn wound healing in diabetes through activation of the HIF1α signaling pathway. *Burns* **2022**, *48*, 1452–1461. [CrossRef] [PubMed]
43. Noushahi, H.A.; Khan, A.H.; Noushahi, U.F.; Hussain, M.; Javed, T.; Zafar, M.; Batool, M.; Ahmed, U.; Liu, K.; Harrison, M.T.; et al. Biosynthetic pathways of triterpenoids and strategies to improve their Biosynthetic Efficiency. *Plant Growth Regul.* **2022**, *97*, 439–454. [CrossRef] [PubMed]
44. Shao, Y.; Dang, M.; Lin, Y.; Xue, F. Evaluation of wound healing activity of plumbagin in diabetic rats. *Life Sci.* **2019**, *231*, 116422. [CrossRef] [PubMed]
45. Chan, W.-K.; Tan, L.T.-H.; Chan, K.-G.; Lee, L.-H.; Goh, B.-H. Nerolidol: A sesquiterpene alcohol with multi-faceted pharmacological and biological activities. *Molecules* **2016**, *21*, 529. [CrossRef]
46. Chen, J.; Qin, S.; Liu, S. Targeting matrix metalloproteases in diabetic wound healing. *Front. Immunol.* **2023**, *14*, 1089001. [CrossRef]
47. Wu, Z.; Zheng, X.; Gong, M.; Li, Y. Myricetin, a potent natural agent for treatment of diabetic skin damage by modulating TIMP/MMPs balance and oxidative stress. *Oncotarget* **2016**, *7*, 71754–71760. [CrossRef]
48. Yan, T.; Cheng, F.; Wei, X.; Huang, Y.; He, J. Biodegradable collagen sponge reinforced with chitosan/calcium pyrophosphate nano flowers for rapid hemostasis. *Carbohydr. Polym.* **2021**, *170*, 271–280. [CrossRef]
49. Tan, W.S.; Arulselvan, P.; Ng, S.-F.; Taib, C.N.M.; Sarian, M.N.; Fakurazi, S. Improvement of diabetic wound healing by topical application of Vicenin-2 hydrocolloid film on Sprague Dawley rats. *BMC Complement. Altern. Med.* **2019**, *19*, 20. [CrossRef]
50. Lwin, O.M.; Giribabu, N.; Kilari, E.K.; Salleh, N. Topical administration of mangiferin promotes healing of the wound of streptozotocin-nicotinamide-induced type-2 diabetic male rats. *J. Dermatol. Treat.* **2020**, *32*, 1039–1048. [CrossRef]
51. Akbik, D.; Ghadiri, M.; Chrzanowski, W.; Rohanizadeh, R. Curcumin as a wound healing agent. *Life Sci.* **2014**, *116*, 1–7. [CrossRef]
52. Huang, Y.W.; Zhu, Q.Q.; Yang, X.Y.; Xu, H.H.; Sun, B.; Wang, X.J.; Sheng, J. Wound healing can be improved by (−)-epigallocatechin gallate through targeting Notch in streptozotocin-induced diabetic mice. *FASEB J.* **2019**, *33*, 953–964. [CrossRef] [PubMed]
53. Li, W.; Kandhare, A.D.; Mukherjee, A.A.; Bodhankar, S.L. Hesperidin, a plant flavonoid accelerated the cutaneous wound healing in streptozotocin-induced diabetic rats: Role of TGF-ß/Smads and Ang-1/Tie-2 signaling pathways. *EXCLI J.* **2018**, *17*, 399–419. [PubMed]
54. Li, S.; Yang, P.; Ding, X.; Zhang, H.; Ding, Y.; Tan, Q. Puerarin improves diabetic wound healing via regulation of macrophage M2 polarization phenotype. *Burn. Trauma* **2022**, *10*, 46. [CrossRef] [PubMed]
55. Zhu, W.; Dong, Y.; Xu, P.; Pan, Q.; Jia, K.; Jin, P.; Zhou, M.; Xu, Y.; Guo, R.; Cheng, B. A composite hydrogel containing resveratrol-laden nanoparticles and platelet-derived extracellular vesicles promotes wound healing in diabetic mice. *Acta Biomater.* **2022**, *154*, 212–230. [CrossRef] [PubMed]

56. Deng, L.; Du, C.; Song, P.; Chen, T.; Rui, S.; Armstrong, D.G.; Deng, W. The Role of Oxidative Stress and Antioxidants in Diabetic Wound Healing. *Oxidative Med. Cell. Longev.* **2021**, *2021*, 8852759. [CrossRef] [PubMed]
57. Ren, J.; Yang, M.; Xu, F.; Chen, J.; Ma, S. Acceleration of wound healing activity with syringic acid in streptozotocin induced diabetic rats. *Life Sci.* **2019**, *233*, 116728. [CrossRef]
58. Guimarães, I.; Baptista-Silva, S.; Pintado, M.; Oliveira, A.L. Polyphenols: A Promising Avenue in Therapeutic Solutions for Wound Care. *Appl. Sci.* **2021**, *11*, 1230. [CrossRef]
59. Yin, M.; Wu, J.; Deng, M.; Wang, P.; Ji, G.; Wang, M.; Zhou, C.; Blum, N.T.; Zhang, W.; Shi, H.; et al. Multifunctional magnesium organic framework-based microneedle patch for accelerating diabetic wound healing. *ACS Nano* **2021**, *15*, 17842–17853. [CrossRef]
60. Zhou, R.; Xiang, C.; Cao, G.; Xu, H.; Zhang, Y.; Yang, H. Berberine accelerated wound healing by restoring TrxR1/JNK in diabetes. *Clin. Sci.* **2021**, *135*, 613–627. [CrossRef]
61. Hsu, C.-Y.; Lin, S.-C.; Wu, Y.-H.; Hu, C.-Y.; Chen, Y.-T.; Chen, Y.-C. The Antimicrobial Effects of Bacterial Cellulose Produced by Komagataeibacter intermedius in Promoting Wound Healing in Diabetic Mice. *Int. J. Mol. Sci.* **2022**, *23*, 5456. [CrossRef]
62. Moradpoor, H.; Mohammadi, H.; Safaei, M.; Mozaffari, H.R.; Sharifi, R.; Gorji, P.; Sulong, A.B.; Muhamad, N.; Ebadi, M. Recent Advances on Bacterial Cellulose-Based Wound Management: Promises and Challenges. *Int. J. Polym. Sci.* **2023**, *2022*, 1214734. [CrossRef]
63. García-Sánchez, M.E.; de Guadalajara, U.; Robledo-Ortiz, J.R.; Jiménez-Palomar, I.; González-Reynoso, O.; González-García, Y. Production of bacterial cellulose by Komagataeibacter xylinus using mango waste as alternative culture medium. *Rev. Mex. Ing. Quim.* **2020**, *19*, 851–865. [CrossRef]
64. Abazari, M.; Akbari, T.; Hasani, M.; Sharifikolouei, E.; Raoufi, M.; Foroumadi, A.; Sharifzadeh, M.; Firoozpour, L.; Khoobi, M. Polysaccharide-based hydrogels containing herbal extracts for wound healing applications. *Carbohydr. Polym.* **2022**, *294*, 119808. [CrossRef] [PubMed]
65. Alven, S.; Aderibigbe, B.A. Chitosan and Cellulose-Based Hydrogels for Wound Management. *Int. J. Mol. Sci.* **2020**, *21*, 9656. [CrossRef] [PubMed]
66. Escárcega-Galaz, A.A.; De La Cruz-Mercado, J.L.; López-Cervantes, J.; Sánchez-Machado, D.I.; Brito-Zurita, O.R.; Ornelas-Aguirre, J.M. Chitosan treatment for skin ulcers associated with diabetes. *Saudi J. Biol. Sci.* **2018**, *25*, 130–135. [CrossRef]
67. Bollyky, P.L.; Falk, B.A.; Long, S.A.; Preisinger, A.; Braun, K.R.; Wu, R.P.; Evanko, S.P.; Buckner, J.H.; Wight, T.N.; Nepom, G.T. CD44 costimulation promotes $FoxP^{3+}$ regulatory T cell persistence and function via production of IL-2, IL-10, and TGF-β. *J. Immunol.* **2009**, *183*, 2232–2241. [CrossRef] [PubMed]
68. Fan, S.-L.; Lin, J.-A.; Chen, S.-Y.; Lin, J.-H.; Lin, H.-T.; Chen, Y.-Y.; Yen, G.-C. Effects of Hsian-tsao (*Mesona procumbens* Hemsl.) extracts and its polysaccharides on the promotion of wound healing under diabetes-like conditions. *Food Funct.* **2021**, *12*, 119–132. [CrossRef]
69. Zhang, Y.; Wang, Y.; Li, Y.; Yang, Y.; Jin, M.; Lin, X.; Zhuang, Z.; Guo, K.; Zhang, T.; Tan, W. Application of Collagen-Based Hydrogel in Skin Wound Healing. *Gels* **2023**, *9*, 185. [CrossRef]
70. Yue, K.; Trujillo-de Santiago, G.; Alvarez, M.M.; Tamayol, A.; Annabi, N.; Khademhosseini, A. Synthesis, properties, and biomedical applications of gelatin methacryloyl (GelMA) hydrogels. *Biomaterials* **2015**, *73*, 254–271. [CrossRef]
71. Zhao, Y.; Wang, X.; Qi, R.; Yuan, H. Recent advances of natural-polymer-based hydrogels for wound antibacterial therapeutics. *Polymers* **2023**, *15*, 3305. [CrossRef]
72. Güiza-Argüello, V.R.; Solarte-David, V.A.; Pinzón-Mora, A.V.; Ávila-Quiroga, J.E.; Becerra-Bayona, S.M. Current Advances in the Development of Hydrogel-Based Wound Dressings for Diabetic Foot Ulcer Treatment. *Polymers* **2022**, *14*, 2764. [CrossRef]
73. Ghosal, K.; Chakraborty, D.; Roychowdhury, V.; Ghosh, S.; Dutta, S. Recent Advancement of Functional Hydrogels toward Diabetic Wound Management. *ACS Omega* **2022**, *7*, 43364–43380. [CrossRef] [PubMed]
74. Wang, T.; Liao, Q.; Wu, Y.; Wang, X.; Fu, C.; Geng, F.; Qu, Y.; Zhang, J. A composite hydrogel loading natural polysaccharides derived from Periplaneta americana herbal residue for diabetic wound healing. *Int. J. Biol. Macromol.* **2020**, *164*, 3846–3857. [CrossRef] [PubMed]
75. El-Samad, L.M.; Hassan, M.A.; Basha, A.A.; El-Ashram, S.; Radwan, E.H.; Aziz, K.K.A.; Tamer, T.M.; Augustyniak, M.; El Wakil, A. Carboxymethyl cellulose/sericin-based hydrogels with intrinsic antibacterial, antioxidant, and anti-inflammatory properties promote re-epithelization of diabetic wounds in rats. *Int. J. Pharm.* **2022**, *629*, 122328. [CrossRef] [PubMed]
76. Pu, Y.; Wang, P.; Yang, R.; Tan, X.; Shi, T.; Ma, J.; Xue, W.; Chi, B. Bio-fabricated nanocomposite hydrogel with ROS scavenging and local oxygenation accelerates diabetic wound healing. *J. Mater. Chem. B* **2022**, *10*, 4083–4095. [CrossRef] [PubMed]
77. Khalid, A.; Madni, A.; Raza, B.; Islam, M.U.; Hassan, A.; Ahmad, F.; Ali, H.; Khan, T.; Wahid, F. Multiwalled carbon nanotubes functionalized bacterial cellulose as an efficient healing material for diabetic wounds. *Int. J. Biol. Macromol.* **2022**, *203*, 256–267. [CrossRef]
78. Shah, S.A.; Sohail, M.; Khan, S.A.; Kousar, M. Improved drug delivery and accelerated diabetic wound healing by chondroitin sulfate grafted alginate-based thermoreversible hydrogels. *Mater. Sci. Eng. C Mater. Biol. Appl.* **2021**, *126*, 112169. [CrossRef]
79. Barbosa, M.A.G.; Paggiaro, A.O.; de Carvalho, V.F.; Isaac, C.; Gemperli, R. Effects of Hydrogel with Enriched Sodium Alginate in Wounds of Diabetic Patients. *Plast. Surg. Nurs.* **2020**, *40*, 110–115. [CrossRef]
80. Bi, Q.; Zhang, Q.; Ma, J.; Xu, M.; Zhang, S.-J.; Qiu, B.-S.; Xia, B.; Gu, H.-F.; Hong, J.-F.; Zhao, C.; et al. Effect of combination therapy with alginate dressing and mouse epidermal growth factor on epidermal stem cells in patients with refractory wounds. *Chin. Med. J. Engl. Chin. Med. J. Engl.* **2012**, *125*, 257–261.

81. Casado-Díaz, A.; La Torre, M.; Priego-Capote F.; Verdú-Soriano, J.; Lázaro-Martínez, J.L.; Rodríguez-Mañas, L.; Berenguer Pérez, M.; Tunez, I. EHO-85: A Multifunctional Amorphous Hydrogel for Wound Healing Containing Olea europaea Leaf Extract: Effects on Wound Microenvironment and Preclinical Evaluation. *J. Clin. Med.* **2022**, *11*, 1229. [CrossRef]
82. Soares, R.D.; Campos, M.G.; Ribeiro, G.P.; Salles, B.C.; Cardoso, N.S.; Ribeiro, J.R.; Souza, R.M.; Leme, K.C.; Soares, C.B.; de Oliveira, C.M. Development of a chitosan hydrogel containing flavonoids extracted from Passiflora edulis leaves and the evaluation of its antioxidant and wound healing properties for the treatment of skin lesions in diabetic mice. *J. Biomed. Mater. Res. A* **2022**, *108*, 654–662. [CrossRef]
83. Ferreira, M.O.G.; Leite, L.L.R.; de Lima, I.S.; Barreto, H.M.; Nunes, L.C.C.; Ribeiro, A.B.; Osajima, J.A.; Filho, E.C.d.S. Chitosan hydrogel in combination with nerolidol for healing wounds. *Carbohydr. Polym.* **2016**, *152*, 409–418. [CrossRef] [PubMed]
84. Shah, S.A.; Sohail, M.; Karperien, M.; Johnbosco, C.; Mahmood, A.; Kousar, M. Chitosan and carboxymethyl cellulose-based 3D multifunctional bioactive hydrogels loaded with nano-curcumin for synergistic diabetic wound repair. *Int. J. Biol. Macromol.* **2023**, *227*, 1203–1220. [CrossRef] [PubMed]
85. Prasathkumar, M.; Sadhasivam, S. Chitosan/Hyaluronic acid/Alginate and an assorted polymers loaded with honey, plant, and marine compounds for progressive wound healing—Know-how. *Int. J. Biol. Macromol.* **2021**, *186*, 656–685. [CrossRef] [PubMed]
86. Velazco, G.; Gonzalez, A.; Ortiz, R. Chitosan films for the diabetic foot treatment. *Avan. Biomed.* **2012**, *1*, 38–41.
87. Available online: https://chitotech.com/page/627/ChitoHeal-Gel (accessed on 10 August 2023).
88. Pan, W.; Qi, X.; Xiang, Y.; You, S.; Cai, E.; Gao, T.; Tong, X.; Hu, R.; Shen, J.; Deng, H. Facile formation of injectable quaternized chitosan/tannic acid hydrogels with antibacterial and ROS scavenging capabilities for diabetic wound healing. *Int. J. Biol. Macromol.* **2022**, *195*, 190–197. [CrossRef]
89. Zeng, X.; Chen, B.; Wang, L.; Sun, Y.; Jin, Z.; Liu, X.; Ouyang, L.; Liao, Y. Chitosan@Puerarin hydrogel for accelerated wound healing in diabetic subjects by miR-29ab1 mediated inflammatory axis suppression. *Bioact. Mater.* **2023**, *19*, 653–665. [CrossRef]
90. Shen, T.; Dai, K.; Yu, Y.; Wang, J.; Liu, C. Sulfated chitosan rescues dysfunctional macrophages and accelerates wound healing in diabetic mice. *Acta Biomater.* **2020**, *117*, 192–203. [CrossRef]
91. Sanaye, P.M.; Mojaveri, M.R.; Ahmadian, R.; Jahromi, M.S.; Bahramsoltani, R. Apigenin and its dermatological applications: A comprehensive review. *Phytochemistry* **2022**, *203*, 113390. [CrossRef]
92. Li, Q.; Liu, K.; Jiang, T.; Ren, S.; Kang, Y.; Li, W.; Yao, H.; Yang, X.; Dai, H.; Chen, Z. Injectable and self-healing chitosan-based hydrogel with MOF-loaded α-lipoic acid promotes diabetic wound healing. *Mater. Sci. Eng. C Mater. Biol. Appl.* **2021**, *131*, 112519. [CrossRef]
93. Hao, Y.; Zhao, W.; Zhang, H.; Zheng, W.; Zhou, Q. Carboxymethyl chitosan-based hydrogels containing fibroblast growth factors for triggering diabetic wound healing. *Carbohydr. Polym.* **2022**, *287*, 119336. [CrossRef]
94. Cifuentes, A.; Gómez-Gil, V.; Ortega, M.A.; Asúnsolo, Á.; Coca, S.; Román, J.S.; Álvarez-Mon, M.; Buján, J.; García-Honduvilla, N. Chitosan hydrogels functionalized with either unfractionated heparin or bemiparin improve diabetic wound healing. *Biomed. Pharmacother.* **2020**, *129*, 110498. [CrossRef]
95. Singh, S.K.; Dwivedi, S.D.; Yadav, K.; Shah, K.; Chauhan, N.S.; Pradhan, M.; Singh, M.R.; Singh, D. Novel Biotherapeutics Targeting Biomolecular and Cellular Approaches in Diabetic Wound Healing. *Biomedicines* **2023**, *11*, 613. [CrossRef] [PubMed]
96. Yang, H.; Song, L.; Sun, B.; Chu, D.; Yang, L.; Li, M.; Li, H.; Dai, Y.; Yu, Z.; Guo, J. Modulation of macrophages by a paeoniflorin-loaded hyaluronic acid-based hydrogel promotes diabetic wound healing. *Mater. Today Bio* **2021**, *12*, 100139. [CrossRef]
97. Hauck, S.; Zager, P.; Halfter, N.; Wandel, E.; Torregrossa, M.; Kakpenova, A.; Rother, S.; Ordieres, M.; Räthel, S.; Berg, A.; et al. Collagen/hyaluronan based hydrogels releasing sulfated hyaluronan improve dermal wound healing in diabetic mice via reducing inflammatory macrophage activity. *Bioact. Mater.* **2021**, *6*, 4342–4359. [CrossRef] [PubMed]
98. Hu, B.; Gao, M.; Boakye-Yiadom, K.O.; Ho, W.; Yu, W.; Xu, X.; Zhang, X.-Q. An intrinsically bioactive hydrogel with on-demand drug release behaviors for diabetic wound healing. *Bioact. Mater.* **2021**, *6*, 4592–4606. [CrossRef]
99. Xu, Z.; Liu, G.; Liu, P.; Hu, Y.; Chen, Y.; Fang, Y.; Sun, G.; Huang, H.; Wu, J. Hyaluronic acid-based glucose-responsive antioxidant hydrogel platform for enhanced diabetic wound repair. *Acta Biomater.* **2022**, *147*, 147–157. [CrossRef] [PubMed]
100. Edmonds, M. European and Australian Apligraf Diabetic Foot Ulcer Study Group Apligraf in the treatment of neuropathic diabetic foot ulcers. *Int. J. Low. Extremity Wounds* **2009**, *8*, 11–18. [CrossRef]
101. Sharma, A.D.B.; Jarman, E.H.B.; Fox, P.M. Scoping Review of Hydrogel Therapies in the Treatment of Diabetic Chronic Wounds. *Plast. Reconstr. Surg. Glob. Open* **2023**, *11*, e4984. [CrossRef]
102. Djavid, G.E.; Tabaie, S.M.; Tajali, S.B.; Totounchi, M.; Farhoud, A.; Fateh, M.; Ghafghazi, M.; Koosha, M.; Taghizadeh, S. Application of a collagen matrix dressing on a neuropathic diabetic foot ulcer: A randomised control trial. *J. Wound Care* **2020**, *29*, S13–S18. [CrossRef]
103. Narisepalli, S.; Salunkhe, S.A.; Chitkara, D.; Mittal, A. Asiaticoside polymeric nanoparticles for effective diabetic wound healing through increased collagen biosynthesis: In-vitro and in-vivo evaluation. *Int. J. Pharm.* **2023**, *631*, 122508. [CrossRef]
104. Liu, J.; Chen, Z.; Wang, J.; Li, R.; Li, T.; Chang, M.; Yan, F.; Wang, Y. Encapsulation of Curcumin Nanoparticles with MMP9-Responsive and Thermos-Sensitive Hydrogel Improves Diabetic Wound Healing. *ACS Appl. Mater. Interfaces* **2018**, *10*, 16315–16326. [CrossRef] [PubMed]

105. Xia, S.; Weng, T.; Jin, R.; Yang, M.; Yu, M.; Zhang, W.; Wang, X.; Han, C. Curcumin-incorporated 3D bioprinting gelatin methacryloyl hydrogel reduces reactive oxygen species-induced adipose-derived stem cell apoptosis and improves implanting survival in diabetic wounds. *Burn. Trauma* **2022**, *10*, tkac001. [CrossRef] [PubMed]
106. Bai, L.; Zhang, X.; Li, X.; Wang, S.; Zhang, Y.; Xu, G. Impact of a Novel Hydrogel with Injectable Platelet-Rich Fibrin in Diabetic Wound Healing. *J. Diabetes Res.* **2023**, *2023*, 7532637. [CrossRef] [PubMed]

gels

MDPI

Article

Temperature-Sensitive Hydrogels as Carriers for Modulated Delivery of Acetaminophen

Snežana Ilić-Stojanović [1,*], Ljubiša Nikolić [1], Vesna Nikolić [1], Ivan Ristić [2], Suzana Cakić [1] and Slobodan D. Petrović [3]

1 Faculty of Technology, University of Niš, Bulevar Oslobodjenja 124, 16000 Leskovac, Serbia; nljubisa@tf.ni.ac.rs (L.N.); nikolicvesna@tf.ni.ac.rs (V.N.); cakics@tf.ni.ac.rs (S.C.)
2 Faculty of Technology, University of Novi Sad, Bulevar Cara Lazara 1, 21000 Novi Sad, Serbia; ivan.ristic@uns.ac.rs
3 Faculty of Technology and Metallurgy, University of Belgrade, 11000 Belgrade, Serbia; sloba@tmf.bg.ac.rs
* Correspondence: snezanai@tf.ni.ac.rs

Abstract: The purposes of this study are the polymerization of temperature-sensitive copolymers based on N-isopropyl acrylamide and 10 mol % of 2-hydroxypropylmethacrylate, characterisations of their thermal, morphological and swelling properties, as well as the analysis of potential application in drug-delivery systems. Acetaminophen, the representative of non-steroidal anti-inflammatory drugs, was used as a model drug in this study. It is a common pain relief drug, which is also used for fever treatment. However, oral administration comes with certain health risks, mainly the overdose and frequent administration of up to four times a day. The goal of applying temperature-sensitive hydrogel is to enable extended administration once a day, depending on the body temperature. The swelling behavior of the obtained poly(*N*-isopropyl acrylamide-*co*-2-hydroxypropylmethacrylate) (p(NIPA/HPMA)) hydrogels and their temperature-sensitivity, kinetics and order of swelling processes at 18 and 38 °C were analyzed. The thermal properties of these hydrogels were observed by the DSC method, and the obtained thermograms showed both melting and glass transitions. The drug delivery system of p(NIPA/HPMA) hydrogels with loaded acetaminophen was analyzed using scanning electron microscopy and Fourier transform infrared spectroscopy methods. Structural analysis of FTIR spectra indicates that non-covalent intermolecular interactions of the type of hydrogen bonds were formed among functional groups of acetaminophen and side-chains of p(NIPA/HPMA) hydrogels. The surface structure of p(NIPA/HPMA) hydrogels after drug loading indicates the acetaminophen presence into the pores of the hydrogel network, and their loading efficiency was higher than 92%. Qualitative and quantitative analysis of acetaminophen, determined by the high-pressure liquid chromatography method, showed that about 90–99% of the loaded amount was released from p(NIPA/HPMA) hydrogels within 24 h. Kinetic parameters of the acetaminophen release under simulated gastrointestinal conditions were determined. Based on obtained results, the drug delivery system of temperature-sensitive p(NIPA/HPMA) hydrogels with loaded acetaminophen could be suitable for additional investigation for modulated drug administration, e.g., for extended drug administration.

Keywords: *N*-isopropyl acrylamide; 2-hydroxypropylmethacrylate; hydrogels; DSC; drug carrier; modulated drug release; acetaminophen

1. Introduction

Temperature-sensitive hydrogels belong to the wide group of stimuli-sensitive polymers, characterized by a cross-linked three-dimensional network obtained from synthetic or natural materials, which have the capability to swell while retaining plenty of liquid, without dissolving [1]. They are well-known as "smart" or "intelligent" hydrogels, because, with the temperature changes in the surroundings, they can respond with changing certain physicochemical characteristics [2]. Possible changes in its properties are, for example, swelling capacity, phase transitions, network structure, surface characteristics, permeability, mechanical properties, shape and volume, and they can turn back to their original state on the condition of elimination of the stimulus. Temperature-sensitive hydrogels are materials that are rationally designed, and they are broadly applied in human medicine and pharmacy [3–5]. The volume phase transition (VPT) in hydrogels was characterized by an immediate transformation in the swelling ratio [6,7]. Based on the volume phase transition and critical solution temperature, temperature-sensitive hydrogels can be divided into positively and negatively thermosensitive [8]. Negatively temperature-sensitive hydrogels possess a lower critical solution temperature (LCST), and they swell in the solution below critical temperature, while the hydrogel contracts above it [9]. The best-known negatively temperature-sensitive polymer is poly(N-isopropylacrylamide), which is widely researched and used to design many thermo-sensitive hydrogel systems [10]. It exhibits LCST in aqueous fluids at 32 °C, which is near the physiological body temperature [11–13]. Monomer *N*-isopropylacrylamide (NIPA) was usually copolymerized with numerous types of other synthetic monomers [14,15]), natural polymers or their combinations, in order to form the new, innovative hydrogels with different desired properties. Hydrogels have been utilized in many pharmaceutical and biomedical applications, for example, in temperature-sensitive modulated drug delivery [16–19], bioanalysis, bioseparations [20,21] and biosensors [22].

Acetaminophen, *N*-(4-hydroxyphenyl)ethanamide, is one of the most widely used pharmaceutical compounds, it belonging to the class of "aniline analgesics" [23,24]. It is a metabolite of phenacetin. A possible route for its synthesis from 4-aminophenol and acetic anhydride is presented on Figure 1 [25].

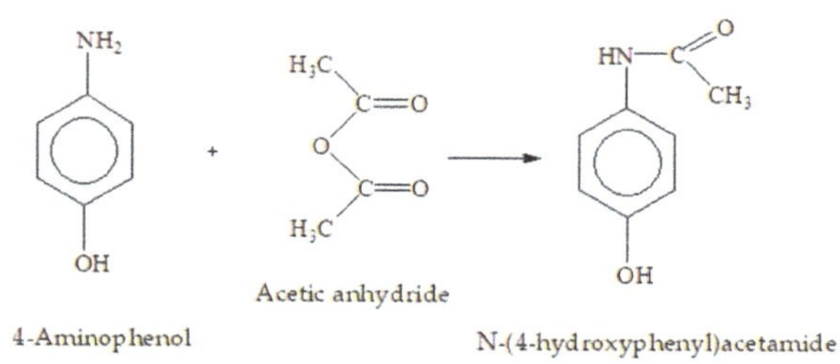

Figure 1. A possible route for acetaminophen synthesis from 4-aminophenol and acetic anhydride.

It shows different levels of analgesia, anti-inflammatory, antipyretic and antiplatelet activity [26–28]. Acetaminophen is a feeble inhibitor of two cyclooxygenase enzyme (COX-1, COX-2). It has no vasoconstrictive effect and is an effective in the treatment of acute, mild or moderate migraine attacks [29]. Its half-life is 1–4 h, and is increased in cases of liver damage and overdose patients [30]. It has been recently used in the treatment of neurodegenerative diseases, such as Alzheimer's disease, protecting the brain's endothelial cells from oxidative stress [31]. N-acetylcysteine can be used as an antidote in cases of acetaminophen overdose [32]. According to the Biopharmaceutical Classification System (BSK/BSC), i.e., the guide for predicting the intestinal absorption of drugs, acetaminophen belongs to class I (highly soluble and highly permeable) medicinal substances [33]. With oral administration, there are certain health risks of overdose and frequent administration of up to four times a day. Orodispersible hydrophilic wax tablets have been formulated to increase the solubility and bioavailability of acetaminophen [34]. Micronization of acetaminophen particles with supercritical carbon(IV)-oxide was also performed, resulting in particles with a finer structure and greater bioavailability [35]. Many researchers investigated the possibility of using intelligent gels as carriers of acetaminophen for its modified/controlled release. Multifunctional copolymer hydrogels of *N*-vinyl-2-pyrrolidone and poly(acrylic acid) were tested as acetaminophen and aspirin carriers, but without analysis of their release mechanisms [36]. Kinetics of swelling and release of acetaminophen from pH- and temperature-sensitive hydrogels based on sodium alginate and NIPA under simulated gastric conditions indicate anomalies in the transport mechanism and a good fit with a first-order kinetic model [37]. The effect of the composition of hydrogels on the swelling process and release of acetaminophen in aqueous media of different pH values from poly(acrylamide-*co*-itaconic acid) was also investigated [38]. Tests have shown that the functional monomer 4-[(4-methacryloyloxy)phenylazo]benzenesulfonic acid with poly(acrylamide) hydrogel provides the affinity of the hydrogel to acetaminophen, whose release can be photoregulated, but without the release mechanism analysis [39]. The acetaminophen release from poly(2-hydroxyethyl methacrylate-*co*-*N*-vinyl-2-pyrrolidone) hydrogels at 37 °C showed a different rate and release mechanisms due to weaker interactions of the polymer with water at pH = 7.2 in the simulated intestinal fluid, SIF, compared with the simulated gastric fluid, SGF, at pH = 1.2 [40]. The mathematical modelling of releasing the acetaminophen from gel based on hydroxypropylcellulose and poly(acrylamide), 25/75 wt %, was analyzed using a genetic programming algorithm [41]. The obtained results showed that a greater amount of acetaminophen was released at pH = 7.38 than at pH = 7. Polyurethane nanocomposite hydrogels with organofillized montmorillonite were applied as matrix for analysis of the acetaminophen release process [42]. The temperature increases at 37 °C speed up the acetaminophen release process compared with the results at 23 °C. Poly(*N*-isopropylacrylamide-acrylamide) membrane was applied in order to study of the mechanism of both vitamin B12 and acetaminophen transport through hydrogels at phosphate buffer solution (pH 7.4) and different temperatures [43]. Biodegradable composites obtained from polyvinyl alcohol in combination with nanofibrillated cellulose and 2,2,6,6-tetramethylpiperidine-*N*-oxyl-oxidized nanofibrillated cellulose showed a release of acetaminophen only about 14% from nanofibrillated cellulose, and about 28% from 2,2,6,6-tetramethylpiperidine-*N*-oxyl-oxidized nanofibrillated cellulose with the occurrence of a burst release from the outer film layer, during 6 days [44].

Based on presented results in available literature, there is no single temperature sensitive carrier which could be suitable for extended release and administration once a day. For this reason, the goal of this study is the characterization of hydrogels poly(*N*-isopropyl acrylamide-*co*-2-hydroxypropylmethacrylate), p(NIPA/HPMA), as potential carriers of acetaminophen. The profiles of drug delivery from corresponding hydrogel network systems were monitored in previous studies for the investigation of possible controlled release of some NSAID, e.g., caffeine [45], piroxicam [46], naproxen [47], phenacetin (as a structural analog of acetaminophen) [48] and ibuprofen [49]. Based on the previous studies, the major hypothesis of the current study was that similar "intelligent" p(NIPA/HPMA) hydrogels will improve the acetaminophen release mechanism with benefits over previous results. Not one comparable investigation mentioning hydrogel networks as carriers for modified acetaminophen release was found in the available publications.

2. Results and Discussion

2.1. Synthesis of p(NIPA/HPMA) Hydrogel

The process of obtaining p(NIPA/HPMA) hydrogels consisted of 10 mol % of comonomer 2-hydroxypropylmethacrylate (HPMA) calculated in relation to the initial NIPA monomer amount and 0.5, 1.0, 1.5, 2.0 and 3.0 mol % of ethyleneglycoldimethacylate (EGDM) as crosslinker. The probe with 0.5 mol % of EGDM crosslinker maintained the liquid structure after synthesis without the need for hydrogel consistency, and, for that reason, it was excluded from further analysis. The other samples of obtained hydrogels were extracted unreacted molecules during the polymerization process and, after that, they were dried in the xerogel state for further analyses, analogous to the procedure described in the author's previous studies [9,46,47,50].

2.2. Differential Scanning Calorimetry

The thermal phase transitions of obtained p(NIPA/HPMA) hydrogels with 10 mol % of HPMA and with 1.0, 2.0 or 3.0 mol % of the ethyleneglycoldimethacylate were detected in differential scanning calorimetry (DSC) thermogram. The first derivative of heat flow in the range of 30–180 °C is shown in Figure 2. It shows two sharp peaks from melting of ordered structures and also the glass transitions with weak intensity.

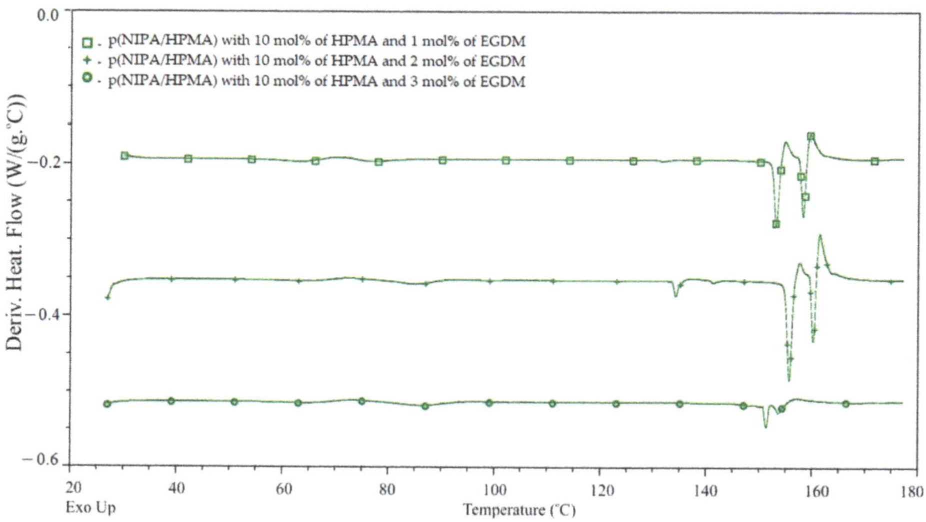

Figure 2. The first derivative of heat flow of the p(NIPA/HPMA) xerogels with 10 mol % of HPMA and 1.0, 2.0 and 3.0 mol % of ethyleneglycoldimethacylate.

Values obtained from DSC thermogram, including the melting temperatures (T_m), the melting enthalpy and the glass transition temperatures (T_g) for the obtained p(NIPA/HPMA) xerogels are presented in Table 1.

Table 1. The melting temperatures (T_m), the melting enthalpy (ΔH_m) and the glass transition temperatures (T_g), for the obtained p(NIPA/HPMA) xerogels with 1.0, 2.0, and 3.0 mol % of EGDM.

Poly(*N*-isopropyl acrylamide-*co*-2-hydroxypropylmethacrylate)	**Melting Transition Temperature, °C**		**Melting Enthalpy, J·g^{-1}**	**Glass Transition Temperature, T_g, °C**		
p(NIPA/HPMA) with	T_{m1}	T_{m2}	ΔH_m	T_{g1}	T_{g2}	T_{g3}
1 mol % of EGDM	153.90	159.07	5.99	63.41	77.53	131.76
2 mol % of EGDM	156.49	160.73	9.85	64.79	85.17	134.16
3 mol % of EGDM	154.16	156.98	2.79	64.23	86.47	-

Data obtained from DSC thermogram for the p(NIPA/HPMA) xerogels with 1, 2 and 3 mol % of ethyleneglycoldimethacylate shows that the melting process happens in two intervals (first from temperature 153.90 °C to 156.49 °C and second from 156.98 °C to 160.73 °C). The melting enthalpy is in the range of 2.79–9.85 J·g^{-1}. When the hydrogel is heating, its crystalline regions become disordered and pass through a viscous liquid phase, which causes transformation of its physical state, characterized as the melting temperature. The hydrogel network structure has different distances between crosslinking nodes, divergent branching and chain length, with various melting temperatures. This is the reason for its broader melting temperature and varied enthalpy. These results also confirmed the suitable miscibility and compatibility of the monomers NIPA and HPMA, as well as structural irregularity [9,44,51]. Synthesized hydrogels are stable up to temperatures of about 150 °C. The common sterilizing temperatures are in the range of 121–132 °C. Obtained hydrogels are convenient for sterilizing and enable biomedical application as drug carriers.

The DSC analysis of p(NIPA/HPMA) xerogels also provides results of the glass transition temperatures. The glass transition is the characteristic process for amorphous regions of the hydrogels during its heating, when it changes from a glassy to a soft state. This state is usually named as "viscous liquid". However, since it has not been melted yet, the temperature range of its occurrence depends on the hydrogel chemical structure. The samples with 2.0 and 3.0 mol % of ethyleneglycoldimethacylate display three glass transition temperatures, while the sample with 3 mol % of ethyleneglycoldimethacylate displays two glass transitions temperatures. The first occurrence of the glass transition temperatures is in the range 63.41–64.79 °C, the second is in the range 77.53–86.47 °C and the third is in the range 134.81–131.76 °C. Some difference in the glass transition temperatures between hydrogels with different molar mass of the crosslinker are visible. These differences are probably consequences of diverse crosslinking density, conditioned from the content of ethyleneglycoldimethacylate, and show similarity to previously published works [46,51].

DSC analysis confirmed the amorphous, semi-crystalline structure of the synthesized p(NIPA/HPMA) hydrogels.

2.3. Swelling Behavior

2.3.1. Temperature Sensitivity Analysis

The swelling behavior of p(NIPA/HPMA) xerogels with 10 mol % of HPMA in distilled water was monitored during heating from 5 to 60 °C to examine their temperature sensitivity. Results of the equilibrium swelling degree, α_e, calculated using Equation (1), for p(NIPA/HPMA) xerogels as a function of the temperature are presented in Figure 3.

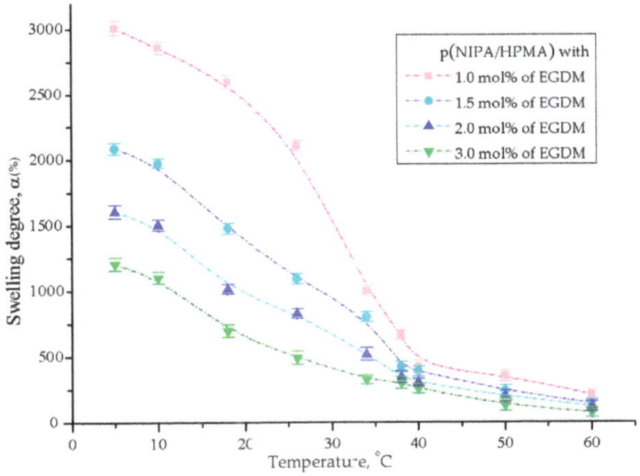

Figure 3. Temperature sensitivity of the p(NIPA/HPMA) xerogels with 10 mol % of HPMA and 1.0, 2.0 and 3.0 mol % of ethyleneglycoldimethacylate. Error bars represent the standard deviation of three replicates.

All p(NIPA/HPMA) hydrogel samples show sensitivity to temperature changes. The greatest swelling degree at all tested temperatures was achieved at lower temperatures (from 5 to about 20 °C) by the sample with 1 mol % of ethyleneglycoldimethacylate [9,46,50]. Significant reduction of the swelling degree happens when temperature increase in the range of 30–38 °C, when hydrogels pass a volume phase transition temperature (VPTT) which is named as the lower critical solution temperature (LCST). When the temperature exceeds 50 °C, these dependencies asymptomatically approach a constant value. As expected, the swelling degree increased with decreasing crosslinking ratio, because the smaller number of crosslinker decreases the network density, increasing the elasticity and mobility of the polymer chains and pore size. At temperatures below LSCT, hydrogels are swollen, soft and transparent. However, at temperatures above the LCST, they deswell, reducing their volume and changing to opaque white and rigid due to the breaking of intermolecular hydrogen bonds [46,51]. The hydrogen bond formation between the amide and hydroxyl groups of hydrogel and water molecules causes both hydration and swelling, while the amide–amide hydrogen bond formation led causse dehydration; therefore, they have a potential role in the phase transition. When the temperature rises above the LCST, intermolecular hydrogen bonds break up, and hydrophobic interactions cause the prevention of the enlargement of hydrogel network and water molecules ejectment becomes dominant [48,52]. The swelling behavior of p(NIPA/HPMA) hydrogels exhibited a resemblance to the results of the comparable hydrogels available in the literature [9,46,47,49,50,53]. The exhibited swelling behavior of synthesized hydrogels is known as negatively thermosensitive, which is very useful for efficient drug absorption at a temperature below the LCST, and drug delivery when temperature rises above the LCST.

2.3.2. Analysis of the p(NIPA/HPMA) Hydrogel Swelling Kinetics

The equilibrium swelling degree, α_e, values for p(NIPA/HPMA) hydrogels at 18 °C and 38 °C in the fluid with pH value 7.4, and kinetic parameters (the diffusion coefficient, diffusion exponent, the correlation coefficient (R^2) and kinetic constant,) calculated with expressions (2)–(6), are presented in Figure 4 and in Table 2.

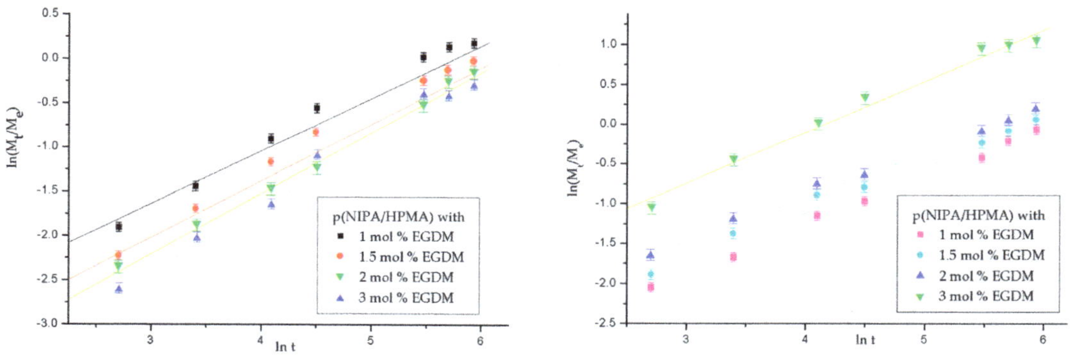

(a) (b)

Figure 4. Kinetic parameters n and k for series of p(NIPA-HPMA) hydrogels with 10 mol % HPMA at: (**a**) 18 °C and (**b**) 38 °C. In each figure, error bars represent the standard deviation of three replicates.

Based on the swelling kinetics results at pH = 7.4 (Figure 4a,b and Table 2) and especially the diffusion exponent, *n*, values, the transport mechanism into the p(NIPA/HPMA) hydrogels samples at both temperatures (18 °C and 38 °C) corresponds to "non-Fickian diffusion" for all samples and contributes to the water-sorption process. The fluid transport into the p(NIPA/HPMA) network exhibited an anomaly in the fluid transport mechanism, i.e., the swelling mechanism was controlled by diffusion of fluid and relaxation of macromolecular chains and did not obey the rules of the Fickian law. The fluid transport is complicated by segmental mobility retardation, molecular relaxation, crystallization, functional interaction among penetrant and macromolecules, etc. Non-Fickian diffusion was characterized by an initial induction time, when the sharp front is established near the film surface, which separates the highly swollen area from a dry, glassy area, and the amount of absorbed fluid increases linearly with time [54]. Comparable results for the diffusion exponent of hydrogels p(NIPA/HPMA) at 5, 25, and 38 °C have been previously published [32,46,47]. Increasing the crosslinker content and enhancing the network density leads to a reduction in the value of n, which means that the swelling degree decreases [47,48].

Table 2. The equilibrium swelling degree (α_e) of p(NIPA/HPMA) hydrogels at 18 °C and pH 7.4 and kinetic parameters of the fluid diffusion (n, k and D).

p(NIPA/HPMA) * with	Equilibrium Swelling Degree, α_e	Diffusion Exponent, n	Kinetic Constant, $k \times 10^2$, $min^{-1/2}$	Linear Correlation Coefficient, R^2	Diffusion Coefficient, D, $cm^2 \cdot min^{-1}$
		18 °C			
1 mol % of EGDM	25.960	0.822	1.998	0.981	8.998×10^{-6}
1.5 mol % of EGDM	14.748	0.761	1.105	0.991	2.703×10^{-6}
2 mol % of EGDM	10.143	0.681	0.989	0.993	0.379×10^{-6}
3 mol % of EGDM	6.943	0.584	0.971	0.996	0.691×10^{-6}
		38 °C			
1 mol % of EGDM	6.662	0.984	3.323	0.996	0.784×10^{-5}
1.5 mol % of EGDM	4.196	0.701	2.601	0.964	0.676×10^{-5}
2 mol % of EGDM	3.429	0.692	2.251	0.981	1.399×10^{-5}
3 mol % of EGDM	2.986	0.611	2.292	0.936	3.102×10^{-5}

* p(NIPA/HPMA)—poly(*N*-isopropyl acrylamide-*co*-2-hydroxypropylmethacrylate) hydrogel.

The kinetic constant, k, values are slightly smaller at 18 °C relative to data at 38 °C due to decrease of the free volume above the LCST, after the process of shrinking of the macromolecular chains [47,50].

The diffusion coefficient, *D*, values (calculated for the initial phase of swelling to 60%) at 18 °C are in the range 0.397×10^{-6}–8.998×10^{-6} cm^2/min. Above the LCST, at 38 °C, the diffusion coefficient values are in the range of 0.676×10^{-5}–3.102×10^{-5} cm^2/min, and indicate faster diffusion and growth in the values for all analyzed samples of p(NIPA/HPMA) hydrogels. The increased values of the *D* (diffusion coefficient) with the increase of temperature, showed known effect of temperature on sigmoidal non-Fickian diffusion process, also described in the literature [46,47].

2.3.3. The Order of the Swelling Reaction

The experimentally obtained and calculated normalized equilibrium swelling ratio values, the rate constant for I-order and II-order and the linear correlation coefficient values, R^2, for p(NIPA/HPMA) hydrogels were obtained by applying Equation (7) for the first order and Equation (8) for the second order of the swelling reaction at 18 and 38 °C (Table 3 and Figures 5 and 6).

Table 3. The experimentally and calculated normalized equilibrium swelling degree and kinetic parameters of the first- and second-order reaction at 18 °C and 38 °C.

p(NIPA/HPMA) * with	Equilibrium Swelling Ratio, α_e (exp)	Equilibrium Swelling Ratio, α_e (I-order)	Rate Constant (I-order), $K \cdot 10^3$, min^{-1}	Linear Correlation Coefficient (I-order), R^2	Equilibrium Swelling Ratio, α_e (II-order)	Rate Constant (II-order), $K \cdot 10^3$, min^{-1}	Linear Correlation Coefficient (II-order), R^2
			18 °C				
1 mol % of EGDM	25.960	27.214	2.35	0.982	26.113	26.286	0.999
1.5 mol % of EGDM	14.748	15.972	3.96	0.971	14.986	5.691	0.999
2 mol % of EGDM	10.143	11.266	3.23	0.934	10.534	13.118	0.999
3 mol % of EGDM	6.943	8.32	3.98	0.961	7.112	4.882	0.999
			38 °C				
1 mol % of EGDM	6.662	8.012	9.36	0.989	6.892	6.141	0.999
1.5 mol % of EGDM	4.196	4.949	7.33	0.991	4.464	6.841	0.999
2 mol % of EGDM	3.429	3.833	7.18	0.996	3.678	5.654	0.999
3 mol % of EGDM	2.986	3.524	8.59	0.991	3.274	4.991	0.998

* p(NIPA/HPMA)—poly(*N*-isopropyl acrylamide-*co*-2-hydroxypropylmethacrylate) hydrogel.

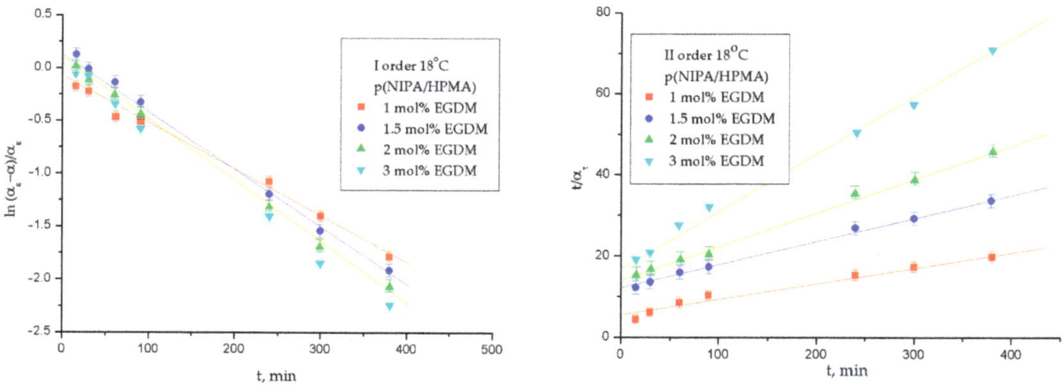

(a) (b)

Figure 5. Determination of the reaction order of the swelling process for series of p(NIPA/HPMA) hydrogels during 6 h at 18 °C (**a**) first order; (**b**) second order. In each figure, error bars represent the standard deviation of three replicates.

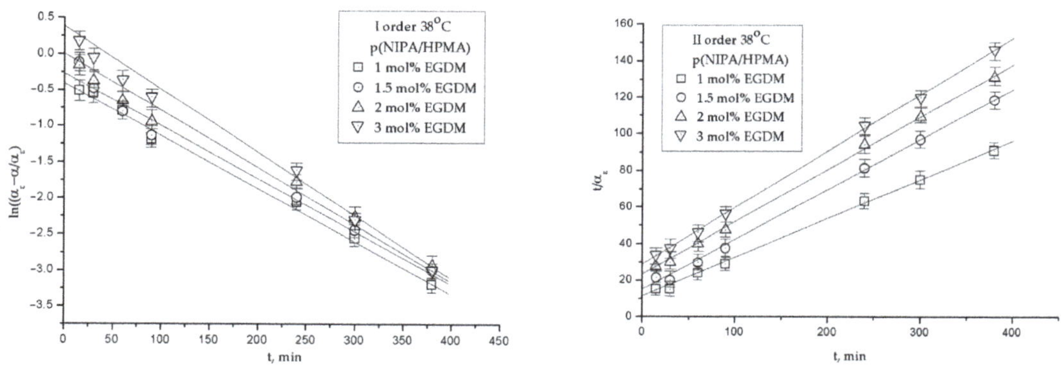

(a) (b)

Figure 6. Determination of the reaction order of the swelling process for series of p(NIPA/HPMA) hydrogels during 6 h at 38 °C: (**a**) first order; (**b**) second order. In each figure, error bars represent the standard deviation of three replicates.

The calculated and experimentally obtained values of normalized equilibrium swelling degree for p(NIPA/HPMA) hydrogels at the temperature of 18 °C were in a concordance (Table 3 and Figure 5). The values of the linear correlation coefficient, R^2, are nearly 1, and designate good agreement of the experimental values with the supposed order of swelling reaction. However, they are more suitable for the second-order reaction.

The calculated and experimentally obtained values of normalized equilibrium swelling degree for series of p(NIPA/HPMA) hydrogels at 38 °C (Table 3, Figure 6) showed similarity with data at the lower temperature (18 °C).

The calculated values of normalized equilibrium swelling degree and experimentally obtained at 38 °C for p(NIPA/HPMA) hydrogels were in agreement. The values of the linear correlation coefficient, R^2, are also close to 1, and indicate good a of the experimental results with the assumed swelling reaction order, but they correspond better to a second-order reaction. The values of the reaction rate constants, *K*, for all samples are within the same order of magnitude (Table 3). It might be said that the swelling process of p(NIPA/HPMA) hydrogels demonstrated a little higher deviation from the first order of the swelling mechanism at both tested temperatures (18 °C, 38 °C). Consequently, it could be concluded that the swelling process of p(NIPA/HPMA) hydrogels obeys the second order of the swelling reaction, which is comparable to the one described in previous work for analogous hydrogels [46].

2.4. Fourier Transform Infrared Spectra

2.4.1. FTIR Spectrum Analysis of Acetaminophen

The FTIR spectrum of acetaminophen is shown in Figure 7, and the wavenumber maximums values of the characteristic absorption bands are given in Table 4. In the FTIR spectrum of acetaminophen, there is an absorption band of medium strength, originating from the stretching vibrations of the amino group of the secondary amide, ν(NH), with an absorption maximum at 3324 cm^{-1}. The broad band with a maximum at 3162 cm^{-1} originates from the stretching vibrations of the phenolic hydroxyl group, ν(Ar-OH). The stretching C-H vibrations of the phenyl group, ν(C-H), give bands of weak intensity with a maximum at 3035 cm^{-1}. The stretching C-H vibrations of the methyl group give bands of weak intensity originating from $\nu_{as}(CH_3)$ with a maximum at 2929 cm^{-1} and $\nu_s(CH_3)$ with a maximum at 2880 cm^{-1}, as well as bending vibrations of medium intensity, $\delta_s(CH_3)$, whose bands appear in the spectrum of acetaminophen at 1370 cm^{-1} and 1328 cm^{-1}. The stretching vibrations of the C=O group from the secondary amide, ν(C=O), are characterized by a strong band appearing in the spectrum with a maximum at 1654 cm^{-1}. The maximum of the "amide band II" from the in-plane bending vibrations of the -NH group, δ(N-H), occurs at 1564 cm^{-1} in the spectrum of acetaminophen. The out-of-plane bending vibrations, γ(C-H), characteristic of para-disubstituted benzene, give a sharp band with a maximum at 837 cm^{-1}. The skeletal vibrations of the C=C bond from the phenyl group in the spectrum give characteristic strong bands with maximum at 1610 cm^{-1} and 1507 cm^{-1} [55].

2.4.2. FTIR Spectrum Analysis of p(NIPA/HPMA) Hydrogel with Loaded Acetaminophen

In the FTIR spectrum of the p(NIPA/HPMA) hydrogel with loaded acetaminophen (Figure 7 and Table 4), the absorption maximum from the stretching vibrations of the -OH group, ν(OH), can be observed at 3436 cm^{-1}, which was moved by 2 units toward lesser wavenumbers relative to the location in the hydrogel. In the FTIR spectrum of the p(NIPA/HPMA) hydrogel with loaded acetaminophen, the band of stretching vibrations of the N-H group, ν(NH), appears at 3326 cm^{-1} and was moved to higher wavenumbers by 7 units compared with the p(NIPA/HPMA) hydrogel, and by 2 units compared with acetaminophen. These shifts and the presence of a broadened band in the part around 3400 cm^{-1} indicate the formation of intermolecular hydrogen bonds among the chains of the p(NIPA/HPMA) hydrogel and acetaminophen NH•••O and OH•••O, which is also confirmed in the previous investigation [56]. In the FTIR spectrum of the p(NIPA/HPMA) hydrogel with loaded acetaminophen, there are no bands from the stretching vibrations of the C-H group, ν(C-H), and -OH from the phenolic group of acetaminophen, ν(Ar-OH), and it is considered that they are covered by the C-H vibrations from the p(NIPA/HPMA). "Amide band I", ν(C=O), appears in the spectrum of the p(NIPA/HPMA) hydrogel with loaded acetaminophen at 1656 cm^{-1} and was moved towards higher wavenumbers by 6 units in relation to the position in the spectrum of the hydrogel, and by 2 units in relation to position in the spectrum of acetaminophen. The maximum shift of the "amide band II", δ(N-H), which occurs at 1560 cm^{-1} in the spectrum of the p(NIPA/HPMA) hydrogel with loaded acetaminophen, by 16 units to higher wavenumbers compared to the position in the spectrum of the hydrogel and by 4 units towards lower wavenumbers compared to acetaminophen, confirms that the -NH group participates in the hydrogen bond construction. The band with the maximum at 1720 cm^{-1} in the spectrum of the p(NIPA/HPMA) hydrogel with loaded acetaminophen originates from the stretching vibrations of the keto group from the ester part of the structure, ν(C=O), and was shifted to lower wavenumbers by 8 units compared with the hydrogel, which indicates that it participated in the formation of intermolecular hydrogen bonds. The maximum of the band of hydroxyl group bending vibrations, δ(OH), in the spectrum of the p(NIPA/HPMA) hydrogel with loaded acetaminophen, occurs at 1453 cm^{-1} and was shifted by 11 units towards higher wavenumbers in relation to the position in the spectrum of acetaminophen, which confirms that the -OH group participates in building a hydrogen bond. The presence of acetaminophen within the p(NIPA/HPMA) hydrogel structure is confirmed by the appearance of characteristic bands' maximum values from the stretching vibrations of the C=C bond from the aromatic structure at 1610 and 1507 cm^{-1}. Small shiftings between p(NIPA/HPMA) copolymer and acetaminophen suggest that the mentioned interactions are of the non-covalent type. The specified shifts of the individual bands maximum values indicate the formed intermolecular interactions of hydrogen bond type between the hydrogel and acetaminophen over hydroxyl-, ester- and amino- groups. These studies are in agreement with available literature data [36,37].

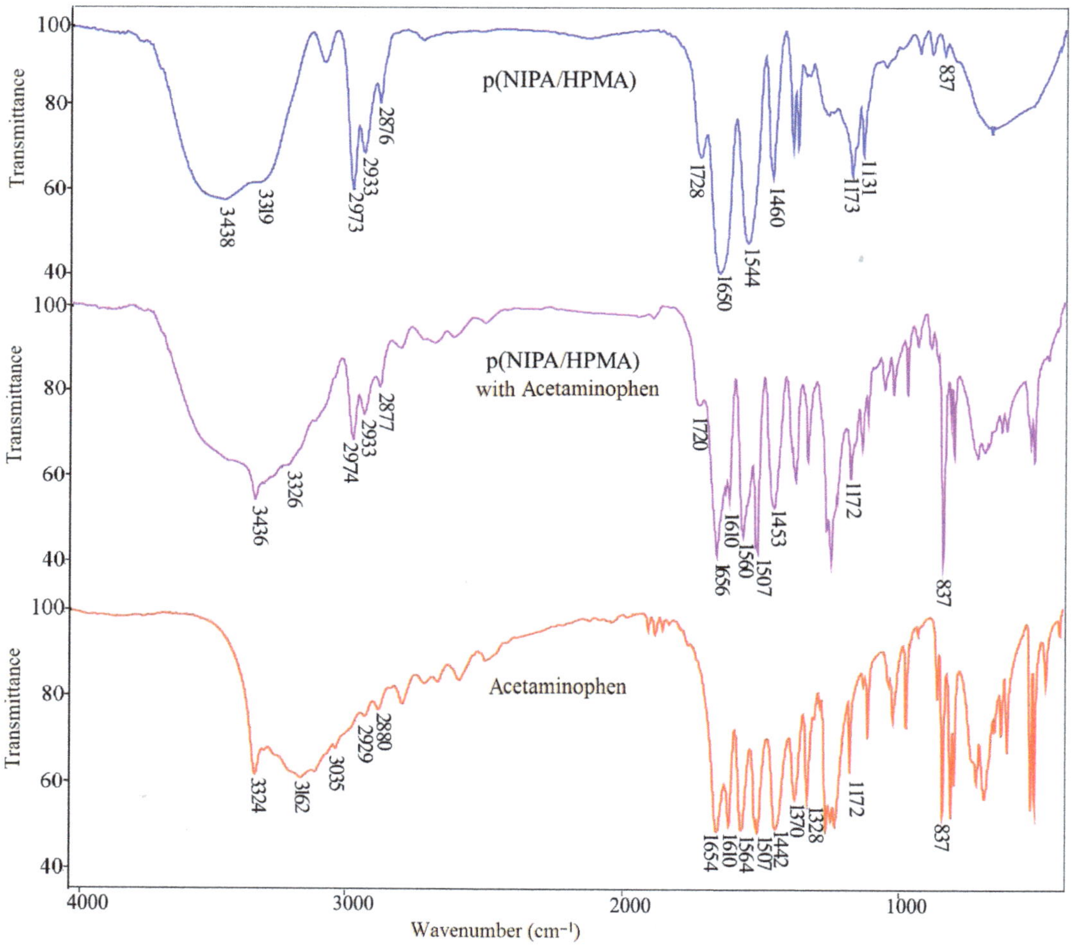

Figure 7. FTIR spectra of: synthesized p(NIPA/HPMA) hydrogel, p(NIPA/HPMA) hydrogel with loaded acetaminophen and acetaminophen.

Table 4. Characteristic absorption band maximum positions for p(NIPA/HPMA), acetaminophen, p(NIPA/HPMA) with loaded acetaminophen and value of wavenumbers shifts after drug incorporation.

Wavenumber of Functional Group, cm^{-1}			Functional Group	Shifts in Relation to the FTIR Spectra, cm^{-1}	
p(NIPA/HPMA) *	**Acetaminophen**	**p(NIPA/HPMA) with Acetaminophen**		**p(NIPA/HPMA)**	**Acetaminophen**
3438		3436	ν(OH)	−2	
3319	3324	3326	ν(NH)	+7	+2
	3162	-	ν(Ar-OH)		-
	3035	-	ν(C-H) Ar		-
2973	2929	2974	$\nu_{as}(CH_3)$	+1	
2933		2934	$\nu_{as}(CH_2)$	+1	
2876	2880	2877	$\nu_s(CH_3)$	+1	−3
1728		1720	ν(C=O) ester	−8	
1650	1654	1656	ν(C=O) amide I	+6	+2
	1610	1610	ν(C=C) Ar		0
1544	1564	1560	δ(N-H) amide II	+16	−4
	1507	1507	ν(C=C) Ar		0
1460, 1387	1442	1453, 1369	δ(OH)	−7, −9	+11
1367		1369	δ(CH)-isopropyl	+2	
1306		-	ν_s(C-N) amide III		
	1370, 1328	1327	$\delta_s(CH_3)$		−1
	1260	1250	ν(C-O)		
	1227	1243	ν(NCH)		+16
1173	1172	1172	ν(CN)	−1	0
1131		1131	ν_s(C-O)	0	
837	837	837	γ(CH)	0	0
	808	797	γ(CH)		−11
674	686	688	γ(OH)	+14	+2

* p(NIPA/HPMA)—poly(*N*-isopropyl acrylamide-*co*-2-hydroxypropylmethacrylate) hydrogel.

The corresponding FTIR spectrum of p(NIPA/HPMA) was analyzed in the previously mentioned paper by the authors [48] and used for this investigation.

The structural variations in the p(NIPA/HPMA) hydrogel with loaded acetaminophen are given as the FTIR spectrum in Figure 7, and the wavenumber maximum shifts of the characteristic absorption bands are given in Table 4.

2.5. Acetaminophen Loading Efficiency into p(NIPA/HPMA) Hydrogels

The loaded amount of acetaminophen into the three-dimensional networks of p(NIPA/HPMA) hydrogels was calculated based on the difference in sample weights before and after the loading process, i.e., its swelling in the acetaminophen solution. The acetaminophen loading efficiency into the p(NIPA/HPMA) hydrogels, obtained by applying Equation (9), is shown in Table 5.

Table 5. The amount of loaded acetaminophen (L_g) into the series of p(NIPA/HPMA) hydrogels and acetaminophen loading efficiency ($\eta_{acetaminophen}$).

p(NIPA/HPMA) * with	Amount of Loaded Acetaminophen, L_g, $mg/g_{xerogel}$	Acetaminophen Loading Efficiency, $\eta_{acetaminophen}$, %
1 mol % of EGDM	480.34	96.06
1.5 mol % of EGDM	460.84	92.16
2 mol % of EGDM	467.63	93.52
3 mol % of EGDM	480.81	96.14

* p(NIPA/HPMA)—poly(*N*-isopropyl acrylamide-*co*-2-hydroxypropylmethacrylate) hydrogel.

The available amount of acetaminophen for loading into each sample of p(NIPA/HPMA) hydrogels was in the range of 92.16–96.06% and confirms excellent loading efficiency.

The potential process of loading acetaminophen into the network structure of p(NIPA/HPMA) hydrogels with the possible formation of non-covalent hydrogen bonds between drug and lateral functional groups from hydrogel, according to FTIR analysis, is schematically presented in Figure 8.

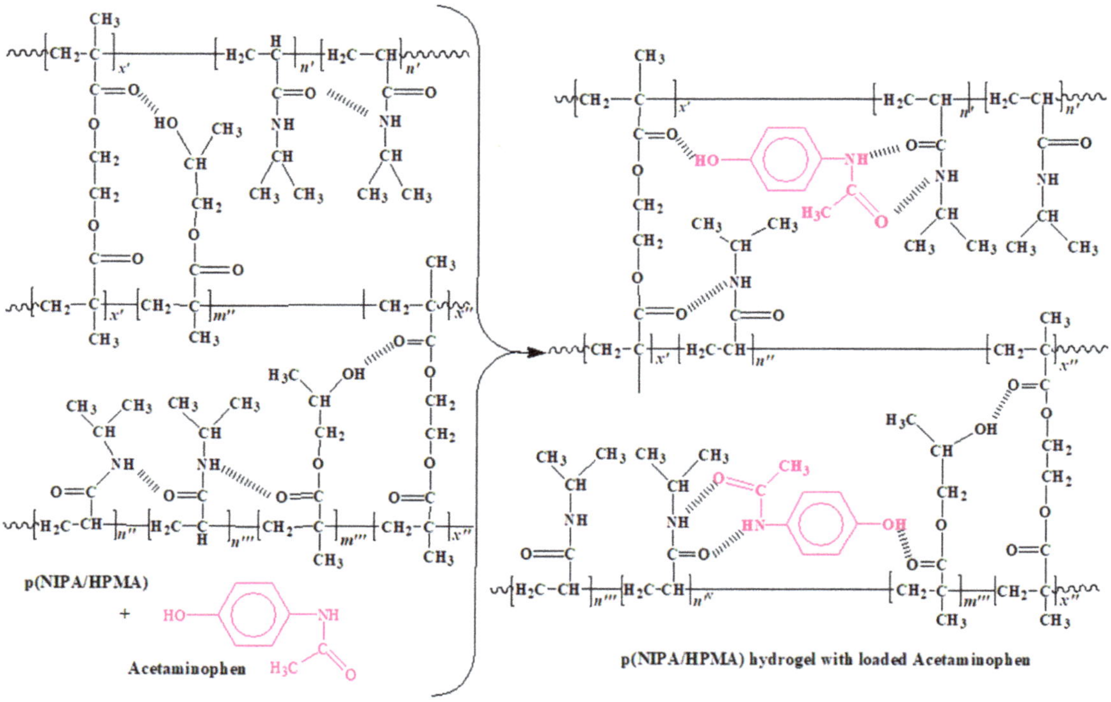

Figure 8. Drug delivery system based on acetaminophen (in the pink color) loaded into the p(NIPA/HPMA) hydrogels (the possible intramolecular interactions are designated).

2.6. Morphology Characterization

SEM micrographs of p(NIPA/HPMA) xerogels without and with loaded acetaminophen are shown in Figure 9. The microstructure of the p(NIPA/HPMA) xerogel after synthesis confirms the effect of copolymerization on gel morphology. It shows numerous micro-cracks and micro-pores. The pore diameters of the p(NIPA/HPMA) xerogel ranged from 0.2–0.80 μm, with an average pore diameter of 0.5 μm (Figure 9a) with interconnecting cracks. The pore distribution on the micrograph (Figure 9a) showed the presence of large pore diameters, which contribute to the drug diffusion process as much as the smaller area of the smaller pores.

The surface structure of p(NIPA/HPMA) with loaded acetaminophen in the xerogel state (Figure 9b) indicates the incorporation of acetaminophen inside the pores because it is significantly different compared with the structure of the empty xerogel (Figure 9a). Crystals of acetaminophen that are observed on the surface show similarity to the ones discussed in the previously published works [35].

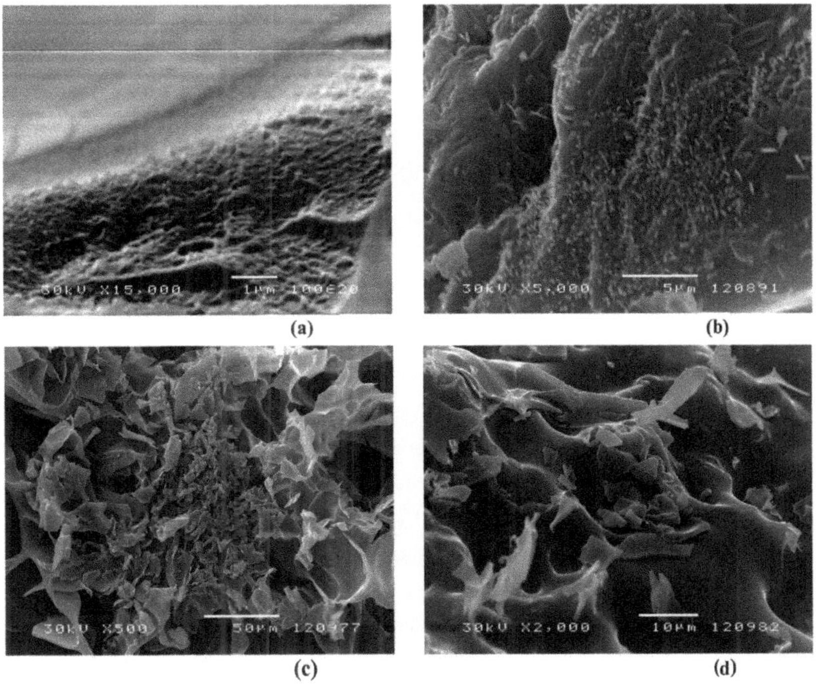

Figure 9. SEM micrographs of: (**a**) the p(NIPA/HPMA) in xerogel state, magnification 15,000× *g*, scale bar 1 μm; (**b**) p(NIPA/HPMA) with loaded acetaminophen in the xerogel state, magnification 5000× *g*, scale bar 5 μm; (**c**) p(NIPA/HPMA) with loaded acetaminophen in the equilibrium swollen state, magnification 500× *g*, scale bar 50 μm, (**d**) p(NIPA/HPMA) with loaded acetaminophen in the equilibrium swollen state, magnification 2000× *g*, scale bar 10 μm.

A clearer representation of the microsurface of p(NIPA/HPMA) hydrogel with loaded acetaminophen (swollen to equilibrium and freeze-dryed) is shown in Figure 9c,d. In these micrographs, the change in the hydrogel network structure is clearly visible, which is caused by the semi-homogeneous arrangement of acetaminophen crystals within the three-dimensional p(NIPA/HPMA) hydrogel network. The surface of p(NIPA/HPMA) with loaded acetaminophen indicates a nonuniformal dispersal of acetaminophen inside the hydrogel network. Certain areas are abundant with acetaminophen crystals that are retained on the three-dimensional microsurface. During the delivery process, they are the first to leave the network structure

Based on the results obtained by the FTIR and SEM methods, as well as the acetaminophen loading efficiency, it can be concluded that acetaminophen has shown excellent loading in the p(NIPA/HPMA) hydrogels, especially in the sample with 1 mol % of crosslinker, thanks to the weak interactions between the electronegative oxygen from -OH and the ester group C=C or nitrogen from the -NH group.

2.7. In Vitro Acetaminophen Delivery from p(NIPA/HPMA) Copolymers

The amount of released acetaminophen from p(NIPA/HPMA) hydrogels was determined by Equation (10), which corresponds to the linear part of the constructed calibration curve. The maximum in the HPLC chromatogram at a retention time Rt = 2.469 min (according determined RP HPLC settings) originates from acetaminophen, and it is presented in Figure 10a.

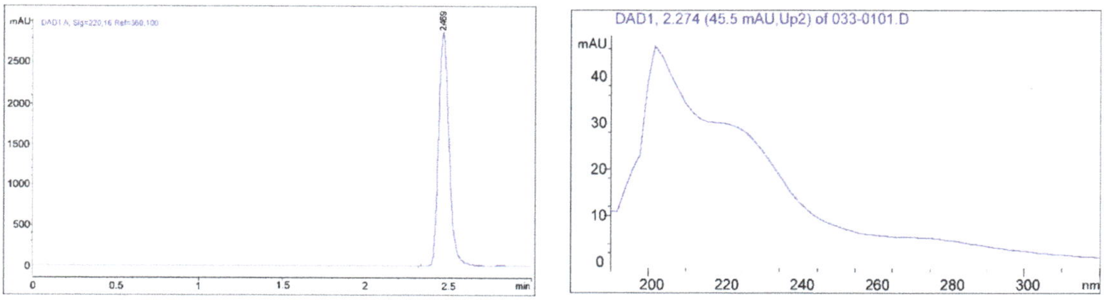

(**a**) (**b**)

Figure 10. Chromatogram of acetaminophen from RP HPLC (**a**) with UV spectrum from DAD detector (**b**).

The UV spectrum of acetaminophen, obtained by recording on a DAD detector, has shown two characteristic absorption maxima (Figure 10b). The primary maximum at λ_{max} = 205 nm originates from the $\pi \rightarrow \pi^*$ transition of C=C bonds from benzene, and the secondary one with a band at λ_{max} = 230 nm is the result of the $\pi \rightarrow \pi^*$ transition from the keto group. The bathochromic shift of the primary and secondary bands is the result of the conjugation of the unbound electrons of the substituent -OH and -NH groups with the π electrons of benzene [57].

In vitro cumulative acetaminophen release from the synthesized p(NIPA/HPMA) hydrogels at 38 °C during 24 h in the acidic fluid (with pH values of 2.2) and in the alkaline fluid (pH = 7.4) are given in Figure 11.

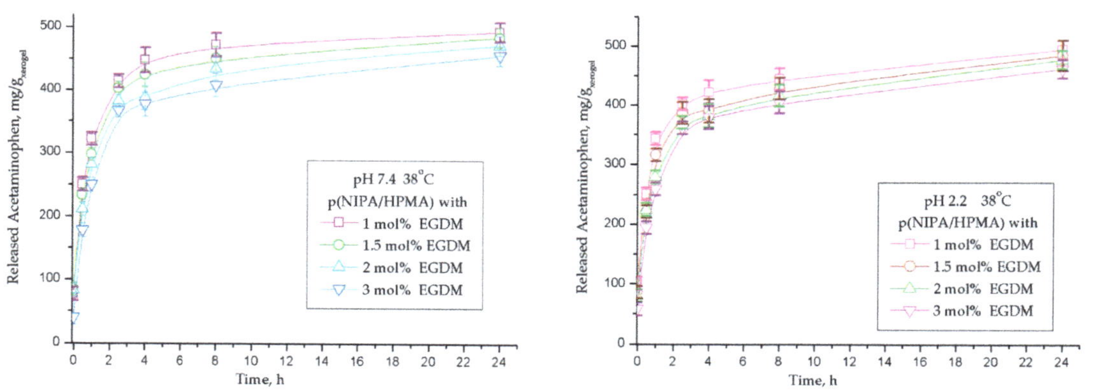

(**a**) (**b**)

Figure 11. In vitro cumulative acetaminophen release from series of p(NIPA/HPMA) hydrogels during 24 h at 38 °C at: (**a**) pH 7.4, (**b**) pH 2.2. In each figure, error bars represent the standard deviation of three replicates.

The content of released acetaminophen in the simulated gastrointestinal conditions (pH = 2.2 and pH = 7.4 at 38 °C) from samples of copolymer p(NIPA/HPMA) hydrogels is shown in Figure 11a,b, respectively. It can be noted that the p(NIPA/HPMA) hydrogel with the lowest crosslinking (sample with 1 mol % of crosslinker ethyleneglycoldimethacylate) released the highest quantity of acetaminophen: 492.77 mg/$g_{xerogel}$, or 98.56% of the loaded quantity at pH = 7.4, and 480.16 mg/$g_{xerogel}$, i.e., 96.03% at pH = 2.2 (Table 6). The smallest acetaminophen quantity was released from the sample with the highest crosslinking density (with 3 mol % of ethyleneglycoldimethacylate): 455.59 mg/$g_{xerogel}$, or 91.12% of the loaded quantity at pH = 7.4 and 451.58 mg/$g_{xerogel}$, i.e., 90.31% at pH = 2.2. The analysis of the obtained results shows a high similarity in the amount of acetaminophen released from all hydrogel samples under conditions with different pH values. Small differences in the quantity of released acetaminophen in an acidic environment compared with the weak alkaline fluid can be attributed to the influence of ionization of the -OH group of acetaminophen, because in the molecular structure of p(NIPA/HPMA) hydrogels, there are no functional groups that could be ionized in aqueous solution. The probable reason is an increase in network density and a smaller internal free volume that can accommodate the drug, which is in agreement with the swelling results. At 38 °C, which is above the LCST (and also the internal physiological body temperature), the p(NIPA/HPMA) hydrogels are in a contracted state because they undergo the phase transition. The intermolecular hydrogen bonds between the drug and the hydrogel are broken at this temperature, and acetaminophen is released from the hydrogel network. The drug release rate from the macromolecular network is dependent on the intermolecular bonds among the acetaminophen and the lateral groups of the hydrogel, which was also shown in the studies of other authors [36,38].

Table 6. Quantity of released acetaminophen from p(NIPA/HPMA) hydrogels (in mg/g and %) and kinetic parameters of diffusion (n, k and D) at 38 °C, pH 2.2 and pH 7.4.

Poly(*N*-isopropyl acrylamide-*co*-2-hydroxypropylmethacrylate) with	Quantity of Released Acetaminophen, mg/$g_{xerogel}$ %		Diffusion Exponent, n	Kinetic Constant, K, $min^{-1/2}$	Linear Correlation Coefficient, R^2	Diffusion Coefficient, D, cm^2/min
			pH 7.40			
1 mol % of EGDM	492.77	98.56	0.297	0.641	0.972	3.79×10^{-3}
1.5 mol % of EGDM	484.11	96.82	0.325	0.694	0.997	3.74×10^{-3}
2 mol % of EGDM	471.59	94.32	0.297	0.756	0.972	4.48×10^{-3}
3 mol % of EGDM	455.58	91.12	0.375	0.639	0.944	3.21×10^{-3}
			pH 2.2			
1 mol % of EGDM	480.16	96.03	0.486	0.747	0.881	4.38×10^{-3}
1.5 mol % of EGDM	474.78	94.96	0.541	0.737	0.846	4.26×10^{-3}
2 mol % of EGDM	465.66	93.12	0.533	0.722	0.961	4.10×10^{-3}
3 mol % of EGDM	451.43	90.32	0.662	0.666	0.961	3.48×10^{-3}

The mechanism of acetaminophen diffusion from p(NIPA/HPMA) hydrogels was assessed by fitting results of experimental release using Equations (4)–(7), and the kinetic parameters (n, k and D) are all presented in Table 6.

The kinetic parameters of the acetaminophen release from the p(NIPA/HPMA) hydrogels at slightly alkaline conditions (at pH 7.4) and only the sample with 1 mol % of crosslinker in acidic fluid (at pH 2.2) display that the mechanism of fluid transport belongs to Fickian diffusion, characterized by a fluid permeation process which is considerably slower compared with the relaxation of the macromolecular chains and is controlled by the diffusion (Table 6). Obtained results of kinetic parameters of the acetaminophen released from these hydrogels at acidic fluid (at pH 2.2) for samples with 1.5, 2 and 3 mol % of crosslinker show that the fluid transport mechanism follows non-Fickian diffusion process. Values of the diffusion coefficient, D, from all analyzed p(NIPA/HPMA) hydrogels were similar and show slight variations between samples.

It is notable that approximately full quantity of loaded acetaminophen (about 90–99%) was delivered during the first 24 h and only 1–10% persisted within the hydrogel's network. This result provides the possibility that the analysed drug delivery system based on acetaminophen and synthesized p(NIPA/HPMA) hydrogels could be suitable for extended release and administration once a day. The average adult dosage for fever and pain may be given depending on the formulation: parenterally every 4 h, orally immediate-release every 4 to 6 h or extended-release orally every 8 h [58,59]. Design of the new drug delivery system could be easier for administration, in comparison with conventional therapy. The advantages of the current study are evident after comparison with the published acetaminophen polymeric carriers mentioned in the introduction, which are presented in Table 7.

Table 7. The comparison of the acetaminophen release from different polymeric carriers.

Polymeric Carrier	Experimental Conditions and Released Acetaminophen	Mechanism of Acetaminophen Diffusion	Reference
Poly(*N*-isopropyl acrylamide-*co*-2-hydroxypropylmethacrylate) with ethyleneglycoldimethacrylate	98.56% at pH = 7.4, and 96.03% at pH = 2.2 during 24 h at 38 °C.	At pH = 7.4—Fickian diffusion. At pH = 2.2—non-Fickian diffusion (gels with 1.5, 2 and 3 mol % of EGDM).	[current study]
Poly(*N*-vinylpyrrolidinone-*co*-acrylic acid) (with 30 wt % AA) and polyethylene glycol 600 dimethacrylate	At pH 2 approx. 24 h pH 6.8 approx. 5 h pH 9 approx. 5 h at 37 °C.	-	[36]
Sodium alginate and *N*-isopropyl acrylamide crosslinked with *N,N'*methylenebisacrylamide	At 37 °C in pH 2.2 for 9 days ~90%.	The first-order kinetic. Non-Fickian (anomalous) diffusion.	[37]
Poly(acrylamide-*co*-itaconic acid) crosslinked with *N,N*-methylenebisacrylamide	At 37 °C during 8 h: at pH = 2.2 ~20–55% pH = 4.5 ~90–99% pH = 6.8 ~90–99%.	Slow drug release under acidic conditions and rapid release at higher pH value. Fickian diffusion.	[38]
Polyacrylamide with 4-[(4-methacryloyloxy) phenylazo] benzenesulfonic acid and *N,N'*-hexylenebismethacrylamide	The photoregulated release at 353 nm for 120 min of irradiation, a total of 83.6% was released in aqueous HEPES buffer pH 7.16.	-	[39]
Poly(2-hydroxyethyl methacrylate-*co*-*N*-vinyl-2-pyrrolidone) with *N,N'*-methylenebisacrylamide	at 37 °C and SGF pH = 1.2, SIF pH = 7.2 ~95% during 240 min (4 h).	The tablet from the polymer NVP3 crumbled within an hour of immersion, resulting in burst release. Non-Fickian (anomalous) diffusion.	[40]
Hydroxypropyl cellulose with polyacrylamide, 25/75 wt %	In deionized water, pH 7 in phosphate buffer, pH 7.38 at 35 °C, 37 °C, 39 °C, during 6 h.	Drug was crystallized on the gel surface. Fickian diffusion.	[41]
Polyurethane nanocomposite hydrogels PU/PEG 4000 with 1% of organofillized montmorillonite (Cloisite® 30B)	In water for 24 h at 23 °C and 37 °C.	Easier release from nanocomposites than from a pure hydrogel matrix. Non-Fickian (anomalous) diffusion.	[42]
Poly(N-isopropylacrylamide-acrylamide) with *N,N*-methylenbisacrylamide, and *N,N,N,N*-tetra-methylethylenediamine	pH= 7.4 at 27 °C, 32 °C, 41 °C, 44 °C ± 0.1 °C	The pore mechanism of drug transport.	[43]
Poly(vinyl alcohol) with nanofibrillated cellulose, (NFC)/PVA, and 2,2,6,6-tetramethylpiperidine-*N*-oxyl-oxidized nanofibrillated cellulose (TNFC)/PVA without any chemical linkers	At 37 °C and phosphate buffer, 14% from (NFC)/PVA and about 28% from (TNFC)/PVA over 144 h (6 days).	Diffusion-controlled and burst release, with small fractions of relaxation-induced and prolonged-diffusional release.	[44]

The obtained results presented in this study show that thermosensitive p(NIPA/HPMA) hydrogels with acetaminophen would be of interest as drug delivery systems for further testing, especially for extended acetaminophen release.

3. Conclusions

Temperature-sensitive p(NIPA/HPMA) hydrogels with 10 mol % of comonomer 2-hydroxypropylmethacrylate and 1, 1.5, 2 and 3 mol % of ethyleneglycoldimethacylate were successfully synthesized. DSC thermograms of p(NIPA/HPMA) hydrogels showed two sharp peaks from melting of crystalline regions (first from temperature 153.90 °C to 156.49 °C and second from 156.98 °C to 160.73 °C). Also, they exhibited more than one peak with weak intensity from glass transition of amorphous regions (first in the range of 63.41–64.79 °C, second in the range of 77.53–86.47 °C and third at 134.81–131.76 °C). DSC results confirmed that p(NIPA/HPMA) hydrogels are convenient for sterilizing and enable biomedical application as drug carriers. All p(NIPA/HPMA) hydrogels show temperature sensitivity, and the exhibited swelling behavior is known as negatively thermo-sensitive. The fluid transport into the p(NIPA/HPMA) network at 18 °C and 38 °C exhibited an anomaly in the fluid transport mechanism, i.e., the swelling dynamics were controlled by both diffusion of fluid and macromolecular chain relaxation, and did not obey the Fickian law. The mathematical modeling of the swelling process indicates that hydrogels follow a second-order kinetics. The morphological and structural characteristics in the p(NIPA/HPMA) hydrogels with loaded acetaminophen were verified using SEM and FTIR methods. The loading efficiency of acetaminophen into the p(NIPA/HPMA) hydrogels is in the range of 92.16–96.06% and confirms excellent loading efficiency. The kinetic parameters of the acetaminophen release from the p(NIPA/HPMA) hydrogels at pH 7.4 and for hydrogel sample with 1 mol % of crosslinker at pH 2.2 correspond to Fickian diffusion, and p(NIPA/HPMA) hydrogels with 1.5, 2 and 3 mol % of crosslinker at pH 2.2 correspond to non-Fickian diffusion. Applied p(NIPA/HPMA) hydrogels could be suitable as acetaminophen carriers due to their superior thermosensitive properties. The in vitro acetaminophen delivery from the temperature-sensitive p(NIPA/HPMA) hydrogels proved that it might be interesting for further studies on extended acetaminophen delivery.

4. Materials and Methods

4.1. Reagents

N-Isopropyl acrylamide 99%, 2-hydroxypropylmethacrylate 96.5% and α,α'-azoisobutyronitrile 98% (Acros Organics, Morris Plains, NJ, USA); methanol 99.9% HPLC grade and ethyleneglycoldimethacylate 97% (Fluka Chemical Corp., Buchs, Switzerland); 4-Acetamidophenol $\geq$ 98% (Acros Organics N.V., Fair Lawn, NJ, USA); potassium bromide, KBr, for IR spectroscopy $\leq$ 100% (Merck KGaA, Darmstadt, DE, USA); methanol p.a. (Unichem, Belgrade, RS, USA); acetone (Centrohem, Belgrade, RS, USA).

4.2. Hydroges Synthesis

Hydrogel networks based of both comonomers *N*-isopropyl acrylamide and 2-hydroxypropylmethacrylate (10 mol % related to the quantity of NIPA monomer) were synthesized using the free radical polymerization method [9,43,44,47]. Ethyleneglycoldimethacylate (0.5, 1.0, 1.5, 2.0 and 3.0 mol % related to the total comonomers mass) was added to crosslinking reactants. Acetone was used to dilute reaction mixture, which was initiated by 2.7 mol % of α,α'-azoisobutyronitrile relative to the total comonomers mass. All samples after homogenization were syringed in glass ampoules, and afterward, they were heat sealed. The polymerization was performed as following: 80 min at 75 °C, 100 min at 80 °C and 22 min at 85 °C, until all unsaturated bonds of α,α'-azoisobutyronitrile were cleaved to produce radicals. Synthesized p(NIPA/HPMA) hydrogels were cut into discs ($d \times l = 5 \times 2$, d, mm, the diameter, and l, mm, the thickness) after separation from the glass ampoules. For the removal of unreacted reactants, hydrogels were extracted using methanol as solvent during 48 h. After that, hydrogels were subsequently immersed in methanol/distilled water solutions 100/0, 75/25, 50/50, 25/75, for a day, and 0/100% for 2 days, and then dehydrated to constant mass to reach the xerogel stage (at 40 °C). Figure 12 shows a flow chart of the polymerization process.

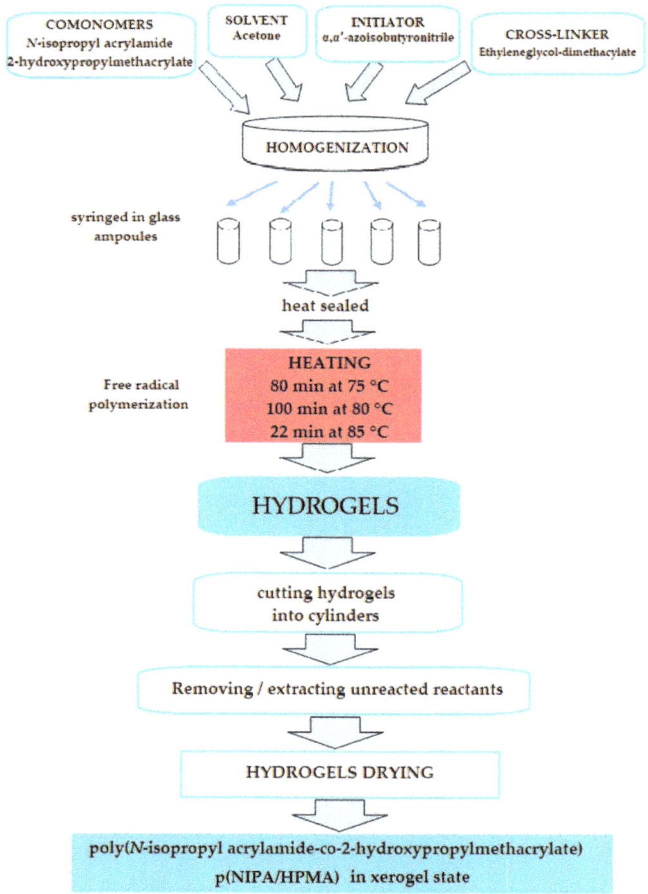

Figure 12. Flow chart of p(NIPA/HPMA) hydrogel polymerization process.

4.3. DSC Method

Thermal characteristics of the p(NIPA/HPMA) copolymers were examined using the differential scanning calorimetry (DSC) method. Hermetically sealed aluminum vessels with 3–5 mg of xerogels in powder state were heated in a nitrogen atmosphere from ambient temperature up to 180 °C (heating dynamics 10 °C/min^{-1}). The sensitivity of the used instrument DSC Q20 (TA instruments, New Castle, DE, USA) was 10 mVcm^{-1}. Standard calibration was accomplished by using indium.

4.4. Swelling Behavior

The swelling process of the p(NIPA/HPMA) hydrogels in the xerogel state was supervised gravimetrically at 18 °C (as the lower recommended room temperature) and 38 °C (as an inner body temperature) [46,47]. All samples were first measured, and after that, immersed in the fluid (pH = 7.4). At determined times, the hydrogels were taken out, their masses were measured, and then they were returned until constant mass was achieved. The percentage of swelling degree, α(%), was determined according to Equation (1) [45] and the equilibrium swelling degree, α_e, was determined according to Equation (2) [9,21,60]:

$$\alpha(\%) = \frac{m_t - m_0}{m_0} \times 100 \quad (1)$$

$$\alpha_e = \frac{m_e - m_0}{m_0} \quad (2)$$

m_t—the mass of the hydrogel swollen at the time t, m_e—the mass of the hydrogel swollen at equilibrium, m_0—the initial mass of xerogel.

4.4.1. Temperature Sensitivity

The temperature sensitivity of p(NIPA/HPMA) hydrogels was supervised in a water bath (Sutjeska, Belgrade, RS, USA) during heating from 5 to 60 °C. The measured p(NIPA/HPMA) xerogels were swollen in distilled water for 24 h at determined temperatures to reach equilibrium, the mass of the swollen hydrogel, m_t, was measured, and after that, the hydrogels were immersed into distilled water. The percentage of swelling degree was evaluated by Equation (1).

4.4.2. Kinetic Analysis

The nature of the fluid diffusion process into the analyzed p(NIPA/HPMA) hydrogel samples, throughout the swelling and acetaminophen release processes, was evaluated using Fick's law (3) in order to calculate the fractional sorption, F, and fitting the experimental results [61,62]:

$$F = \frac{M_t}{M_e} = k \times t^{n} \quad (3)$$

M_t—content of the absorbed liquid at the time t, M_e—content of the absorbed liquid in the equilibrium, k, $min^{1/n}$—the characteristic kinetic constant, n—the diffusion exponent (indicative of the transport mechanism). If the value of n < 1/2, the mechanism of fluid transport obeys the Fickian diffusion; a value of 1/2 < n < 1 designates sigmoidal (anomalous), or non-Fickian diffusion characterized by the water-sorption process. If the value of n = 1/2, the diffusion grade is significantly lesser than the relaxation of the macromolecular chains (Case I), while n = 1 points to the diffusion of fluid being significantly faster in relation to the polymer chains relaxation (Type II, Case II). If the value of n > 1, the relaxation of polymer chains controls swelling, and this case is named Super Case II, Type III, or Case III.

By logarithmization of Equation (3), Equation (4) is obtained [61,62]:

$$\ln F = \ln\frac{M_t}{M_e} = \ln k + n \cdot \ln t \quad (4)$$

The values of the exponents n and k are calculated from the slope and intercept of the linear relationship between ln$F$ and ln$t$. The diffusion coefficient D is calculated for the initial phase of swelling to 60% from Equation (5) [61,62]:

$$\frac{M_t}{M_e} = 4 \cdot \left(\frac{Dt}{\pi l^2}\right)^{0.5} \quad (5)$$

l—the dried sample thickness (cm), and D—the diffusion coefficient (cm^2/min). By taking the logarithm of Equation (4), Equation (6) is obtained (the linear relationship between ln(M_t/M_e) and ln$t$):

$$\ln\frac{M_t}{M_e} = \left(\frac{4D^{0.5}}{\pi^{0.5} l}\right) + \frac{1}{2}\ln t \quad (6)$$

The diffusion coefficient is calculated from the intercept of the linear relationship between ln(M_t/M_e) and ln$t$, where t is the time for which the hydrogel absorbs half of the total amount of fluid [61,62].

4.4.3. The Order of the Swelling Reaction

The order of the swelling process was calculated using the results of swelling kinetics, for synthesized p(NIPA/HPMA) hydrogels [60,63,64]. The normalized swelling ratio at time t is calculated applying the following expression (7):

$$\ln \frac{\alpha_e - \alpha_t}{\alpha_e} = -Kt \tag{7}$$

If the swelling process obeys first-order kinetics, this dependence (7) is linear with a slope of −K. Should this not happen, it is than checked whether the swelling process follows a second-order reaction or not. The swelling kinetics obey a second-order reaction described by Robinson–Schott's Equation (8) [63–65]:

$$\frac{t}{\alpha_t} = \frac{1}{K{\alpha_e}^2} + \frac{t}{\alpha_e} \tag{8}$$

The constructed plot of t/α_e versus t, according to the expression (8), should be linear. The normalized equilibrium swelling ratio, the second-order rate constant as an intercept of $1/K\alpha_e$ and the linear correlation coefficient are calculated using the experimental results by expressions 7 and 8.

4.5. Acetaminophen Loading into the p(NIPA/HPMA)

Acetaminophen solution, 40 mg/cm^3, was prepared by dissolving the drug in a solvent mixture of methanol/distilled water, 80/20 v/v, at a room temperature. With the aim of loading the drug into the p(NIPA/HPMA) hydrogels, measured samples (at about 0.10 g) in the xerogel state were swollen in the prepared acetaminophen solution for 48 h at 5 °C (stored in refrigerator) to reach equilibrium state. The acetaminophen maximum available amount for loading into the p(NIPA/HPMA) hydrogel was 500 mg/$g_{xerogel}$. Swollen p(NIPA/HPMA) hydrogels with loaded acetaminophen were washed using distilled water to remove excess acetaminophen, then dried to remove water and methanol, and measured in the xerogel state. Acetaminophen loading efficiency, $\eta(\%)$, was calculated using Equation (9) [43–45]:

$$\eta(\%) = \frac{L_g}{L_u} \times 100 \tag{9}$$

L_g (mg/$g_{xerogel}$)—the acetaminophen loaded mass into p(NIPA/HPMA) hydrogel sample,

L_u (mg/$g_{xerogel}$)—the acetaminophen maximum available mass for loading into p(NIPA/HPMA) hydrogel sample.

4.6. In Vitro Acetaminophen Release Study

Acetaminophen release from the p(NIPA/HPMA) carrier with loaded drug was studied in vitro under simulated physiological conditions. The p(NIPA/HPMA) hydrogel samples with loaded acetaminophen were overflowed with 7 cm^3 of alkaline or acidic fluids (either sodium hydroxide at pH 7.4 or hydrochloric acid at pH 2.2) and thermostated at 38 °C. Acetaminophen release was monitored during 24 h. For RP HPLC analysis, aliquot parts of the flud with released acetaminophen (100 µL) were sampled in fixed times, and methanol was added and then filtered for RP HPLC analysis through a cellulose membrane filter (0.45 µm). The HPLC Agilent 1100 Series with diode-array detector, DAD 1200 Series (Agilent Technologies, Santa Clara, CA, USA) and column Supelcosil LC-18-DB, Hypersil GOLD 25 cm × 4.6 mm, 5 µm particle size (Supelco, Sigma-Aldrich Chemie GmbH, Taufkirchen, Germany) devices were used at a working column temperature of 25 °C, with methanol for HPLC as eluent (flow 1 cm^3/min), 20 µL injected volumes of each sample, at detector wavelength 220 nm for the acetaminophen.

Aimed at the calibration curve plot construction, first a set of standard solutions containing known concentrations of acetaminophen analyte were made ready for use.

Then, they were filtered over a cellulose membrane filter (0.45 μm) for RP HPLC analysis. Agilent ChemStation software was used for analysis of each recorded chromatograms. The linear part of the calibration curve ranged from 0.005 to 0.800 mg/cm^3 with the linear correlation coefficient R^2 = 0.9994, and the quantity of acetaminophen released from p(NIPA/HPMA) hydrogels was calculated applying Equation (10):

$$c_{\text{acetaminophen}} = \frac{A - 165.6}{14{,}756.4} \tag{10}$$

where $c_{\text{acetaminophen}}$, mg/cm^3 is the acetaminophen amount and A, mAUs, is the peak area.

4.7. Characterization

4.7.1. FTIR Method

Acetaminophen, both samples in the xerogel state of the obtained pristine p(NIPA/HPMA) copolymer and p(NIPA/HPMA) with entrapped acetaminophen, were turned into powder form with an Amalgamator (WIG-L-Bug, Dentsply RINN, a Division of Dentsply International Inc., Smile Way, York, PA, USA). All samples in the powder state (0.9 mg) with the 150 mg of KBr were vacuumed and pressed under 200 MPa to create transparent tablets. FTIR spectra were scanned (16 scans) using an FTIR spectrophotometer, Bomem Hartmann & Braun MB-series (Hartmann & Braun, Baptiste, CA, USA), at the resolution of 2 cm^{-1} in the near IR range (4000–400 cm^{-1} wavenumbers). Win-Bomem Easy software was used for FTIR spectra analysis.

4.7.2. Freeze-Drying of Hydrogels

Freeze-drying of synthesized p(NIPA/HPMA) hydrogels without and with loaded acetaminophen was carried out on a LH Leybold, Lyovac GT2 device (Frenkendorf, Switzerland). The p(NIPA/HPMA) hydrogels swollen to equilibria were firstly rapidly frozen at −22 °C over 24 h. Then, absorbed fluid was removed at −30 °C at a high vacuum (0.05 kPa) over 14 h in the sublimation subphase. In the next subphase, isothermal desorption, the hydrogels were kept at 20 °C through 6 h at the same vacuum (0.05 kPa). Freeze-dried p(NIPA/HPMA) hydrogels without and with loaded acetaminophen were packed and kept in the refrigerator.

4.7.3. Scanning Electron Microscopy

The morphology of selected p(NIPA/HPMA) hydrogels freeze-dried in the equilibrium swollen state, pure and with acetaminophen loaded, was observed by scanning electron microscopy. Firstly, all samples were covered with a gold/palladium alloy (85/15) under vacuum conditions with a sprayer, JEOL Fine Coat JFC 1100E Ion Sputter (JEOL Co., Tokyo, Japan), and, after that, scanned on a JEOL Scanning Electron Microscope JSM-5300 (JEOL Ltd., Tokyo, Japan).

4.8. Statistical Analysis

The data obtained in the experiments of swelling and in vitro acetaminophen release were evaluated using one-way analysis of variance (ANOVA) using the SPSS statistical package. Statistical differences results ($p \leq 0.05$) were considered significant. All results were determined to be within the 95% confidence level for reproducibility.

5. Patents

Granted patent RS53220B: Ilić-Stojanović, S.; Nikolić, Lj.; Nikolić, V.; Petrović, S.D.; Stanković, M. Process for synthesis of thermosensitive hydrogels and pharmaceutical applications, The Intellectual Property Office of the Republic of Serbia, https://worldwide.espacenet.com/patent/search/family/046025219/publication/RS53220B?q=RS53220B (accessed on 1 April 2023).

Author Contributions: Conceptualization, S.I.-S., L.N., V.N. and S.D.P.; methodology, S.I.-S., L.N., V.N., I.R., S.C. and S.D.P.; validation, S.I.-S., L.N., V.N., I.R., S.C. and S.D.P.; investigation, S.I.-S., L.N., V.N., I.R. and S.C.; writing—original draft preparation, S.I.-S., L.N., I.R. and V.N.; writing—review and editing, S.I.-S., L.N., V.N., I.R., S.C. and S.D.P. All authors have read and agreed to the published version of the manuscript.

Funding: This research received no external funding.

Institutional Review Board Statement: Not applicable.

Informed Consent Statement: Not applicable.

Data Availability Statement: Not applicable.

Acknowledgments: The authors wish to express their gratitude to the Ministry of Science, Technological Development and Innovation of the Republic of Serbia, Programs for financing scientific research works, ev. no. 451-03-47/2023-01/200133, University of Niš, Faculty of Technology and ev. no. 451-03-47/2023-01/200134, University of Novi Sad, Faculty of Technology.

Conflicts of Interest: The authors declare no conflict of interest.

References

1. Koetting, M.C.; Peters, J.T.; Steichen, S.D.; Peppas, N.A. Stimulus-responsive hydrogels: Theory, modern advances, and applications. *Mater. Sci. Eng. R Rep.* **2015**, *93*, 1–49. [CrossRef]
2. Chaterji, S.; Kwon, I.K.; Park, K. Smart polymeric gels: Redefining the limits of biomedical devices. *Prog. Polym. Sci.* **2007**, *32*, 1083–1122. [CrossRef]
3. Zarrintaj, P.; Jouyandeh, M.; Ganjali, M.R.; Hadavand, B.S.; Mozafari, M.; Sheiko, S.S.; Vatankhah-Varnoosfaderani, M.; Gutiérrez, T.J.; Saeb, M.R. Thermo-sensitive polymers in medicine: A review. *Eur. Polym. J.* **2019**, *117*, 402–423. [CrossRef]
4. Vanparijs, N.; Nuhn, L.; De Geest, B.G. Transiently thermoresponsive polymers and their applications in biomedicine. *Chem. Soc. Rev.* **2017**, *46*, 1193–1239. [CrossRef] [PubMed]
5. Nagase, K.; Yamato, M.; Kanazawa, H.; Okano, T. Poly(N-isopropylacrylamide)-based thermoresponsive surfaces provide new types of biomedical applications. *Biomaterials* **2018**, *153*, 27–48. [CrossRef] [PubMed]
6. Dušek, K.; Dušková-Smrčková, M. Volume phase transition in gels: Its discovery and development. *Gels* **2020**, *6*, 22. [CrossRef] [PubMed]
7. Siegel, R.A. *Responsive Gels: Volume Transitions I.*; Dusek, K., Ed.; Advanced in Polymer Science; Springer-Verlag: Berlin/Heidelberg, Germany, 1993; Volume 109, pp. 233–267.
8. Barrett, C.J.; Mamiya, J.I.; Yager, K.G.; Ikeda, T. Photo-mechanical effects in azobenzene-containing soft materials. *Soft Matter* **2007**, *3*, 1249–1261. [CrossRef]
9. Ilić-Stojanović, S. *Synthesis and Characterization of Negatively Thermosensitive Hydrogels*; LAP LAMBERT Academic Publishing: Saarbrücken, Germany, 2015; ISBN 978-3-659-47484-2.
10. Lizundia, E.; Meaurio, E.; Laza, J.M.; Vilas, J.L.; Isidro, L.L. Study of the chain microstructure effects on the resulting thermal properties of poly (L-lactide)/poly (N-isopropylacrylamide) biomedical materials. *Mater. Sci. Eng. C* **2015**, *50*, 97–106. [CrossRef]
11. Heskins, M.; Guillet, J.E. Solution properties of poly(N-isopropylacrylamide). *J. Macromol. Sci.—Chem.* **1968**, *2*, 1441–1455. [CrossRef]
12. Bawa, P.; Pillay, V.; Choonara, Y.E.; Du Toit, L.C. Stimuli-responsive polymers and their applications in drug delivery. *Biomed. Mater.* **2009**, *4*, 022001. [CrossRef]
13. Coughlan, D.C.; Quilty, F.P.; Corrigan, O.I. Effect of drug physicochemical properties on swelling/deswelling kinetics and pulsatile drug release from thermoresponsive poly(N-isopropyl acrylamide) hydrogels. *J. Control. Release* **2004**, *98*, 97–114. [CrossRef]
14. Urošević, M.Z.; Nikolić, L.B.; Ilić-Stojanović, S.; Zdravković, A.; Nikolić, V.D. Synthesis and characterization of poly(N-isopropylmethacrylamide-*co*-N-isopropyl acrylamide) copolymers. *Hem. Ind.* **2020**, *74*, 103–117. [CrossRef]
15. Urošević, M.Z.; Nikolić, L.B.; Ilić-Stojanović, S.S.; Nikolić, V.D.; Petrović, S.M.; Zdravković, A.S. Hydrogels based on N-isopropylmethacrylamide and N-isopropyl acrylamide. *Adv. Technol.* **2018**, *7*, 79–91. [CrossRef]
16. Gheysoori, P.; Paydayesh, A.; Jafari, M.; Peidayesh, H. Thermoresponsive nanocomposite hydrogels based on Gelatin/poly (N–isopropylacrylamide)(PNIPAM) for controlled drug delivery. *Eur. Polym. J.* **2023**, *186*, 111846. [CrossRef]
17. Chung, J.E.; Yokoyama, M.; Yamato, M.; Aoyagi, T.; Sakurai, Y.; Okano, T. Thermo-responsive drug delivery from polymeric micelles constructed using block copolymers of poly(N-isopropylacrylamide) and poly (butylmethacrylate). *J. Control. Release* **1999**, *62*, 115–127. [CrossRef] [PubMed]
18. Radu, I.C.; Mirica, A.C.I.; Hudita, A.; Tanasa, E.; Iovu, H.; Zaharia, C.; Galateanu, B. Thermosensitive Behavior Defines the Features of Poly(N-isopropylacrylamide)/Magnetite Nanoparticles for Cancer Management. *Appl. Sci.* **2023**, *13*, 4870. [CrossRef]
19. Huang, Y.C.; Zeng, Y.J.; Lin, Y.W.; Tai, H.C.; Don, T.M. In Situ Encapsulation of Camptothecin by Self-Assembly of Poly (acrylic acid)-b-Poly (N-Isopropylacrylamide) and Chitosan for Controlled Drug Delivery. *Polymers* **2023**, *15*, 2463. [CrossRef] [PubMed]

20. Nagase, K.; Okano, T. Thermoresponsive-polymer-based materials for temperature-modulated bioanalysis and bioseparations. *J. Mater. Chem. B* **2016**, *4*, 6381–6397. [CrossRef]
21. Cai, W.; Anderson, E.C.; Gupta, R.B. Separation of lignin from aqueous mixtures by ionic and nonionic temperature-sensitive hydrogels. *Ind. Eng. Chem. Res.* **2001**, *40*, 2283–2288. [CrossRef]
22. Ciarleglio, G.; Toto, E.; Santonicola, M.G. Conductive and Thermo-Responsive Composite Hydrogels with Poly (N-isopropylacrylamide) and Carbon Nanotubes Fabricated by Two-Step Photopolymerization. *Polymers* **2023**, *15*, 1022. [CrossRef]
23. 4-Acetamidophenol, 98% Material Safety Data Sheet. Available online: https://fscimage.fishersci.com/msds/00229.htm (accessed on 12 March 2023).
24. USP. *U.S. Pharmacopeia National Formulary 2018: USP 41 NF 36*; United States Pharmacopeial Convention: Rockville, MD, USA, 2018.
25. Čeković, Ž. *Organske Sinteze: Reakcije i Metode*; Zavod za Udžbenike i Nastavna Sredstva: Belgrade, Serbia, 2006. (In Serbian)
26. Bertolini, A.; Ferrari, A.; Ottani, A.; Guerzoni, S.; Tacchi, R.; Leone, S. Paracetamol: New vistas of an old drug. *CNS Drug Rev.* **2006**, *12*, 250–275. [CrossRef]
27. Knights, K.; Rowland, A.; Darroch, S.; Bushell, M. *Pharmacology for Health Professionals*, 6th ed.; Elsevier: Melbourne, Australia, 2023.
28. Byrant, B.; Knights, K.; Salerno, E. *Pharmacology for Health Professionals*; Elsevier: Melbourne, Australia, 2003; p. 270.
29. Diener, H.C.; Tfelt-Hansen, P.; Dahlöf, C.; Láinez, M.J.; Sandrini, G.; Wang, S.J.; Neto, W.; Vijapurkar, U.; Doyle, A.; Jacobs, D. MIGR-003 Study Group. Topiramate in migraine prophylaxis: Results from a placebo–controlled trial with propranolol as an active control. *J. Neurol.* **2004**, *251*, 943–950. [PubMed]
30. Qin, N.; Zhang, S.P.; Reitz, T.L.; Mei, J.M.; Flores, C.M. Cloning, expression, and functional characterization of human cyclooxygenase-1 splicing variants: Evidence for intron 1 retention. *J. Pharmacol. Exp. Ther.* **2005**, *315*, 1298–1305. [CrossRef] [PubMed]
31. Tripathy, D.; Grammas, P. Acetaminophen protects brain endothelial cells against oxidative stress. *Microvasc. Res.* **2009**, *77*, 289–296. [CrossRef] [PubMed]
32. Shayani-Jam, H.; Nematollahi, D. Electrochemical evidences in oxidation of acetaminophen in the presence of glutathione and N-acetylcysteine. *Chem. Comm.* **2010**, *46*, 409–411. [CrossRef] [PubMed]
33. Folkers, G.; Waterbeemd, H.; Lennernäs, H.; Artursson, P.; Mannhold, R.; Kubinyi, H. *Drug Bioavailability: Estimation of Solubility, Permeability, Absorption and Bioavailability (Methods and Principles in Medicinal Chemistry)*; Wiley-VCH, Verlag GmbH & Co. KgaA: Weinheim, Germany, 2003.
34. Abdelbary, G.; Prinderre, P.; Eouani, C.; Joachim, J.; Reynier, J.P.; Piccerelle, P.H. The preparation of orally disintegrating tablets using a hydrophilic waxy binder. *Int. J. Pharm.* **2004**, *278*, 423–433. [CrossRef]
35. Li, G.; Chu, J.; Song, E.S.; Row, K.H.; Lee, K.H.; Lee, Y.W. Crystallization of acetaminophen micro-particle using supercritical carbon dioxide. *Korean J. Chem. Eng.* **2006**, *23*, 482–487. [CrossRef]
36. Devine, D.M.; Devery, S.M.; Lyons, J.G.; Geever, L.M.; Kennedy, J.E.; Higginbotham, C.L. Multifunctional polyvinylpyrrolidinone-polyacrylic acid copolymer hydrogels for biomedical applications. *Int. J. Pharm.* **2006**, *326*, 50–59. [CrossRef]
37. Dumitriu, R.P.; Oprea, A.M.; Vasile, C. A drug delivery system based on stimuli-responsive alginate/N-isopropylacryl amide hydrogel. *Cellul. Chem. Technol.* **2009**, *43*, 251–262.
38. Stanojević, M.; Kalagasidis-Krušić, M.; Filipović, J.; Parojčić, J.; Stupar, M. An investigation into the influence of hydrogel composition on swelling behavior and drug release from poly(acrylamide-*co*-itaconic acid) hydrogels in various media. *Drug Deliv.* **2006**, *13*, 1–7. [CrossRef]
39. Gong, C.; Wong, K.L.; Lam, M.H. Photoresponsive molecularly imprinted hydrogels for the photoregulated release and uptake of pharmaceuticals in the aqueous media. *Chem. Mater.* **2008**, *20*, 1353–1358. [CrossRef]
40. Subramanian, K.G.; Vijayakumar, V. Characterization of crosslinked poly(2-hydroxyethyl methacrylate-co-n-vinyl-2-pyrrolidone) as a carrier for controlled drug delivery. *J. Pharm. Res.* **2011**, *4*, 743–747.
41. Díaz-Guerrero, A.M.; Castillo-Miranda, C.A.; Castro-Guerrero, C.F.; Peraza-Vázquez, H.; Morales-Cepeda, A.B.; Peña-Delgado, A.F. Mathematical modelling of acetaminophen release in HPC/PAAm hydrogel: Synthesis and application. *Int. J. Polym. Sci.* **2019**, *2019*, 9306459. [CrossRef]
42. Miotke, M.; Strankowska, J.; Kwela, J.; Strankowski, M.; Józefowicz, M. Transport of paracetamol in swellable and relaxing polyurethane nanocomposite hydrogels. *Polym. Bull.* **2020**, *77*, 483–499. [CrossRef]
43. Elham, K.; Omid, R.; Farshad, F.; Afshin, J.; Farnaz Sadat Mirzazadeh, T. Preparation and investigation of poly [N-isopropylacrylamide-acrylamide] membranes in temperature responsive drug delivery. *Iran. J. Basic Med. Sci.* **2010**, *13*, 102–110.
44. O'Donnell, K.L.; Oporto-Velásquez, G.S.; Comolli, N. Evaluation of Acetaminophen Release from Biodegradable Poly(Vinyl Alcohol) (PVA) and Nanocellulose Films Using a Multiphase Release Mechanism. *Nanomaterials* **2020**, *10*, 301. [CrossRef]
45. Nizam El-Din, H.M. Characterization and caffeine release properties of N-isopropyl acrylamide/hydroxypropyl methacrylate, copolymer hydrogel synthesized by gamma radiation. *J. Appl. Polym. Sci.* **2011**, *119*, 577–585. [CrossRef]
46. Ilić-Stojanović, S.; Nikolić, L.; Nikolić, V.; Ristić, I.; Cakić, S.; Petrović, S.D. Biomedical applications of thermosensitive hydrogels for controlled/modulated piroxicam delivery. *Gels* **2023**, *9*, 70. [CrossRef]

47. Ilić-Stojanović, S.; Nikolić, L.; Nikolić, V.; Petrović, S.; Oro, V.; Mitić, Ž.; Najman, S. Semi-Crystalline copolymer hydrogels as smart drug carriers: In vitro thermo-responsive naproxen release study. *Pharmaceutics* **2021**, *13*, 158. [CrossRef]
48. Ilić-Stojanović, S.; Nikolić, L.; Nikolić, V.; Milić, J.; Stamenković, J.; Nikolić, G.M.; Petrović, S.D.; Kapor, A. Potential application of thermosensitive hydrogels for controlled release of phenacetin. *Chem. Ind.* **2012**, *66*, 831–839. [CrossRef]
49. Ilić-Stojanović, S.S.; Nikolić, L.B.; Nikolić, V.D.; Milić, J.R.; Stamenković, J.; Nikolić, G.M.; Petrović, S.D. Synthesis and characterization of thermosensitive hydrogels and the investigation of modified release of ibuprofen. *Hem. Ind.* **2013**, *67*, 901–912. [CrossRef]
50. Ilić-Stojanović, S.; Nikolić, L.B.; Nikolić, V.; Stanković, M.; Stamenković, J.; Mladenović-Ranisavljević, I.; Petrović, S. Influence of monomer and crosslinker molar ratio on the swelling behaviour of thermosensitive hydrogels. *Chem. Ind. Chem. Eng.* **2012**, *18*, 1–9. [CrossRef]
51. Ilić-Stojanović, S.; Nikolić, L.; Nikolić, V.; Ristić, I.; Budinski-Simendić, J.; Kapor, A.; Nikolić, G.M. The structure characterization of thermosensitive poly(N-isopropyl acrylamide-*co*-2-hydroxypropylmethacrylate) hydrogel. *Polym. Int.* **2013**, *63*, 973–981. [CrossRef]
52. Khare, A.R.; Peppas, N.A. Swelling/deswelling of anionic copolymer gels. *Biomaterials* **1995**, *16*, 559–567. [CrossRef]
53. Guirguis, O.W.; Moselhey, M.T.H. Thermal and structural studies of poly(vinyl alcohol) and hydroxypropyl cellulose blends. *Nat. Sci.* **2012**, *4*, 57–67. [CrossRef]
54. De Kee, D.; Liu, Q.; Hinestroza, J. Viscoelastic (non-Fickian) diffusion. *Can. J. Chem. Eng.* **2005**, *83*, 913–929. [CrossRef]
55. Habiba, U.; Alam, A.; Rahman, S.; Shamim, S.U.D.; Piya, A.A. IR spectra of paracetamol. *Bangladesh J. Sci. Ind. Res.* **2021**, *56*, 255–262. [CrossRef]
56. Boldyreva, E.V. High-pressure studies of the anisotropy of structural distortion of molecular crystals. *J. Mol. Struct.* **2003**, *647*, 159–179. [CrossRef]
57. Milosavljević, S.M. *Strukturne Instrumentalne Metode*; Univerzitet u Beogradu, Hemijski Fakultet: Beograd, Serbia, 1994. (In Serbian)
58. What Is Extended-Release Medication? | PainScale. Available online: www.painscale.com/article/what-is-extended-release-medication, (accessed on 20 April 2023).
59. Acetaminophen Dosage. Available online: www.drugs.com/dosage/acetaminophen.html (accessed on 20 April 2023).
60. Diez-Pena, E.; Quijada-Garrido, I.; Barrales-Rienda, J.M. Hydrogen-bonding effects on the dynamic swelling of P(N-iPAAm-co-MAA) copolymers. A case of autocatalytic swelling kinetics. *Macromolecules* **2002**, *35*, 8882–8888. [CrossRef]
61. Ritger, P.L.; Peppas, N.A. A simple equation for description of solute release II. Fickian and anomalous release from swellable devices. *J. Control. Release* **1987**, *5*, 37–42. [CrossRef]
62. Bajpai, S.K. Swelling–deswelling behavior of poly(acrylamide-*co*-maleic acid) hydrogels. *J. Appl. Polym. Sci.* **2001**, *80*, 2782–2789. [CrossRef]
63. Schott, H.J. Swelling kinetics of polymers. *Macromol. Chem. Phys.* **1992**, *B31*, 1–9. [CrossRef]
64. Schott, H. Kinetics of swelling of polymers and their gels. *J. Pharm Sci.* **1992**, *81*, 467–470. [CrossRef] [PubMed]
65. Krušić, M.K.; Filipović, J. Copolymer hydrogels based on N-isopropylacrylamide and itaconic acid. *Polymer* **2006**, *47*, 148–155. [CrossRef]

gels

Article

Hydrogel-Transformable Antioxidant Poly-γ-Glutamic Acid/Polyethyleneimine Hemostatic Powder for Efficient Wound Hemostasis

Xiang Li [1,†], Wenli Han [2,†], Gao He [1], Jiahao Yang [2], Jing Li [2], Hongxia Ma [1,*] and Shige Wang [2,*]

1 School of Public Health, Center for Global Health, Nanjing Medical University, Nanjing 211100, China
2 School of Materials and Chemistry, University of Shanghai for Science and Technology, No. 516 Jungong Road, Shanghai 200093, China
* Correspondence: hongxiama@njmu.edu.cn (H.M.); sgwang@usst.edu.cn (S.W.)
† These authors contributed equally to this work.

Abstract: Hemostatic powder, which can absorb large amounts of water and tends to produce repeated hydration with tissue, has been clinically proven as an ideal engineering material for treating wounds and tissues. We herein designed a polypeptide-based hemostatic powder. A water-soluble polypeptide, γ-polyglutamic acid (γ-PGA), was mixed with the polyethyleneimine (PEI), N-hydroxysuccinimide, and 1-(3-dimethylaminopropyl)-3-ethylcarbodiimide. The solution of these polymers was lyophilized to harvest the γ-PGA/PEI powder (PP hemostatic powder). When deposited on a bleeding wound, the PP hemostatic powder can quickly absorb a large amount of blood and interstitial fluid, concentrate coagulation factors, coagulate blood cells, and eventually form a stable mechanical hydrogel. The wound bleeding time of the PP hemostatic powder group was 1.8 ± 0.4 min, significantly lower than that of the commercial chitosan hemostatic powder group (2.8 ± 0.4 min). The PP hemostatic powder was endowed with antioxidant capacity by introducing protocatechuic aldehyde, which can effectively inhibit inflammation and promote wound healing. Therefore, via preparation through a facile lyophilization method, the PP hemostatic powder is expected to find a wide application prospect as a qualified hemostatic powder.

Keywords: hemostatic powder; hydrogel; γ-polyglutamic acid; polyethyleneimine; wound hemostasis

1. Introduction

Wound hemostasis is a complex physiological process that relies on multiple factors such as biomolecules, signaling pathways, cell populations cytokines, etc. [1]. In recent years, with the continuous development of wound-management techniques, wound dressings used for hemostasis have attracted worldwide attention [2]. Hydrogels represent three-dimensional networks of hydrophilic polymers cross-linked by chemical and physical bonds that can absorb and retain large amounts of water or biological fluids [3,4]. As soft and moist materials, hydrogels usually have good biocompatibility, strong adhesion, high stretchability, and good self-healing properties [5–8]. Hydrogel bio-adhesives have a variety of unique advantages, such as good biosafety, degradability, similarity to the natural extracellular matrix, hydrophilicity, moisture retention, flexibility, etc. [9–12]. Moreover, equipped with diverse functional groups, certain hydrogels can establish covalent or non-covalent bonds with tissue surfaces, becoming one of the most promising candidates for medical wound dressings [13–16]. In recent years, the hemostatic properties of hydrogels have been extensively studied [17–20]. For example, Tang and co-workers prepared hemostatic hydrogels based on carboxymethyl chitosan, 2,3,4-trihydroxybenzaldehyde, peptidyl repressing enzyme, and 4-arm poly (ethylene glycol) aldehyde, which showed a promising future for the clinical treatment of full-thickness wounds [21].

Another commonly used hemostatic dressing is hemostatic powder, which has been clinically proven as an ideal engineering material for treating wounds and tissue [22].

Hemostatic powder can not only provide a moist environment for wound healing but also form a protective barrier with blood at the wound interface. The hemostatic powder absorbs large amounts of water and tends to produce repeated hydration, drawing moisture from the tissue into the dressing when in contact with tissue [19]. Meanwhile, the excess water on the surface of the hemostatic powder material can be used as a lubricant to reduce tissue adhesion. In addition, the hemostatic powder is injectable, can be closely adhered to the uneven wound surface, and can effectively prevent wound inflammation [3]. More importantly, the hemostatic powder can bind with anti-bacterial and anti-inflammatory functions to promote the formation of new blood vessels and tissues and help epithelial cells grow to repair wounds [23].

The design of polypeptide-based hydrogel bio-adhesives with intelligent drug-controlled release properties and stimulus responsiveness for chronic wounds is one of the current research hotspots [24]. However, polypeptide-based hydrogel bio-adhesives have limited fluid uptake capacity and are not suitable for highly exudative wounds. Improper application could lead to the excessive accumulation of tissue exudate at the wound site and macerate surrounding healthy tissue, thereby delaying healing and increasing treatment costs. Considering the unique features of hemostatic powder, we designed a composite hemostatic powder based on a mixture of γ-polyglutamic acid (γ-PGA), polyethyleneimine (PEI), N-hydroxysuccinimide (NHS), and 1-(3-dimethylaminopropyl)-3-ethylcarbodiimide (EDC) using a facile lyophilization method. γ-PGA has good hydrophilicity and is able to quickly absorb blood and interstitial fluid when deposited on a bleeding wound and eventually form a stable mechanical hydrogel. Furthermore, the γ-PGA/PEI powder (PP hemostatic powder) was endowed with antioxidant capacity by introducing protocatechuic aldehyde (PCA), which can effectively inhibit inflammation and promote wound healing. This study provides a facile method to prepare multifunctional hemostatic powder, which is anticipated to promote the research and clinical transformation of wound dressing materials.

2. Results and Discussion

2.1. Preparation and Characterization of the PP Hemostatic Powder

The PP hemostatic powder was prepared using a simple lyophilization method. In the powder, γ-PGA has excellent hydrophilicity to quickly absorb blood and then form a hydrogel with PEI under the activation of EDC/NHS (Figure 1). Upon the introduction of PCA, the PP hemostatic powder was endowed with antioxidant capacity, which can effectively inhibit inflammation and promote wound healing.

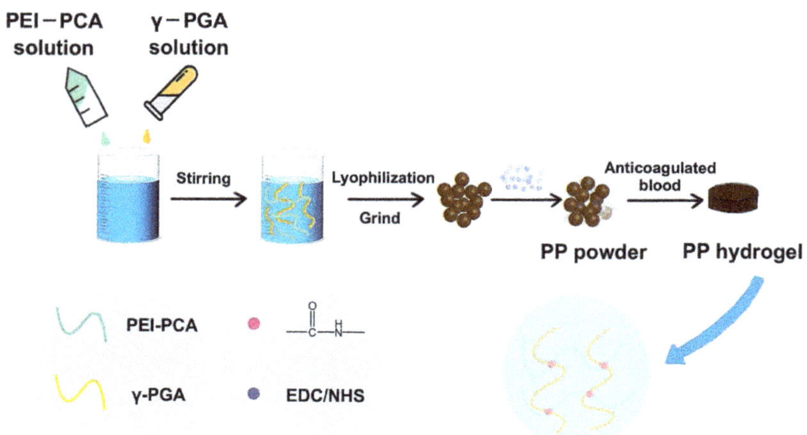

Figure 1. Preparation of the PP hemostatic powder and PP hemostatic powder-derived hydrogel.

The PEI-PCA was analyzed by FTIR. As shown in Figure 2a, comparing the FTIR spectra of PEI and PEI-PCA, it can be seen that PEI-PCA has a new sharp peak at 1643 cm^{-1}, and the amino peak in the range of 3100–3500 cm^{-1} becomes weaker, suggesting the formation of Schiff base (C=N) imine bonds. The mixed aqueous solution of PEI-PCA and γ-PGA was lyophilized and mixed with EDC/NHS to prepare the PP hemostatic powder (Figure 2b). The PP hemostatic powder will rapidly expand to form a hydrogel after contact with anticoagulated blood or water. For example, anticoagulant blood was added to heart-shaped PP hemostatic powder, and a heart-shaped PP hydrogel was obtained (Figure 2c). The microstructure of the PP hemostatic powder and PP hemostatic powder-derived hydrogel was studied using SEM. When the PP hemostatic powder became a PP hemostatic powder-derived hydrogel after absorbing water, the hydrogel exhibited a three-dimensional porous structure (Figure 2d). This three-dimensional porous structure can effectively absorb body fluids exuded from the wound, keep the wound area clean and moist, and facilitate the transportation of cell nutrients, thereby promoting hemostasis and wound healing.

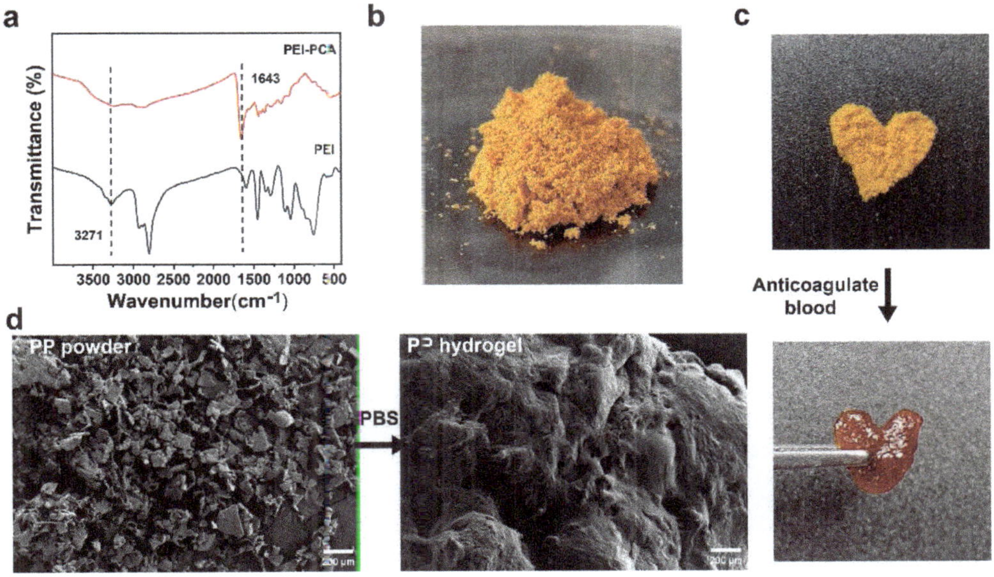

Figure 2. (**a**) Infrared spectra of PEI and PEI-PCA compounds; (**b**) macroscopic image of PP hemostatic powder; (**c**) macroscopic image of PP hemostatic powder absorbing anticoagulant blood to form a hydrogel; (**d**) SEM image of the PP hemostatic powder and PP hemostatic powder-derived hydrogel after absorbing PBS.

2.2. Mechanical and Rheological Properties

As a hemostatic material, the derived hydrogel formed after PP hemostatic powder absorbs body fluids should have strong mechanical properties to serve as a physical barrier to prevent wound bleeding [25,26]. Therefore, we studied the mechanical properties of PP hemostatic powder-derived hydrogels and the gelation time of the PP hemostatic powder. As a hemostatic material, it needs to have the ability to withstand pressure and avoid failure. In order to verify the compression resistance of PP hemostatic powder after absorbing body fluids, compression tests were conducted (Figure 3a,b). As shown in Figure 3c, in the compression experiment, the maximum compressive strength of the PP hemostatic powder-derived hydrogel was 2010.4 $\pm$ 20.1 kPa. The gelation time of the PP hemostatic powder was explored through rheological experiments. After adding PBS to the PP hemostatic powder, the G' curve intersected the G'' curve, and then the G' was higher than the G'', which shows that a stable hydrogel was formed through cross-linking (Figure 3d). The results showed that the G′ of the PP hemostatic powder-derived hydrogel was about three times that of G″, and the storage modulus and loss modulus of the formed hydrogel were respectively 3.3 ± 0.1 Pa and 1.2 ± 0.2 Pa (Figure 3e). Meanwhile, the rheological curve indicated that the PP hemostatic powder can achieve the transformation from powder to hydrogel within 9 s (Figure 3f). In general, the PP hemostatic powder can quickly absorb water to form a hydrogel with excellent mechanical properties. It is worth mentioning that the CHP forms a high viscosity block, which can quickly adhere to and seal the wound. However, it cannot form a typical hydrogel. Therefore, its mechanical and rheological properties were not detected (Figure 3c,f).

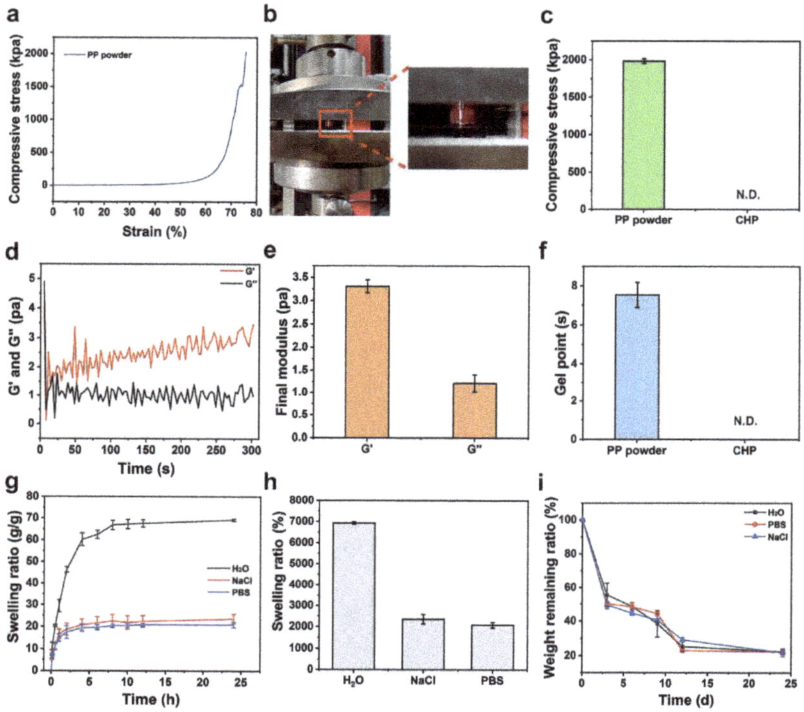

Figure 3. (**a**) Compressive stress–strain curve of the PP hemostatic powder-derived hydrogel; (**b**) schematic diagram of the compression experiment with the PP hemostatic powder-derived hydrogel; (**c**) final compressive strength of the PP hemostatic powder-derived hydrogel and CHP (N.D. = not detectable); (**d**) dynamic change diagram of G′ and G″ after adding PBS to the PP hemostatic powder; (**e**) final G′ and G″ after adding PBS to the PP hemostatic powder; (**f**) gelling time of the PP hemostatic powder and CHP N.D. = not detectable); (**g**) swelling kinetics curve of PP hemostatic powder-derived hydrogel in H_2O, PBS, and NaCl solutions for 24 h; (**h**) swelling ratio of PP hemostatic powder-derived hydrogel in H_2O, PBS, and NaCl solutions for 24 h; (**i**) degradation kinetics curve of PP hemostatic powder-derived hydrogel in H_2O, PBS, and NaCl solutions for 24 h.

2.3. Swelling and Degradation

The good degradation ability of a hemostatic material can avoid secondary damage to the wound [27]. A hemostatic dressing that can effectively absorb tissue fluid and blood will not only increase the concentration of coagulation components at the bleeding site and promote coagulation but also keep the wound moist. Therefore, we performed a swelling experiment to verify the absorption performance of the PP hemostatic powder-derived hydrogel. As shown in Figure 3g,h, the hydrogel reached an equilibrium state after swelling in H_2O, PBS, and 0.9% NaCl solution for 24 h. The equilibrium swelling ratio in water was the highest, reaching 6918 ± 55.4%. The degree of swelling in PBS and 0.9% NaCl solution was lower than that in H_2O, which may be due to the presence of salt ions and the higher osmotic pressure of PBS and 0.9% NaCl solution. Meanwhile, the degradation performance of the PP hemostatic powder-derived hydrogel was evaluated. As shown in Figure 3i, the degradation rate of the derived hydrogel was high in the early stage and gradually decreased in the later stage. On day 24, the remaining mass ratios of the derived hydrogel in H_2O, PBS, and 0.9% NaCl solutions were 22.4 ± 1.7%, 22.2 ± 1.4%, and 21.9 ± 2.4%, respectively, indicating that the PP hemostatic powder-derived hydrogel has good in vitro degradation properties. The degradation can be attributed to the hydrolysis of amide bonds in the hydrogel formed after the water absorption. It is worth mentioning that the CHP cannot form a typical hydrogel. Therefore, its swelling and degradation properties were not studied.

2.4. Antioxidant Properties

Chronic inflammation at the wound site generates large amounts of reactive oxygen species, which may lead to oxidative stress and prevent angiogenesis and extracellular matrix remodeling [28]. Therefore, a hemostatic material with antioxidant properties will find diverse applications as a wound dressing. The antioxidant properties of PP hemostatic powder were explored through oxygen free radical and nitrogen free radical scavenging experiments. When the PP hemostatic powder concentration reaches 50 mg/mL, the colors of ABTS·, DPPH·, and PTIO· working solutions all fade away and become almost transparent and colorless. The UV–visible spectrum also proves that the characteristic peak absorbance is close to 0 (Figure 4a–c). As the PP hemostatic powder concentration increases, the scavenging rate becomes higher. The maximum scavenging rate for ABTS· is 99.1 ± 0.2% (Figure 4d, $p < 0.01$ versus 15 mg/mL; $p < 0.001$ versus 5 mg/mL), while the maximum scavenging rates for DPPH· and PTIO· are 75.6 ± 2.6% and 89.3 ± 0.3%, respectively (Figure 4e, $p < 0.05$ versus 30 mg/mL, $p < 0.05$ versus 15 mg/mL, $p < 0.001$ versus 5 mg/mL; Figure 4f, $p < 0.01$ versus 15 mg/mL, $p < 0.001$ versus 5 mg/mL). Such antioxidant properties can be attributed to the introduction of PCA with catechol groups. With the proven antioxidant capabilities, the PP power can prevent the excessive expression of free radicals at the wound site, inhibit inflammation, and thereby promote wound healing.

2.5. In Vitro Safety

An ideal hemostatic material should have no or a small amount of hemolysis when in contact with bleeding wounds. An in vitro hemolysis test was used to evaluate the blood compatibility of the PP hemostatic powder. As shown in Figure 5a,b, after incubation with PP hemostatic powder at 37 °C for 2 h, the supernatants of all centrifuged samples showed a transparency similar to that of the negative control group, without obvious cell disruption. However, the blood treated with H_2O in the positive control group turned bright red due to hemolysis. The hemolysis ratios of PP hemostatic powders at different concentrations (5, 15, 30, and 50 mg/mL) were all less than 5%, with values of 0.2 ± 0.5%, 2.5 ± 0.3%, 2.7 ± 1.3%, and 3.0 ± 1.6%, respectively, and were considered to represent the safety level of hemostatic materials. These results proved that PP hemostatic powder has good blood compatibility and is a safe hemostatic powder.

2.6. In Vitro Hemostasis Experiments

In order to explore the coagulation ability of PP hemostatic powder, an in vitro whole-blood coagulation experiment was conducted. The whole-blood coagulation kinetics of PP hemostatic powder and commercial CHP powder were evaluated by measuring the absorbance of the supernatant of PP hemostatic powder- and commercial CHP powder-treated blood. As shown in Figure 5c, the PP hemostatic powder group adsorbed some red blood cells initially, and the absorbance value was much lower than that of the CHP group. The absorbance of the PP hemostatic powder group decreased significantly at 1 min until the absorbance was close to 0 at 6 min. The absorbance value representing the hemoglobin concentration decreased as the blood began to coagulate, and the rapid decrease corresponded to a high coagulation rate. Figure 5d records the blood coagulation conditions of PP hemostatic powder, CHP, and blank groups at different time points (0.5, 1, 2, 3, 4, and 6 min). It can be seen that the blood in the blank group and CHP group was diffused red in the water, while the supernatant in the PP group remained transparent after 1 min. Blood coagulation is a dynamic process in which coagulation factors are effectively activated and ultimately convert fibrinogen into fibrin. During the coagulation process, red blood cells are captured by the irregular pores of the PP hemostatic powder, and a large amount of fibrin aggregates to form a fibrin matrix, which captures more blood cells [29,30]. Finally, the blood clot formed by PP hemostatic powder showed a dark red color composed of blood cells and fibrin. As shown in Figure 5e, both the blank group and the CHP group showed higher blood coagulation indices, indicating poor hemostatic performance. The blood coagulation index value of the PP hemostatic powder group was significantly lower than those of the blank group and CHP group, only 3.4 ± 0.6% (*** $p < 0.01$, versus CHP; *** $p < 0.001$, versus blank), indicating a better in vitro hemostatic performance. To further explore the blood absorption ability of the PP hemostatic powder, a blood absorption experiment was conducted. The experimental operation is shown in Figure 6a. After 30 s, the blood absorption rate of the PP hemostatic powder was as high as 373.1 ± 38.8%, which shows that it has strong blood absorption ability and is a potential hemostatic material. Because the CHP cannot form a typical hydrogel, the blood uptake ratio of CHP was not studied (Figure 6b).

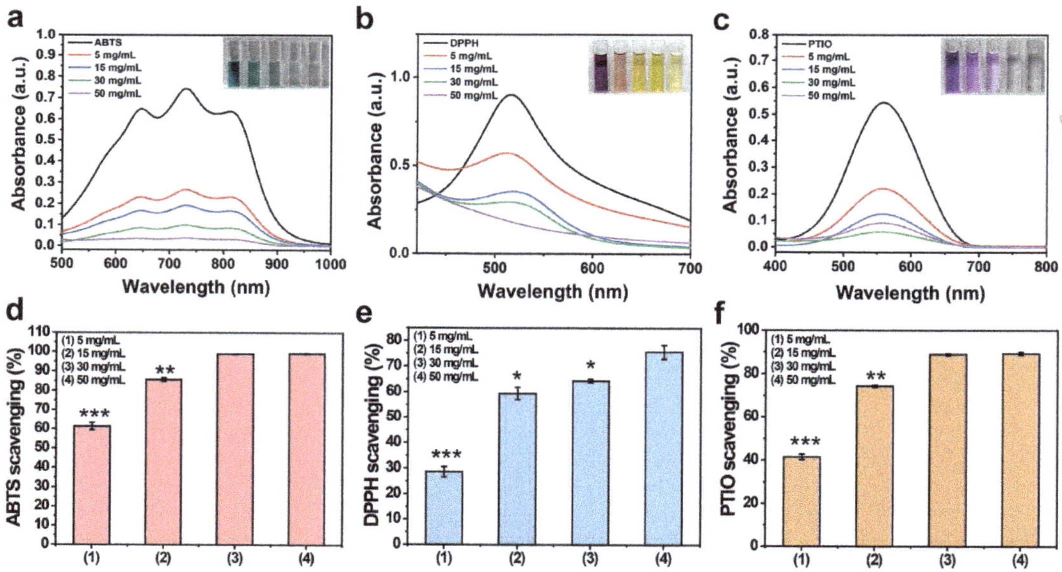

Figure 4. (**a**) UV–visible spectrum of ABTS· after being scavenged by PP hemostatic powder at different concentrations for 3 min; (**b**) UV–visible spectrum of DPPH· after being cleaned by PP hemostatic powder at different concentrations for 30 min; (**c**) PTIO· UV–visible spectrum after being cleaned by PP hemostatic powder at different concentrations for 2 h; (**d**) the clearance rate of ABTS· by PP hemostatic powder at different concentrations; (**e**) the clearance rate of DPPH· by PP hemostatic powder at different concentrations; (**f**) the scavenging rate of PTIO· by PP hemostatic powder at different concentrations. p * < 0.05, p ** < 0.01, p *** < 0.001.

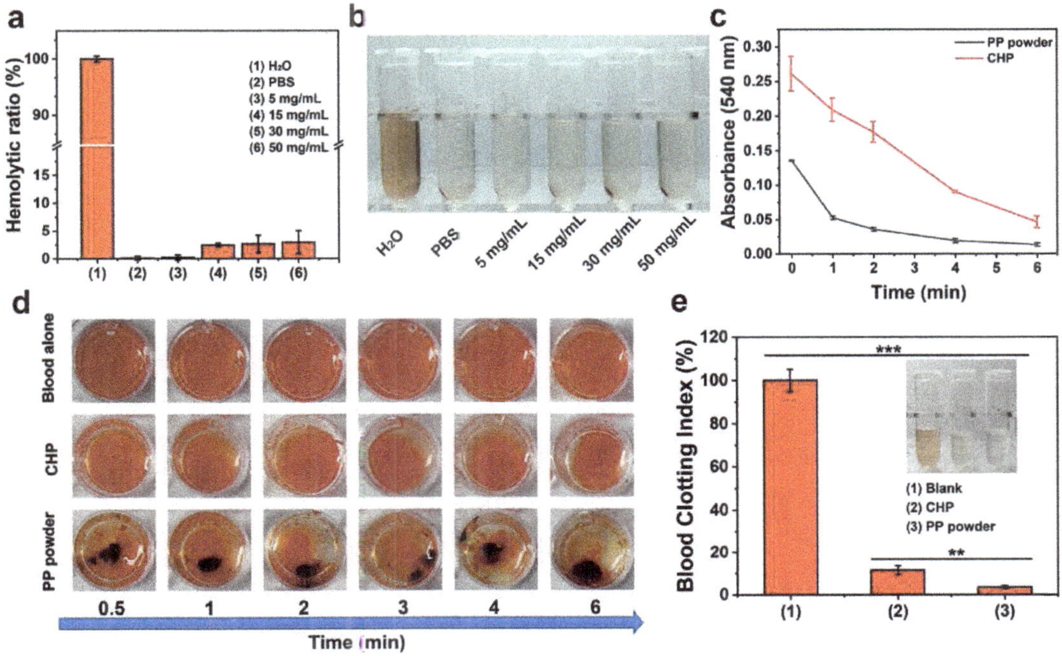

Figure 5. (**a**) Hemolysis ratios of PBS, H_2O, and PP hemostatic powder at different concentrations; (**b**) photos of centrifuged blood cells treated with PBS, H_2O, and PP hemostatic powder at different concentrations; (**c**) in vitro whole-blood coagulation kinetics curves of commercial CHP and PP hemostatic powders; (**d**) images of blood coagulation after different treatments; (**e**) blood coagulation index of full blood and blood treated with commercial CHP and PP hemostatic powders. p ** < 0.01, p *** < 0.001.

2.7. In Vivo Hemostasis Experiments

The in vivo hemostatic properties of the PP hemostatic powder were evaluated through the rat tail-docking model. After treatments, the bleeding time of the PP hemostatic powder group was 1.8 ± 0 4 min (* $p < 0.05$, versus CHP; *** $p < 0.001$, versus blank), while the bleeding time of the CHP group was 2.8 ± 0.4 min. However, the untreated blank control group needed 7.7 ± 0.7 min to stop bleeding (Figure 6c). The schematic diagram of the tail-docking experiment can be seen in Figure 6e. As shown in Figure 6d,f, without any hemostatic treatment, the blood loss of tail-docked rats was 2.3 ± 0.3 g within 10 min. Although CHP can absorb blood, it cannot form a stable physical barrier and dissolve in the blood, resulting in a blood loss of 0.6 ± 0.1 g. The PP hemostatic powder can absorb a large amount of blood, concentrate coagulation factors, gather blood cells and platelets, and achieve the lowest blood loss. Therefore, the blood loss in wounds treated with the PP hemostatic powder was only 0.06 ± 0.01 g (*** $p < 0.001$, versus CHP; *** $p < 0.001$, versus blank). These results indicate that the PP hemostatic powder has good in vivo hemostatic ability and is suitable for complex bleeding sites with irregular geometries.

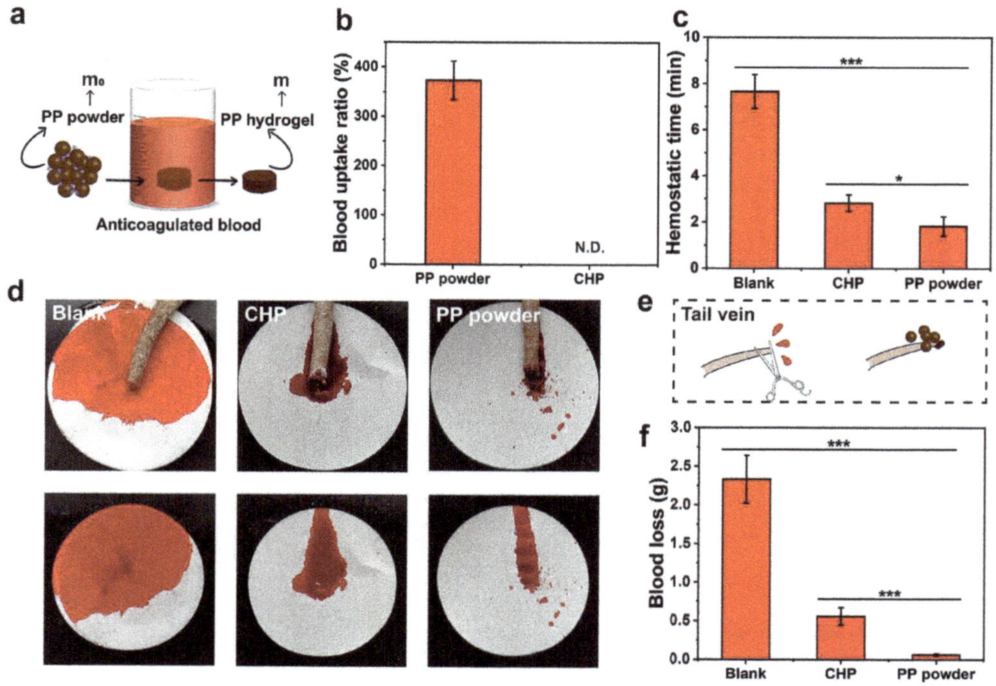

Figure 6. (**a**) Schematic diagram of measuring the blood absorption ratio of the PP hemostatic powder (m and m_0 are the weight of the PP hemostatic powder and the hemostatic powder-derived hydrogel); (**b**) blood uptake ratio of PP hemostatic powder and CHP (N.D. = not detected); (**c**) tail-vein bleeding time of SD rats in the blank group, CHP group, and PP hemostatic powder group; (**d**) corresponding to (**c**) photo of tail-vein hemostasis; (**e**) schematic diagram of tail-vein hemostasis with PP hemostatic powder; (**f**) tail-vein blood loss of SD rats in the blank group, CHP group, and PP hemostatic powder group. p * < 0.05, p *** < 0.001.

3. Conclusions

In summary, we successfully synthesized a multifunctional PP hemostatic powder that can resist inflammation and hemostasis, which is expected to contribute to wound healing. When the PP hemostatic powder is deposited on a bleeding wound, it can quickly absorb a large amount of blood and interstitial fluid, concentrate coagulation factors, coagulate blood cells, and quickly form (<9 s) a stable and strong mechanical hydrogel (maximum compressive strength of 2010.4 ± 20.1 kPa). In addition, due to the presence of PCA, PP hemostatic powder is endowed with antioxidant capacity, which can effectively inhibit inflammation and promote wound healing. PP hemostatic powder also showed good coagulation and hemostasis performance in vitro. The rat-tail venous hemorrhage model showed that its hemostatic performance was better than that of the commercial product CHP. Therefore, we believe that PP hemostatic powder is expected to become a new hemostatic material and wound dressing and has a wide application prospect.

4. Materials and Methods

4.1. Materials

PEI (Mw: 70 kDA) aqueous solution (30 wt.%) was purchased from Thermo Fisher (Shanghai, China). PCA (≥98%) was purchased from Shanghai Yuanye Biotechnology Co., Ltd. (Shanghai, China). γ-PGA (Mw: 700 kDa) was purchased from Shanghai Yika Biotechnology Co., Ltd. (Shanghai, China). Calcium ($CaCl_2$), EDC, NHS, the (2,2′-azino-bis(3-ethylbenzothiazoline-6-sulfonic acid radical (ABTS·), 1,1-diphenyl-2-picrylhydrazyl free radicals (DPPH·), and 2-phenyl-4,4,5,5-tetramethylimidazoline-1-oxyl 3-oxide (PTIO·) were purchased from Shanghai Aladdin Reagent Co., Ltd. (Shanghai, China). Commercial chitosan hemostatic powder (CHP) was bought from Qingdao Biotemed Biomaterials Co., Ltd.

4.2. Preparation of the PP Hemostatic Powder

A 1 mL volume of PEI was diluted in 20 mL of deionized water and stirred evenly at room temperature. Then, 0.075 g of PCA was added to the above solution and stirred at room temperature for 4 h to obtain a solution. The reaction solution was dialyzed in deionized water for 3 days and lyophilized to obtain PEI-PCA powder. To prepare the PP hemostatic powder, PEI-PCA (1 wt.%, 4 mL, in distilled water) and γ-PGA (1 wt.%, 4 mL, in distilled water) solutions were mixed, and 16 mL of deionized water was added thereinto. Then, the mixed solution was placed in a freeze dryer for 72 h for lyophilization. Finally, 0.04 g EDC and 0.04 g NHS were added thereinto and grinded using a grinder for 5 min to obtain PP hemostatic powder. The PP hemostatic powder was stored in a 4 °C refrigerator for later use.

4.3. Characterization

The chemical structure of PEI-PCA was first analyzed by Fourier transform infrared spectroscopy (FTIR). Then, PP hemostatic powder was mixed with PBS to prepare PP hydrogel (PP hemostatic powder-derived hydrogel). After lyophilization, the morphology of the PP hemostatic powder was observed under a scanning electron microscope (SEM). In addition, a hydrogel formed from the PP hemostatic powder was recorded during the full absorption of mouse blood.

4.4. Mechanical Testing

In order to verify the compressive resistance of the hydrogel formed from the PP hemostatic powder, a general material testing machine (Zwick Roell Z2.5 TH with 2.5 kN sensor) for compressive stress–strain measurement was used. In the experiment, the PP hydrogel was prepared in a cylindrical shape with a diameter of 10 mm and a depth of 4 mm. At a predetermined compressive strain rate of 1 mm/min, the hydrogel was compressed to 80% of the initial height to measure its compressive properties, with a strain range of 0 to 90%. All these tests were performed three times in parallel.

In order to study the hydrogel formation process with the PP hemostatic powder, a parallel plate (P20 TiL, diameter 20 mm) rotational rheometer (MARS III HAAKE) was used to conduct dynamic rheological studies. Briefly, the PP hemostatic powder was placed on parallel plates preheated to 37 °C. Subsequently, PBS was added dropwise to the powder. After the powder and PBS were evenly mixed, a time-sweep test was performed at 37 °C (simulated body temperature) with a constant frequency of 1 HZ and a strain of 1%. The storage modulus G' and loss modulus G'' were analyzed. Meanwhile, when the G' curve exceeded the G'' curve, the gelation time point was determined.

4.5. Study of Swelling Behaviors

In this section, a swelling experiment was used to measure the swelling rate of the PP hemostatic powder-derived hydrogel. The freeze-dried hydrogel blocks (0.3 g) were placed into 25 mL H_2O, PBS, and 0.9% NaCl solution (n = 3), and incubated in a 37 °C constant-temperature incubator. At predetermined time intervals, the hydrogel was removed from the solution, and excess water on the surface of the hydrogel was absorbed with filter paper and then weighed to plot the swelling kinetics curve. The swelling ratio was calculated when the hydrogel reached swelling equilibrium using the calculation formula:

$$\text{Swelling ratio (\%)} = \frac{W_t - W_0}{W_0} \times 100\%$$

where W_0 and W_t are the initial weight of the hydrogel and the weight after swelling at different time points, respectively.

4.6. In Vitro Degradation

In vitro degradation experiments were conducted to verify the biodegradability of the PP hemostatic powder. The prepared PP hemostatic powder-derived hydrogel was freeze dried and incubated with 10 mL of H_2O, PBS, and 0.9% NaCl solution containing lysozyme (10^4 U/mL, n = 3) in a constant-temperature shaker for 24 days (37 °C, 120 rpm). All solutions were replaced with fresh solutions containing lysozyme every 2 days. On days 3, 6, 9, 12, and 24, hydrogel samples were taken out, the surface was washed with deionized water, and then the sample were lyophilized and weighed. The degradation efficiency of the PP hemostatic powder-derived hydrogel was calculated and expressed as weight remaining (%). The calculation formula is:

$$\text{Weight remaining (\%)} = \frac{W_t}{W_0} \times 100\%$$

where W_0 and W_t are the initial weight of the lyophilized hydrogel and the dry weight of the remaining hydrogel after degradation at different time points, respectively.

4.7. Antioxidant Properties

In order to evaluate the antioxidant properties of PP hemostatic powder, ABTS-, DPPH·-, and PTIO·-scavenging experiments with PP hemostatic powder-derived hydrogels at different concentrations were conducted [31,32].

4.7.1. ABTS· Scavenging Experiment

The overall antioxidant capacity of the PP hemostatic powder-derived hydrogel was determined through an ABTS· free radical scavenging experiment. Firstly, the ABTS· working solution was prepared and diluted with PBS to the absorbance of 0.70 ± 0.02 (wavelength = 734 nm) [33]. PP hemostatic powder-derived hydrogels (weight: 15, 45, 90, and 150 mg) were placed into 3 mL of diluted ABTS· working solution and let stand in the dark for 3 min at room temperature. The sample without hydrogel was used as a blank. Using a UV–visible near-infrared spectrometer, the absorbance of the mixture from 500 to 1000 nm was measured. In addition, a camera was used to record macroscopic images of the color changes in the mixed solution during the experiment. The ABTS·-scavenging efficiency was calculated as follows:

$$\text{ABTS·scavenging (\%)} = \frac{A_0 - A_t}{A_0} \times 100\%$$

where A_t and A_0 are the absorbance of the experimental group and the absorbance of the blank group, respectively.

4.7.2. DPPH· Scavenging Experiment

The ability of the PP hemostatic powder-derived hydrogel to scavenge nitrogen free radicals was determined using a DPPH· scavenging experiment. The DPPH· working solution was prepared and diluted with ethanol solution to an absorbance value of 0.6–1.0 at 517 nm [34]. Then, PP hemostatic powder-derived hydrogels with different concentrations (weight: 15, 45, 90, and 150 mg) were added to 3 mL of diluted DPPH· working solution and let stand in the dark at room temperature for 30 min. The sample without hydrogel was used as a blank. We then used a UV–visible near-infrared spectrometer to measure the absorbance of the mixture and scan the wavelength information of the mixture from 400 to 700 nm. Meanwhile, a camera was used to record macroscopic images of the color changes in the mixed solution during the experiment. The calculation formula for DPPH·-scavenging efficiency is as follows:

$$\text{DPPH·scavenging } (\%) = \frac{A_0 - A_t}{A_0} \times 100\%$$

where A_t and A_0 are the absorbance of the experimental group and the absorbance of the blank group, respectively.

4.7.3. PTIO· Scavenging Experiment

The oxygen free radical-scavenging ability was studied via the PTIO· scavenging method. Firstly, the PTIO· working solution was diluted with deionized water to the absorbance of 0.2–0.6 at 557 nm [35]. Then, PP hemostatic powder-derived hydrogel with different concentrations (weight: 15, 45, 90, and 150 mg) were added to 3 mL of diluted PTIO· working solution and placed in a 37 °C constant-temperature incubator for 2 h. The sample without hydrogel was used as a blank. We used a UV–visible near-infrared spectrometer to measure the absorbance of the mixture from 400 to 800 nm. At the same time, a camera was used to record macroscopic images of the color changes in the mixed solution during the experiment. The calculation formula for PTIO·-scavenging efficiency is as follows:

$$\text{PTIO·scavenging } (\%) = \frac{A_0 - A_t}{A_0} \times 100\%$$

where A_t and A_0 are the absorbance of the experimental group and the absorbance of the blank group respectively.

4.8. Blood Compatibility

In order to verify the biocompatibility of PP hemostatic powder, blood compatibility experiments were conducted using Kunming mouse blood. Mouse blood was centrifuged (3000 rpm, 5 min) and washed three times with PBS to collect mouse red blood cells. We diluted the obtained red blood cells to 2% with PBS and stored them in the refrigerator for later use. Then, we added 5, 15, 30, and 50 mg PP hemostatic powder-derived hydrogels to the mixed solution of 2.4 mL of PBS and 0.6 mL of red blood cells, respectively. Another two groups were defined as the control group, that is, the red blood cell suspension was treated with PBS (negative control) and H_2O (positive control). Then, we transferred all the above samples to a 37 °C constant-temperature incubator and cultured them for 2 h. Finally, all samples were centrifuged (3000 rpm, 5 min) to extract the supernatant. Using a UV–visible near-infrared spectrometer to detect the absorbance of the supernatant at a wavelength of 541 nm, we were then able to calculate the hemolysis ratio according to the formula:

$$\text{Hemolytic ratio}(\%) = \frac{A_x - A_n}{A_t - A_n} \times 100\%$$

where A_x is the absorbance of the supernatant co-cultured with PP hemostatic powder, A_n is the absorbance of the negative control, and A_t is the absorbance of the positive control.

4.9. In Vitro Hemostasis Test

A 20 μL volume of anticoagulated mouse whole blood was dropped into 10 mg of PP hemostatic powder, and then 1 μL of $CaCl_2$ solution (0.1 mol/mL) was added to initiate blood coagulation. Similarly, the above-mixed blood was added to CHP in the control group. In this experiment, no materials were added to the blank group, and the mixed blood coagulated naturally. The three groups of materials were transferred to a 37 °C constant-temperature incubator and taken out at preset time points (0.5, 1, 3, 4, and 6 min). By slowly adding deionized water (5 mL) to rinse the samples to allow dissociation, the red blood cells were broken up and pictures were taken. The absorbance of the supernatant at a wavelength of 541 nm was detected by a UV–visible near-infrared spectrometer, and the whole-blood coagulation kinetics curve was drawn. The blood coagulation index was calculated according to the formula:

$$\text{Blood clotting index (\%)} = \frac{A_s}{A_0} \times 100\%$$

where A_s is the absorbance of the supernatant of the material group, and A_0 is the absorbance of the supernatant of the blank group.

4.10. In Vivo Hemostasis Study

To study blood absorption, PP hemostatic powder with weight m_0 was added to 2 mL of anticoagulated whole blood at 37 °C for 30 s [36,37]. Then, the PP hemostatic powder-derived hydrogel was taken out, and excess Kuming mouse blood was wiped off with filter paper. The weight was recorded as m. The blood absorption ratio was calculated as follows (m and m_0 are the weight of the PP hemostatic powder and the hemostatic powder-derived hydrogel, respectively):

$$\text{Blood uptake ratio (\%)} = \frac{m}{m_0} \times 100\%$$

An SD rat tail-docking model was used to evaluate the hemostatic potential of PP hemostatic powder. All experimental protocols and animal handling were conducted following the regulations of the Animal Investigation Ethics Committee of Changhai Hospital. Rats were randomly divided into 3 groups (n = 3) and anesthetized with an intraperitoneal injection of chloral hydrate (4%). The experimental group used PP hemostatic powder for hemostasis, the control group used CHP for hemostasis, and the blank group did not undergo any hemostasis treatment. A weighed filter paper was placed at the bleeding site 8 cm from the tail of the rat to absorb blood. After 3 min, photos of the wound were recorded. Finally, the filter paper was weighed again to calculate the blood loss. The hemostasis times of the blank group, control group, and PP hemostatic powder group were measured respectively. All experiments were repeated three times.

4.11. Statistical Analysis

All data are expressed as mean ± standard deviation of at least triplicate samples. Differences between independent groups were analyzed using multiple t-tests (p * < 0.05, p ** < 0.01, p *** < 0.001).

Author Contributions: Conceptualization, X.L., W.H., H.M. and S.W.; Data curation, X.L. and W.H.; Formal analysis, J.Y.; Investigation, X.L.; Methodology, X.L., W.H., G.H., H.M. and S.W.; Project administration, X.L., W.H., H.M. and S.W.; Resources, H.M. and S.W.; Software, X.L., W.H. and J.Y.; Supervision, J.L., H.M. and S.W.; Validation, G.H. and J.Y.; Visualization, W.H. and G.H.; Writing—original draft, X.L. and W.H.; Writing—review & editing, J.L. and S.W. All authors have read and agreed to the published version of the manuscript.

Funding: This research received no external funding.

Institutional Review Board Statement: All animal procedures in this study were approved by the Ethics Committee of the First Affiliated Hospital of Naval Medical University (CHFC(A.E)2023-017) and the Fourth Affiliated Hospital of Nanjing Medical University (SYXK(Su)2019-0028).

Informed Consent Statement: Not applicable.

Data Availability Statement: The data presented in this study are openly available in article.

Acknowledgments: The authors thank Jiulong Zhao of the First Affiliated Hospital of Naval Medical University for his support in animal experiments.

Conflicts of Interest: The authors declare no conflict of interest.

References

1. Cui, T.; Yu, J.; Wang, C.; Chen, S.; Li, Q.; Guo, K.; Qing, R.; Wang, G.; Ren, J. Micro-Gel Ensembles for Accelerated Healing of Chronic Wound via pH Regulation. *Adv. Sci.* **2022**, *9*, e2201254. [CrossRef] [PubMed]
2. Nie, L.; Wei, Q.; Sun, M.; Ding, P.; Wang, L.; Sun, Y.; Ding, X.; Okoro, O.V.; Jiang, G.; Shavandi, A. Injectable, self-healing, transparent, and antibacterial hydrogels based on chitosan and dextran for wound dressings. *Int. J. Biol. Macromol.* **2023**, *233*, 123494. [CrossRef] [PubMed]
3. Altenschmidt, L.; Sanchez-Paradinas, S.; Lubkemann, F.; Zambo, D.; Abdelmonem, A.M.; Bradtmuller, H.; Masood, A.; Morales, I.; de la Presa, P.; Knebel, A.; et al. Aerogelation of Polymer-Coated Photoluminescent, Plasmonic, and Magnetic Nanoparticles for Biosensing Applications. *ACS Appl. Nano Mater* **2021**, *4*, 6678–6688. [CrossRef]
4. Long, M.; Zhang, B.; Peng, S.; Liao, J.; Zhang, Y.; Wang, J.; Wang, M.; Qin, B.; Huang, J.; Huang, J.; et al. Interactions between two-dimensional nanoclay and blood cells in hemostasis. *Mater. Sci. Eng. C* **2019**, *105*, 110081. [CrossRef] [PubMed]
5. Dai, Z.; Zhang, Y.; Chen, C.; Yu, F.; Tian, J.; Cai, H.; Jiang, X.; Zhang, L.; Zhang, W. An Antifouling and Antimicrobial Zwitterionic Nanocomposite Hydrogel Dressing for Enhanced Wound Healing. *ACS Biomater. Sci. Eng.* **2021**, *7*, 1621–1630. [CrossRef]
6. Liu, C.; Liu, C.; Liu, Z.; Shi, Z.; Liu, S.; Wang, X.; Wang, X.; Huang, F. Injectable thermogelling bioadhesive chitosan-based hydrogels for efficient hemostasis. *Int. J. Biol. Macromol.* **2023**, *224*, 1091–1100. [CrossRef]
7. Hu, B.; Ouyang, Y.; Zhao, T.; Wang, Z.; Yan, Q.; Qian, Q.; Wang, W.; Wang, S. Antioxidant Hydrogels: Antioxidant Mechanisms, Design Strategies, and Applications in the Treatment of Oxidative Stress-Related Diseases. *Adv. Healthc. Mater.* **2024**. [CrossRef]
8. Xiao, L.; Hui, F.; Tian, T.; Yan, R.; Xin, J.; Zhao, X.; Jiang, Y.; Zhang, Z.; Kuang, Y.; Li, N.; et al. A Novel Conductive Antibacterial Nanocomposite Hydrogel Dressing for Healing of Severely Infected Wounds. *Front. Chem.* **2021**, *9*, 787886. [CrossRef]
9. Gan, D.; Huang, Z.; Wang, X.; Jiang, L.; Wang, C.; Zhu, M.; Ren, F.; Fang, L.; Wang, K.; Xie, C.; et al. Graphene Oxide-Templated Conductive and Redox-Active Nanosheets Incorporated Hydrogels for Adhesive Bioelectronics. *Adv. Funct. Mater.* **2019**, *30*, 1907678. [CrossRef]
10. Jing, X.; Mi, H.-Y.; Lin, Y.-J.; Enriquez, E.; Peng, X.-F.; Turng, L.-S. Highly Stretchable and Biocompatible Strain Sensors Based on Mussel-Inspired Super-Adhesive Self-Healing Hydrogels for Human Motion Monitoring. *ACS Appl. Mater. Interfaces* **2018**, *10*, 20897–20909. [CrossRef]
11. Luo, X.; Ao, F.; Huo, Q.; Liu, Y.; Wang, X.; Zhang, H.; Yang, M.; Ma, Y.; Liu, X. Skin-inspired injectable adhesive gelatin/HA biocomposite hydrogel for hemostasis and full-thickness dermal wound healing. *Biomater. Adv.* **2022**, *139*, 35882139. [CrossRef] [PubMed]
12. Zhou, L.; Dai, C.; Fan, L.; Jiang, Y.; Liu, C.; Zhou, Z.; Guan, P.; Tian, Y.; Xing, J.; Li, X.; et al. Injectable Self-Healing Natural Biopolymer-Based Hydrogel Adhesive with Thermoresponsive Reversible Adhesion for Minimally Invasive Surgery. *Adv. Funct. Mater.* **2021**, *31*, 2007457. [CrossRef]
13. Prasad, A.S.; Wilson, J.; Thomas, L.V. Designer injectable matrices of photocrosslinkable carboxymethyl cellulose methacrylate based hydrogels as cell carriers for gel type autologous chondrocyte implantation (GACI). *Int. J. Biol. Macromol.* **2023**, *224*, 465–482. [CrossRef]
14. Kim, H.; Kim, J.; Choi, J.; Park, Y.; Ki, C. Characterization of silk hydrogel formed with hydrolyzed silk fibroin-methacrylate via photopolymerization. *Polymer* **2018**, *153*, 232–240. [CrossRef]
15. Wu, C.; Jiao, Q.; Wang, C.; Zheng, Y.; Pan, X.; Zhong, W.; Xu, K. Nanofibrillar peptide hydrogels for self-delivery of lonidamine and synergistic photodynamic therapy. *Acta Biomater.* **2023**, *155*, 139–153. [CrossRef] [PubMed]
16. Zeng, L.; Chang, Y.; Wu, Y.; Yang, J.; Xu, J.-F.; Zhang, X. Charge-reversal surfactant antibiotic material for reducing microbial corrosion in petroleum exploitation and transportation. *Sci. Adv.* **2020**, *6*, eaba7524. [CrossRef] [PubMed]
17. Oleyaei, S.A.; Razavi, S.M.A.; Mikkonen, K.S. Novel nanobiocomposite hydrogels based on sage seed gum-laponite: Physico-chemical and rheological characterization. *Carbohydr. Polym.* **2018**, *192*, 282–290. [CrossRef] [PubMed]
18. Martinez-Garcia, F.D.; van Dongen, J.A.; Burgess, J.K.; Harmsen, M.C. Matrix Metalloproteases from Adipose Tissue-Derived Stromal Cells Are Spatiotemporally Regulated by Hydrogel Mechanics in a 3D Microenvironment. *Bioengineering* **2022**, *9*, 340. [CrossRef]
19. Ahn, D.; Sun, J.; Han, S.; Lee, J.; Jeong, S.; Cha, S.; Noh, S.; Choi, H.; Ren, B.; Yoon, H.; et al. Controllable Physical Synergized Triboelectricity, Shape Memory, Self-Healing, and Optical Sensing with Rollable Form Factor by Zn cluster. *Adv. Sci.* **2022**, *9*, e2200441. [CrossRef]
20. Pei, X.; Wang, J.; Cong, Y.; Fu, J. Recent progress in polymer hydrogel bioadhesives. *J. Polym. Sci.* **2021**, *59*, 1312–1337. [CrossRef]
21. Zhou, Q.; Zhou, X.; Mo, Z.; Zeng, Z.; Wang, Z.; Cai, Z.; Luo, L.; Ding, Q.; Li, H.; Tang, S. A PEG-CMC-THB-PRTM hydrogel with antibacterial and hemostatic properties for promoting wound healing. *Int. J. Biol. Macromol.* **2023**, *224*, 370–379. [CrossRef] [PubMed]
22. Jung, D.H.; Park, C.H.; Choi, S.I.; Kim, H.R.; Lee, M.; Moon, H.S.; Park, J.C. Comparison of a Polysaccharide Hemostatic Powder and Conventional Therapy for Peptic Ulcer Bleeding. *Clin. Gastroenterol. Hepatol.* **2023**, *21*, 2844–2853. [CrossRef] [PubMed]
23. Peng, X.; Xu, X.; Deng, Y.; Xie, X.; Xu, L.; Xu, X.; Yuan, W.; Yang, B.; Yang, X.; Xia, X.; et al. Ultrafast Self-Gelling and Wet Adhesive Powder for Acute Hemostasis and Wound Healing. *Adv. Funct. Mater.* **2021**, *31*, 2102583. [CrossRef]
24. Zhang, W.; Liu, W.; Long, L.; He, S.; Wang, Z.; Liu, Y.; Yang, L.; Chen, N.; Hu, C.; Wang, Y. Responsive multifunctional hydrogels emulating the chronic wounds healing cascade for skin repair. *J. Control. Release* **2023**, *354*, 821–834. [CrossRef] [PubMed]

25. Zhang, M.; Li, S.; Yuan, X.; Zhao, J.; Hou, X. An in situ catechol functionalized ε-polylysine/polyacrylamide hydrogel formed by hydrogen bonding recombination with high mechanical property for hemostasis. *Int. J. Biol. Macromol.* **2021**, *191*, 714–726.
26. Zheng, Y.; Shariati, K.; Ghovvati, M.; Vo, S.; Origer, N.; Imahori, T.; Kaneko, N.; Annabi, N. Hemostatic patch with ultra-strengthened mechanical properties for efficient adhesion to wet surfaces. *Biomaterials* **2023**, *301*, 122240. [CrossRef]
27. Xia, L.; Wang, S.; Jiang, Z.; Chi, J.; Yu, S.; Li, H.; Zhang, Y.; Li, L.; Zhou, C.; Liu, W.; et al. Hemostatic performance of chitosan-based hydrogel and its study on biodistribution and biodegradability in rats. *Carbohydr. Polym.* **2021**, *264*, 117965. [CrossRef]
28. Zhou, C.; Xu, R.; Han, X.; Tong, L.; Xiong, L.; Liang, J.; Sun, Y.; Zhang, X.; Fan, Y. Protocatechuic acid-mediated injectable antioxidant hydrogels facilitate wound healing. *Composites* **2022**, *250 Pt B*, 110451. [CrossRef]
29. Li, X.F.; Lu, P.; Jia, H.R.; Li, G.; Zhu, B.; Wang, X.; Wu, F.G. Emerging materials for hemostasis. *Coord. Chem. Rev.* **2022**, *475*, 214823. [CrossRef]
30. Guo, B.; Dong, R.; Liang, Y.; Li, M. Haemostatic materials for wound healing applications. *Nat. Rev. Chem.* **2021**, *5*, 773–791. [CrossRef]
31. Puertas-Bartolomé, M.; Benito-Garzón, L.; Fung, S.; Kohn, J.; Vázquez-Lasa, B.; Román, J.S. Bioadhesive functional hydrogels: Controlled release of catechol species with antioxidant and antiinflammatory behavior. *Mater. Sci. Eng. C Mater. Biol. Appl.* **2019**, *105*, 110040. [CrossRef] [PubMed]
32. Liu, W.; Chen, M.; Luo, M.; Li, T.; Hu, C.; Xie, C.; Li, S.; Leng, T.; Tian, J.; Xu, P. Bioactive glass ions hydrogels with antiinflammation antioxidant capacity for treating inflammation-related diseases. *Mater. Des.* **2023**, *227*, 111669. [CrossRef]
33. Dulong, V.; Kouassi, M.-C.; Labat, B.; Le Cerf, D.; Picton, L. Antioxidant properties and bioactivity of Carboxymethyl pullulan grafted with ferulic acid and of their hydrogels obtained by enzymatic reaction. *Food Chem.* **2018**, *262*, 21–29. [CrossRef] [PubMed]
34. Komeri, R.; Thankam, F.G.; Muthu, J. Free radical scavenging injectable hydrogels for regenerative therapy. *Biomater. Adv.* **2016**, *71*, 100–110. [CrossRef]
35. Li, X. 2-Phenyl-4,4,5,5-tetramethylimidazoline-1-oxyl 3-Oxide (PTIO•) Radical Scavenging: A New and Simple Antioxidant Assay In Vitro. *J. Agric. Food Chem.* **2017**, *65*, 6288–6297. [CrossRef]
36. Tan, Z.; Li, X.; Yu, C.; Yao, M.; Zhao, Z.; Guo, B.; Liang, L.; Wei, Y.; Yao, F.; Zhang, H.; et al. A self-gelling powder based on polyacrylic acid/polyacrylamide/quaternate chitosan for rapid hemostasis. *Int. J. Biol. Macromol.* **2023**, *232*, 123449. [CrossRef]
37. Wang, Z.; Lyu, T.; Xie, Q.; Zhang, Y.; Sun, H.; Wan, Y.; Tian, Y. Shape-adapted self-gelation hydrogel powder for high-performance hemostasis and wound healing. *Appl. Mater. Today* **2023**, *35*, 101948. [CrossRef]

gels

MDPI

Article

Clinical Perspectives on the Injectability of Cross-Linked Hyaluronic Acid Dermal Fillers: A Standardized Methodology for Commercial Product Benchmarking with Inter-Injector Assessments

Patrick Micheels [1,*], Alexandre Porcello [2,3], Thierry Bezzola [4], Daniel Perrenoud [5], Pierre Quinodoz [6], Yogeshvar Kalia [2,3], Eric Allémann [2,3], Alexis Laurent [7,8,9,†] and Olivier Jordan [2,3,†]

1 Private Medical Practice, CH-1224 Chêne-Bougeries, Switzerland
2 School of Pharmaceutical Sciences, University of Geneva, CH-1211 Geneva, Switzerland; alexandre.porcello@unige.ch (A.P.); yogi.kalia@unige.ch (Y.K.); eric.allemann@unige.ch (E.A.); olivier.jordan@unige.ch (O.J.)
3 Institute of Pharmaceutical Sciences of Western Switzerland, University of Geneva, CH-1211 Geneva, Switzerland
4 Private Medical Practice, CH-1204 Geneva, Switzerland; tbezzola@gmail.com
5 Private Medical Practice, CH-1006 Lausanne, Switzerland; drperrenoud@gmail.com
6 Hôpital de la Tour, CH-1217 Meyrin, Switzerland; pierre.quinodoz@latour.ch
7 Regenerative Therapy Unit, Lausanne University Hospital, University of Lausanne, CH-1015 Lausanne, Switzerland; alexis.laurent@unil.ch
8 Manufacturing Department, TEC-PHARMA SA, CH-1038 Bercher, Switzerland
9 Manufacturing Department, LAM Biotechnologies SA, CH-1066 Epalinges, Switzerland
* Correspondence: patrickscab@bluewin.ch; Tel.: +41-22-347-11-13
† These authors contributed equally to this work.

Abstract: The injectability of cross-linked hyaluronic acid (HA) dermal fillers is influenced by polymer concentration, polymer cross-linking type and degree, the presence of lidocaine or other functional excipients, types of syringes, and injection techniques. Finished product injectability constitutes a critical quality attribute for clinical injectors, as it strongly influences product applicability and ease of use in aesthetic medicine. While injectable product extrusion force specifications are provided by the respective device manufacturers, the qualitative informative value of such datasets is low for injectors wishing to compare product brands and technologies from an injectability standpoint. Therefore, the present study comparatively assessed 28 cross-linked HA dermal fillers (JUVÉDERM®, Restylane®, BELOTERO®, TEOSYAL RHA®, and STYLAGE® brands) using various injectability benchmarking setups for enhanced clinical-oriented relevance. Manual product injections were performed by three specialized and experienced clinicians, whereas automatic product extrusion was performed using a Texture Analyzer instrument. The various hydrogel products were injected into ex vivo human skin and into SimSkin® cutaneous equivalents to appropriately account for injection-related counterpressure. The injectability results revealed important variability between and within product brands, with a strong influence of the local anesthetic lidocaine, HA contents, and needle gauge size. Critical appraisals of the investigated products were performed, notably from manufacturing process-based and clinical ease of application-based standpoints, centered on respective experimental injectability quality levels. Generally, it was confirmed that each HA-based dermal filler product requires specific expertise for optimal injection, mainly due to differing viscoelastic characteristics and injectability attributes. Overall, the present study set forth evidence-based and clinical-oriented rationale elements confirming the importance for injectors to work with injectable products with which they are experienced and comfortable to optimize clinical results.

Keywords: aesthetic medicine; cross-linked hyaluronic acid; dermal fillers; ex vivo skin; hydrogel systems; injectability; lidocaine; medical device; needle gauge; product benchmarking

1. Introduction

Dermal filler injections are nonsurgical cosmetic procedures that are used to plump up wrinkles, smooth cutaneous lines and/or creases and restore fine volumes in the face [1]. Hyaluronic acid (HA)-based dermal fillers stand out at the upper end of the quality spectrum [2]. This is due to their documented safety, efficacy, and reversibility, with more than 2.6 million procedures performed in the USA alone in 2020 [3,4]. Modern HA-based dermal fillers each exhibit specific properties owing to different cross-linking technologies, product compositions, and manufacturing processes [5–8]. The primary objective of manufacturer-specific formulation and production technologies is to adjust the finished product's mechanical or biophysical properties. Such attributes may be finely tuned to closely mimic the biological tissues surrounding the administration site, enabling positive post-injection physiological outcomes [9–11]. Despite the invasive nature of dermal filler administration via localized injection, reported complication rates are low in the hands of medical professionals [12–14].

Among the important biophysical attributes of HA-based dermal fillers, system viscoelasticity, cohesivity, water uptake, and injectability are often cited [6–8,14,15]. Dynamic effects pertaining to product stability and biodistribution are also important from a product development standpoint [16–18]. In addition, specific clinical-oriented product characterization should consider the behavior of the hydrogel system in the primary packaging elements during administration. Throughout the injection process, the hydrogel within the syringes undergoes a variety of stresses and forces. In further detail, the gels are exposed to shear stress, along with vertical compression or elongation forces, resulting in material deformation. Therein, under minimal stress, homogeneous dermal fillers tend to approach the behavior of purely gel-like substances. As the shear stress intensifies, the system exhibits fluid-like behaviors and starts to flow, enabling extrusion through the needle [14]. Of note, the flow resistance is then influenced by external parameters, namely the cutaneous layer in which the gel is implanted, which generates counterpressure. Therein, the subcutaneous tissue generally presents less resistance, as it presents lobules between which HA-based gels can permeate without damaging the cells. Conversely, the dermis is predominantly fibrous, and product distribution and dispersion are influenced by gel cohesivity attributes [17,19].

Upon reviewing the clinical practices around HA-based dermal filler injections in aesthetic medicine, high diversity may be observed. Notably, while some injections are performed in the dermis, others are performed directly into the subcutaneous tissue [20]. Furthermore, the injection techniques vary among practitioners, ranging from closely spaced point-by-point injections to retro-tracing and bolus injections. The latter is especially used for subcutaneous tissue supplementation and in proximity of osseous matter for volumetric correction. Additionally, injectors may use the thumb pad, the thumb's interphalangeal joint, or the thenar eminence to actuate the syringe. This diversity is best exemplified among mesotherapy practitioners [21].

Overall, the extent to which HA-based dermal fillers flow largely depends on their manufacturing conditions, composition, the method and speed of injection, and the injection site. The diversity in available products and the multiple parameters influencing clinical outcomes highlight the importance of evidence-based dermal filler product selection for specific aesthetic applications. As regards product quality attributes, an ideal dermal filler should be easy to inject, with homogeneous (i.e., within one product unit and between product lots) and consistent product behavior upon administration. This ensures optimal performance by conferring the highest level of control over the product and the administration process, which in turn contributes to minimizing administration-related discomfort [14]. The latter may be mitigated by specific formulation means, such as the incorporation of local anesthetics (e.g., lidocaine). Therein, the ancillary pharmacological action of lidocaine may address the administration-related pain stimuli and early patient local reactions.

Since the commercial introduction of non-cohesive HA-based gels in 1994, the practicing co-authors have personally experienced fluctuations in the pressure exerted on dermal filler syringes, particularly during injections using the "Blanching technique" [22–24]. Of note, product manufacturers specify forces of ejection and extrusion curves, which are obtained in the laboratory with hydrogel extrusion in the air. Such data are of low translational relevance, as biological tissues oppose counterpressure upon clinical injection, which depends on the anatomical site and injection depth. Furthermore, cutaneous structures present varying epidermal and dermal proportions, where the dermis is characterized by inhomogeneous density or degrees of dermatoporosis [17,21,25–29].

Instructions for use generally specify that cross-linked HA-based dermal fillers should be injected into the superficial, medium, and deep dermis, in the hypodermis, or in the subperiosteal space [30–34]. Regular personal feedback about perceived in-use variations, either in product viscosity (i.e., both intra- and inter-syringes or lots) or in the required injection force, have been shared with product manufacturers (i.e., Patrick Micheels, unpublished communications, 2000–2023). It is notable that products from the BELOTERO® brand (i.e., original presentation before the addition of lidocaine) could be injected into the superficial reticular dermis [23,24]. After having previously experimentally compared product viscoelastic properties with or without lidocaine, a new area of investigation was designed around the assessment of the pressures exerted on the plunger rods of various cross-linked HA-based dermal fillers present on the European and North American markets [26,27].

Therefore, the purpose of this study was to comparatively assess 28 different cross-linked HA-based dermal fillers (i.e., from the JUVÉDERM®, Restylane®, BELOTERO®, TEOSYAL RHA®, and STYLAGE® brands), using various injectability benchmarking setups. The objective of the study was to obtain robust datasets on the injectability attributes of the considered commercial products for enhanced clinical-oriented relevance as compared to available extrusion force information. The novelty and originality of the study consisted of the adopted methodology (i.e., multifaceted and complementary injectability attribute characterization with high translational relevance) and the vast scope of commercial products ($n = 28$), which were benchmarked. The study investigated the primary hypothesis that intra-product and inter-product variability exists as regards injectability when using appropriate cutaneous scaffolds. The study investigated the secondary hypothesis that inter-injector variability exists as regards product injectability. The study also investigated the secondary hypothesis that the incorporation of lidocaine significantly impacts product injectability from quantitative and qualitative standpoints. Overall, this study covered evidence-based and clinical-oriented rationale elements supporting the use of well-characterized and high-quality dermal filler products for clinical proficiency maximization and clinical result systematic enhancement.

2. Results and Discussion

2.1. Technical Benchmarking of Dermal Filler Product Parameters and Specificities

The modern landscape of HA-based dermal fillers is dense and populated by arrays of diverse injectable products. For the needs of the present study, 28 different commercially available and clinically implemented HA-based dermal filler products were retained and compared in terms of injectability. The predetermined methodology for injectable product selection and inclusion in the study was specified. Namely, widely adopted cross-linked HA-based dermal filler products, commercialized on the European and/or USA markets, were retained. The specific product inclusion criteria in the experimental study were as follows:

- Injectable dermal filler products based on BDDE-cross-linked HA;
- Products from well-established manufacturers with a brand presence on the market >10 years;
- Commercial dermal filler products with a CE mark and/or FDA approval (i.e., medical devices);

- Products indicated for injection in the face for managing fine to medium wrinkles and folds, and for volumizing purposes;
- Products well known and regularly injected by one or more of the practicing co-authors;
- Products available on the European and/or North American markets, with frequent clinical application in Switzerland (i.e., area of practice of the co-authors).

It may be noted here that alternative commercial products conformed to the specified inclusion criteria but were not included in the present study based on co-author consensus. In detail, a compromise was made between a relatively large product panel (i.e., 28 products) and the vast extent of the technical investigations carried out herein on the same products. The 28 products, selected among those proposed by five different industrial manufacturers under well-established brand names, were included in the study (Table 1).

Table 1. Listing of the five manufacturers from which the 28 cross-linked HA-based commercial dermal filler products were obtained. The study focused on well-established brands from highly reputable laboratories with long-term clinical track records and global product commercialization history. CE, European mark of conformity; FDA, US Food and Drug Administration; HA, hyaluronic acid; USA, United States of America.

Parameters	Cross-Linked HA-Based Dermal Filler Product Manufacturers [1]				
Company Name	Allergan Aesthetics (Subsidiary of AbbVie Inc.)	Galderma SA	Merz Aesthetics (Subsidiary of Merz Group)	Teoxane SA	Laboratoires VIVACY
Company Headquarters	Irvine, CA, USA	Zug, Switzerland	Frankfurt am Main, Germany	Geneva, Switzerland	Paris, France
Product Brand of Interest	JUVÉDERM®	Restylane®	BELOTERO®	TEOSYAL RHA®	STYLAGE®
Product Brand Launch Year	2000	1996	2005	2004	2008
Product Types	Class III medical device	Class III medical device	Class III medical device	Class III medical device	Class III medical device
Product Approvals	CE-marked; FDA-approved	CE-marked; FDA-approved	CE-marked; FDA-approved	CE-marked; FDA-approved	CE-marked

[1] Laboratories specified as product legal manufacturers.

Prior to the experimental benchmarking of the retained medical devices (MD) with a focus on injectability attributes, the relevant ad hoc technical documentation was gathered for an initial product technical specification comparison (Table 2).

Table 2. Comparative overview of the technical attributes and characteristics of the cross-linked HA-based commercial dermal filler products included in the study. Within each product brand, formulation-based and clinical indication specificities were found to differentiate the retained injectable products. G, gauge; HA, hyaluronic acid.

Product Brand and Name [1]	Specified Product Clinical Uses	Needle Gauge (G) [2]	HA Concentration [3]	Cross-Linked HA	Contains Lidocaine	Cross-Linking Technology [4]
JUVÉDERM® VOLBELLA®	Fine lines; tear through	30 G	15 mg/mL	Yes	Yes	VYCROSS®
JUVÉDERM® VOLIFT®	Medium fold; lips	30 G	17.5 mg/mL	Yes	Yes	VYCROSS®
JUVÉDERM® VOLUMA®	Volumizer	27 G	20 mg/mL	Yes	Yes	VYCROSS®
JUVÉDERM® VOLUX®	Cheeks; temples; jaw line; chin volumizer	27 G	25 mg/mL	Yes	Yes	VYCROSS®

Table 2. *Cont.*

Product Brand and Name [1]	Specified Product Clinical Uses	Needle Gauge (G) [2]	HA Concentration [3]	Cross-Linked HA	Contains Lidocaine	Cross-Linking Technology [4]
JUVÉDERM® Ultra 2	Medium lines; lip border	30 G	24 mg/mL	Yes	Yes	HYLACROSS®
JUVÉDERM® Ultra 3	Deep folds; lip volume	27 G	24 mg/mL	Yes	Yes	HYLACROSS®
Restylane®	Medium lines	30 G	20 mg/mL	Yes	No	NASHA®
Restylane®	Medium lines	29 G	20 mg/mL	Yes	No	NASHA®
Restylane® Lido	Medium lines	30 G	20 mg/mL	Yes	Yes	NASHA®
Restylane® Lido	Medium lines	29 G	20 mg/mL	Yes	Yes	NASHA®
Restylane® Lyft	Deep folds; volumizer	27 G	20 mg/mL	Yes	No	NASHA®
Restylane® Lyft Lido	Deep folds; volumizer	27 G	20 mg/mL	Yes	Yes	NASHA®
BELOTERO® Soft	Fine lines	30 G	20 mg/mL	Yes	No	CPM®
BELOTERO® Soft +	Fine lines	30 G	20 mg/mL	Yes	Yes	CPM®
BELOTERO® Balance	Medium lines; lip border	30 G	22.5 mg/mL	Yes	No	CPM®
BELOTERO® Balance +	Medium lines; lip border	30 G	22.5 mg/mL	Yes	Yes	CPM®
BELOTERO® Intense	Deep folds; lip volumizer	27 G	25.5 mg/mL	Yes	No	CPM®
BELOTERO® Intense +	Deep folds; lip volumizer	27 G	25.5 mg/mL	Yes	Yes	CPM®
BELOTERO® Volume	Volumizer	30 G	26 mg/mL	Yes	No	CPM®
BELOTERO® Volume +	Volumizer	30 G	26 mg/mL	Yes	Yes	CPM®
TEOSYAL RHA® 1	Fine lines	30 G	15 mg/mL	Yes (mix)	Yes	Preserved Network®
TEOSYAL RHA® 2	Medium folds; lip contour	30 G	23 mg/mL	Yes (mix)	Yes	Preserved Network®
TEOSYAL RHA® 3	Deep folds; lip volumizer	27 G	23 mg/mL	Yes (mix)	Yes	Preserved Network®
TEOSYAL RHA® 4	Volumizer	27 G	23 mg/mL	Yes (mix)	Yes	Preserved Network®
TEOSYAL Ultra Deep	Strong volumizer	25 G	25 mg/g	Yes	Yes	Teosyal PureSense
STYLAGE® S	Fine lines	30 G	16 mg/g	Yes	Yes	IPN-Like® + mannitol
STYLAGE® M	Medium folds; lip contour	30 G	20 mg/g	Yes	Yes	IPN-Like® + mannitol
STYLAGE® L	Deep folds; lip volumizer	27 G	24 mg/g	Yes	Yes	IPN-Like® + mannitol
STYLAGE® XL	Volumizer	27 G	26 mg/g	Yes	Yes	IPN-Like® + mannitol
STYLAGE® XXL	Cheeks; temples; jawline; chin volumizer	27 G	21 mg/g	Yes	No	IPN-Like® + mannitol

[1] All of the data used for the cross-linked HA-based hydrogel product technical comparison work were compiled from manufacturer-provided information. [2] Needles as provided by manufacturers in product secondary packaging or as specified by the manufacturer for a given product. [3] It is notable that while the HA contents of the various products are specified, details on manufacturer-specific cross-linking technologies and the degrees of system cross-linking constitute trade secrets. [4] Various degrees of polymer cross-linking frequently exist within most product brands.

Furthermore, the retained products were briefly compared in terms of physical, chemical, and rheological attributes for optimal technical description prior to focused injectability attribute determination. The relevant data were gathered from manufacturer-provided documentation or from the scientific literature and are presented in Table S1. While important differences were outlined in terms of rheological properties between the investigated products, major technical and chemical similarities were noted (e.g., BDDE cross-linking, Table S1). The presence of lidocaine in the majority of the investigated dermal fillers justified the specific experimental focus on this component in the subsequent assays (Table 2). Furthermore, the technical investigations around the impact of lidocaine on product injectability attributes were devised based on existing (i.e., incomplete) knowledge of the effects of lidocaine on hydrogel system viscoelasticity [5,6,8].

Despite strong similarities in clinical indications and ingredient compositions (i.e., injection-grade HA), the compared dermal filler products are characterized by specific formulation-related attributes (Table 2). In particular, manufacturer-specific cross-linking technologies constitute one of the major factors of differentiation between product brands, as they notably influence hydrogel biophysical attributes, in vivo product behavior, and in vivo product degradation behavior. However, the exact mechanism for a given cross-linking technology is considered a trade secret. Of note, product manufacturers are required to provide the concentration of HA but not the degree of cross-linking (Table 2).

Interestingly, the degree of cross-linking can influence various critical physicochemical parameters of a product, including its rheology and injectability. Namely, the more a hydrogel system is cross-linked, the higher its elasticity and viscosity will be, making it more challenging to inject. Since the exact degrees of cross-linking are unknown, it is difficult to predict injectability attributes within a given product selection. Within the same range, products can vary in their HA concentration and degree of cross-linking despite using the same cross-linking technology. Of note, 1,4-butanediol diglycidyl ether (BDDE) is the cross-linker used within the HA-based dermal fillers included in the study (Table 2). Notably, BDDE was the first cross-linker to be used in a commercially available dermal filler. Furthermore, this cross-linking type is the one used in most clinical studies, making it the default gold standard [35–38]. Notwithstanding the high diversity in product offerings outlined in the study, all of the retained product brands and formulation technologies were confirmed to be of current relevance in the field of aesthetic dermatology (Table 2). Specifically, the listed products are available and routinely clinically applied in the accredited clinical centers of the practicing co-authors in Western Switzerland.

2.2. Comparative Manual Injectability Evaluation in Ex Vivo Human Skin and in SimSkin® Cutaneous Equivalents

In order to reduce the biological variability linked to the use of human skin for the subsequent large-scale comparative hydrogel product injectability studies, a standardized cutaneous equivalent was procured. For the initial validation of the SimSkin® in vitro injectability setup, comparative quantitative analyses were performed against ex vivo human skin for the assessment of potential technical equivalence. A wrinkle-filling product (i.e., TEOSYAL RHA® 2, Teoxane, Geneva, Switzerland) was retained for the experiments and was manually and sequentially injected point-by-point into the superficial to medium dermis of the ex vivo scaffolds and in the superficial polymer layer of the in vitro model. The results were expressed as mean forces of injection and as maximum peak forces of injection (Figure 1).

For both the in vitro and the ex vivo setup, the needle was inserted tangentially into the skin plane (i.e., at an angle $< 10°$). The results revealed no statistically significant differences between injectors for the mean force and the maximum peak force of injection in the ex vivo skin group (Figure 1A,B). Furthermore, no statistically significant differences were observed for injector 1 (PM) between the ex vivo and the in vitro groups (Figure 1A,B). Conversely, the results obtained for injector 2 (TB) showed statistically significant differences between the ex vivo and the in vitro groups, with a slight decrease in mean forces

(−20%) and maximum peak forces (−15%) of injection when TEOSYAL RHA® 2 was injected into SimSkin® (Figure 1A,B). Such different behavior between the injectors was attributed mostly to anatomical specificities of the thumb of the injectors and of the ex vivo skin samples, confirming the need for setup standardization and force measurement normalization. Notably, no very statistically significant or extremely statistically significant differences (i.e., p-values < 0.01) were observed between the groups and between the injectors (Figure 1). While the equivalence between the ex vivo and in vitro skin substrates could not be experimentally validated, the need for setup standardization and the absence of very statistically significant differences warranted the further use of SimSkin® cutaneous equivalents for all subsequent injectability studies. Specifically, the recorded standard deviation values in the SimSkin® setup ranged from almost zero to 20% in the experimental trials, which was assessed as satisfactory given the inter- and intra-sample variability of ex vivo cutaneous models.

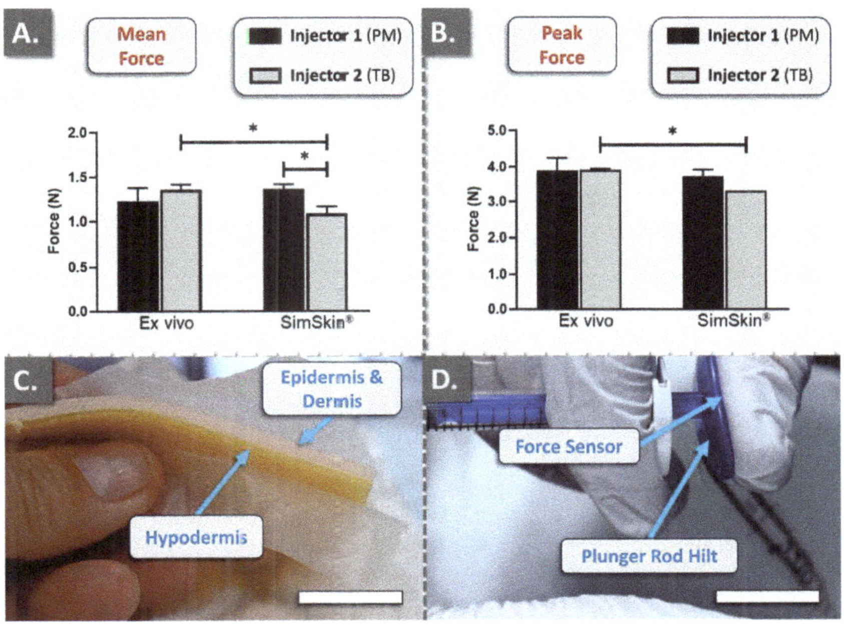

Figure 1. Results of technical equivalence studies for standardized manual hydrogel injectability evaluation, comparing ex vivo human skin and the artificial SimSkin® cutaneous equivalent. The quantitative measurements were performed using the TEOSYAL RHA® 2 product and the FlexiForce® dynamometric sensor attached to the syringe plunger rod hilt. The injections were performed by two clinicians using a point-by-point administration method. (**A**) Quantitative results of the mean injection force. (**B**) Quantitative results of the maximum peak force. Experiments were performed in triplicate, and the results were presented as average values assorted to corresponding standard deviations as error bars. Statistically significant differences (i.e., * or p-value < 0.05) were found between the average values. Detailed results of the statistical analyses are presented in Table S2. (**C**) Profile view of the SimSkin® cutaneous equivalent. Scale bar = 10 mm. (**D**) Profile view of the manual injection setup featuring the FlexiForce® dynamometric sensor, positioned beneath the thumb of the injector on the engaged plunger rod hilt. Scale bar = 15 mm. PM, Patrick Micheels; TB, Thierry Bezzola.

2.3. *Inter-Injector Variability Assessment and Influence of Lidocaine on Product Manual Injectability*

In order to robustly assess the influence of lidocaine on dermal filler product injectability attributes, several BELOTERO® and Restylane® products (i.e., variants with and without lidocaine) were injected by three clinicians into SimSkin® cutaneous equivalents, using a point-by-point injection technique (Figure 2).

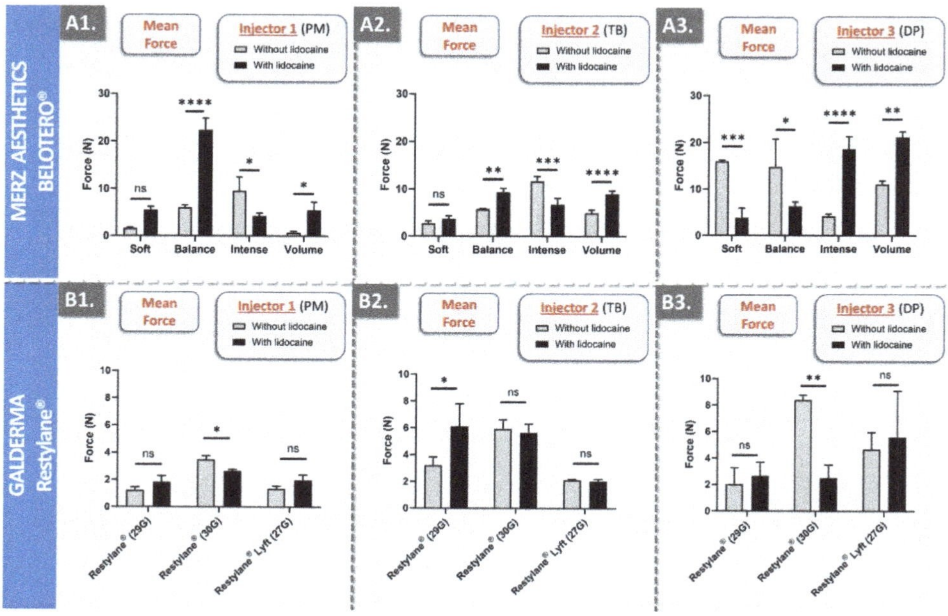

Figure 2. Results of in vitro manual injectability studies for the assessment of inter-injector variability and quantification of the influence of lidocaine on product injectability. The quantitative measurements were performed using the SimSkin® cutaneous equivalent and the FlexiForce® dynamometric sensor. The injections were performed using a point-by-point administration method. (**A1–A3**) Comparison of mean injection forces for BELOTERO® products (i.e., with and without lidocaine) between the three injectors. (**B1–B3**) Comparison of mean injection forces for Restylane® products (i.e., with and without lidocaine) between the three injectors. Experiments were performed in triplicate, and results were presented as average values assorted to corresponding standard deviations as error bars. Statistically significant differences (i.e., * or p-value < 0.05), very significant differences (i.e., ** or 0.001 < p-value < 0.01), or extremely significant differences (i.e., *** or 0.0001 < p-value < 0.001; **** or p-value < 0.0001) were found between the average values. Detailed results of the statistical analyses are presented in Table S3. DP, Daniel Perrenoud; G, gauge; ns, non-significant; PM, Patrick Micheels; TB, Thierry Bezzola.

For all injections, the needle was inserted tangentially into the skin plane (i.e., at an angle < 10°). As concerns the injectability of the dermal fillers from the BELOTERO® brand, the results revealed important inter-product and inter-injector diversity (Figure 2(A1–A3)). While lidocaine presence in the hydrogel system generally resulted in significant injectability attribute modulation, no clear trend was outlined across the board. Of note, the behavior of the Balance, Intense, and Volume products with lidocaine for point-by-point injections was found to be significantly different compared to the product variant without lidocaine. While mean forces of the Balance product were significantly higher with lidocaine for injectors PM and TB, they were found to be inferior for injector DP (Figure 2(A1–A3)). Of further note, the presence of lidocaine in the Volume product systematically resulted in higher mean forces for all three injectors (Figure 2(A1–A3)). It was generally observed that the spread between minimal and maximal force values within the product brand was injector-specific (e.g., low overall spread for injector TB, high spread for injector DP; Figure 2(A1–A3)). Finally, it was observed that the recorded force values vary depending on the injection speed and the area of the thumb used to apply pressure on the plunger rod hilt (Table S4). Specifically, higher values were recorded when the pressure was applied through the P1-P2 joint of the thumb, as opposed to the thumb pad (Figure S1).

As concerns the injectability of the dermal fillers from the Restylane® brand, the result trends were comparable to those of the BELOTERO® brand in terms of the influence of lidocaine presence, intra-brand behavior, and inter-injector behavior (Figure 2). A major difference between the two brands was noted in terms of force scale, where mean values were found to be inferior across the board for the Restylane® products (Figure 2). Additionally, the force differential between product variants (i.e., with or without lidocaine) was generally found to be lower for the Restylane® products as compared to the BELOTERO® products (i.e., lower levels of statistical significance; Figure 2). Specifically considering the Restylane® brand, a trend of higher mean injection force values was outlined for higher-gauge needles, as expected (Figure 2(B1–B3)). From a formulation viewpoint, it is notable that the NASHA® technology used in the manufacture of Restylane® gels is different from the CPM® technology of BELOTERO® gels (Table 2). Furthermore, Restylane® products share a constant concentration of 20 mg/mL HA, while BELOTERO® products contain HA concentrations ranging from 20 mg/mL to 26 mg/mL (Table 2). No correlation was found between the concentration of the HA-based polymer and the mean force of injection of the product. Additionally, according to the literature, the degree of substitution of Restylane® is around 1.2%, whereas no information can easily be found for the CPM® technology [39–41].

The viscosity and the cohesivity (i.e., measurement of maximum normal forces) of Restylane® products were lower than the viscosities of BELOTERO® products, except for the Soft variant, which was found to be congruent with the respective mean forces of injection, especially for injectors 1 and 2 [42,43]. Such quantitative elements may partly explain the lower overall mean injection force values recorded for the Restylane® products (Figure 2 and Table S5). Finally, the lowest overall injection force value spread for Restylane® products was identified for injector PM (Figure 2(B1)). Importantly, it was recorded that injector PM routinely administers Restylane® products, whereas injectors TB and DP do not. Therefore, it may be assessed that a learning phase exists for products that are not routinely administered (i.e., even by specialized physicians) and that practice and experience may result in diminished administration-related variability. Moreover, the investigational panel of three injectors is diverse, consisting of a general practitioner, a surgeon, and a dermatologist. These three injectors do not share the same injection habits. Injector 3 typically injects very small volumes and may prick a patient's face up to a hundred times. However, this injector administers dermal fillers to an average of 10 patients per week. Injectors 1 and 2, on the other hand, treat well over 10 patients per week with fillers. It is also noteworthy that injector 1 tends to inject at a slower pace compared to injectors 2 and 3.

As concerns the injectability of the dermal fillers from the STYLAGE® brand, the results displayed the lowest overall intra-brand and inter-injector variability (Figure 3(A1–A3) and Table S6).

Similar to the results of the Restylane® product brand, the lowest intra-brand spread in force values was recorded for the injector PM (Figure 3(A1)). Furthermore, the lowest overall mean injection force values for the STYLAGE® brand were also recorded for injector PM (Figure 3(A1–A3)). Specifically, STYLAGE® S required the greatest mean injection force for injector PM, with 1.68 N. STYLAGE® M, XL, and XXL necessitated intermediate forces of 1.11 N, 1.38 N, and 1.18 N, respectively, demonstrating marginal variations within the brand. The lowest required force was recorded for STYLAGE® L, averaging at 0.78 N (Figure 3(A1)). Of note, STYLAGE® XXL is the only product variant without lidocaine (Figure 3(A1–A3) and Table 2). While STYLAGE® products differ primarily in their intended uses, needle size, and HA concentration, they all employ the same IPN-Like® cross-linking technology and mannitol as an integrated antioxidant (Table 2) [44,45]. The relevant literature shows that STYLAGE® M displays an injection force close to 10 N at a speed of 13 mm·min^{-1}, which represents an appropriate level for manual injection [46]. The analysis of the experimental datasets revealed that injectors 2 and 3 recorded slightly higher forces, whereas injector 1 recorded slightly lower forces. Notwithstanding, all values were found to fall within a comparable range, and the lower results for injector 1 could be linked to slower speeds of injection in comparison with injectors 2 and 3 [46].

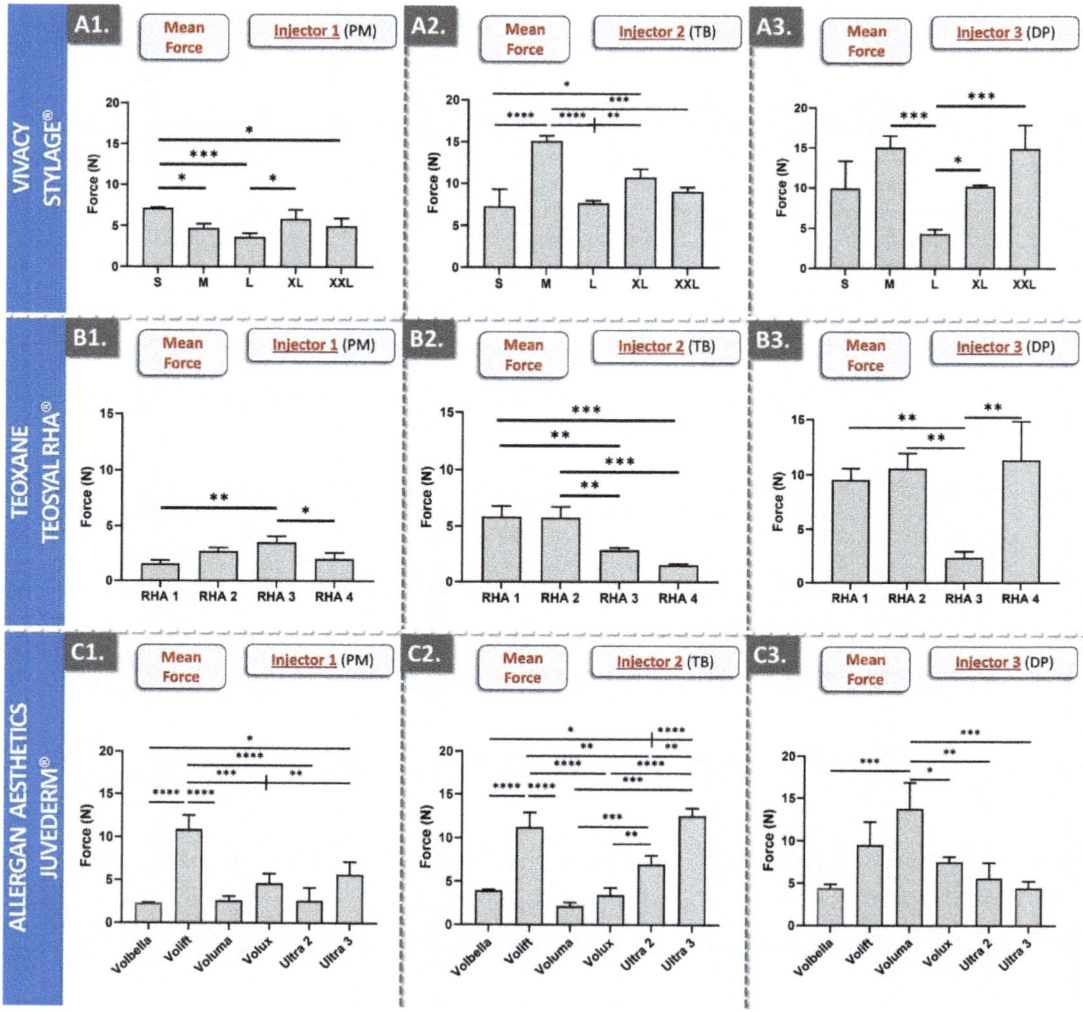

Figure 3. Results of in vitro manual injectability studies for the assessment of inter-injector variability over a diverse panel of dermal filler products. The quantitative measurements were performed using the SimSkin® cutaneous equivalent and the FlexiForce® dynamometric sensor. (**A1–A3**) Comparison of mean injection forces for STYLAGE® products between the three injectors. (**B1–B3**) Comparison of mean injection forces for TEOSYAL RHA® products between the three injectors. (**C1–C3**) Comparison of mean injection forces for JUVÉDERM® products between the three injectors. Experiments were performed in triplicate, and the results were presented as average values assorted to corresponding standard deviations as error bars. Statistically significant differences (i.e., * or p-value < 0.05), very significant differences (i.e., ** or 0.001 < p-value < 0.01), or extremely significant differences (i.e., *** or 0.0001 < p-value < 0.001; **** or p-value < 0.0001) were found between the average values. Detailed results of the statistical analyses are presented in Table S3. DP, Daniel Perrenoud; PM, Patrick Micheels; TB, Thierry Bezzola.

As concerns the injectability of the dermal fillers from the TEOSYAL RHA® brand, the results displayed intra-brand injection force variability and inter-injector force scale differences (Figure 3(B1–B3) and Table S7). Specifically, brand overall mean injection forces were found to be relatively low for injector PM, intermediate for injector TB, and high for injector DP (Figure 3(B1–B3)). The TEOSYAL RHA® products all employ the same Preserved Network® cross-linking technology and contain lidocaine, differing primarily in their intended uses, needle size, and HA concentration (Table 2) [34,47]. From a rheological viewpoint, RHA® 1 exhibits the lowest values, RHA® 2 and 3 score in the same range, and RHA® 4 scores the highest in value [47]. For injector PM, the exerted force increased from RHA® 1 to RHA® 3 (i.e., from 0.37 N to 0.83 N) and then decreased for RHA® 4 (i.e., 0.49 N; Figure 3(B1)). Such patterns may be rationally explained by the specific HA concentrations and the intended product uses (Table 2). In detail, products designed for treating deep folds and for adding volume (e.g., RHA® 3) typically comport a higher HA concentration, thus requiring superior injection forces. RHA® 3 exhibited lower values than those of RHA® 2 for injectors 2 and 3, and this was explained by the fact that RHA® 2 and RHA® 3 show rheological values in the same range, but where RHA® 3 was injected with a 27 G needle (i.e., while RHA® 2 was used with a 30 G needle) [47]. Interestingly, the brand-wide trend of injection forces was inverted for injector TB as compared to injector PM, where RHA® 1 and RHA® 2 scored the highest (Figure 3(B1,B2)). Such results clearly indicated differences in the specific injection techniques used by each injector.

As concerns the injectability of the dermal fillers from the JUVÉDERM® brand, the results outlined high intra-brand variability but the lowest overall inter-injector variability (Figure 3(C1–C3) and Table S8). In detail, the results generally indicated again that injectors TB and DP typically apply more force than injector PM (i.e., with the exception of the Volux and Voluma products, when comparing injectors TB and PM, Figure 3(C1–C3)). As for the other investigated product brands, the recorded differences in the required injection forces are most likely a composite effect of several variables, including the HA concentration, needle gauge, and the specific cross-linking technology (Figures 2 and 3, Table 2). Interestingly, Volbella and Volift are the two products that were assessed as being consistent in terms of injection force for the three injectors, corresponding to the products with the lower HA concentrations within the JUVÉDERM® brand. From a rheological viewpoint, concerning the hydrogels using the VYCROSS® technology, Volbella displayed the lowest injectability values and Volux the highest. Of note, the Hylacross® technology used in the JUVÉDERM® brand shows higher values of tan delta (i.e., ratio of storage modulus/loss modulus), suggesting a lower cross-linking percentage in comparison with the VYCROSS® technology [43]. With regard to the important inter-injector differences that were recorded across the retained product panel, most may be attributed to the specific product administration technique and the respective experience levels of any given injector with a specific product. Importantly, such findings underscore the importance of conjoint injector and product selection, as different combinations were shown herein to significantly influence the required injection force (Figures 2 and 3). As the latter may potentially impact the ease of product injection, such choices may prove to be determinants in the overall patient experience [48,49].

2.4. Automated In Vitro Product Injectability Assessment: Comparative Injection Curves for Standardized Dermal Filler Product Benchmarking

While the previous section (i.e., manual product injections in SimSkin® substrates) enabled to underscore inter-injector diversity and significant impacts of lidocaine presence on product injectability attributes, the assessments of intra-product injectability variability were limited (Tables S3–S7). Specifically, while the injection force was measured during a complete hydrogel unit extrusion cycle (i.e., emptying of the product syringe), the obtained force profile records only enabled to transcribe mean injection forces and peak injection forces (Figures 2 and 3, Tables S3–S7). In order to experimentally investigate the main hypothesis of the study (i.e., intra-product injectability variability), automated in vitro product injectability assessments were performed. This setup enabled normalization of the injection speed for each product and eliminated the potential bias of injector experience with one product brand but not with another.

The first part of the automated injectability study enabled us to assess once more the impact of lidocaine presence on product injectability attributes [50,51]. The results, which were, respectively, obtained with BELOTERO® and Restylane® variants (i.e., with and without lidocaine), revealed either an absence of statistically significant difference in mean plateau injection forces (i.e., Soft, Intense, and Volume products) or extremely statistically significant differences between the respective variants (Figure 4A,B).

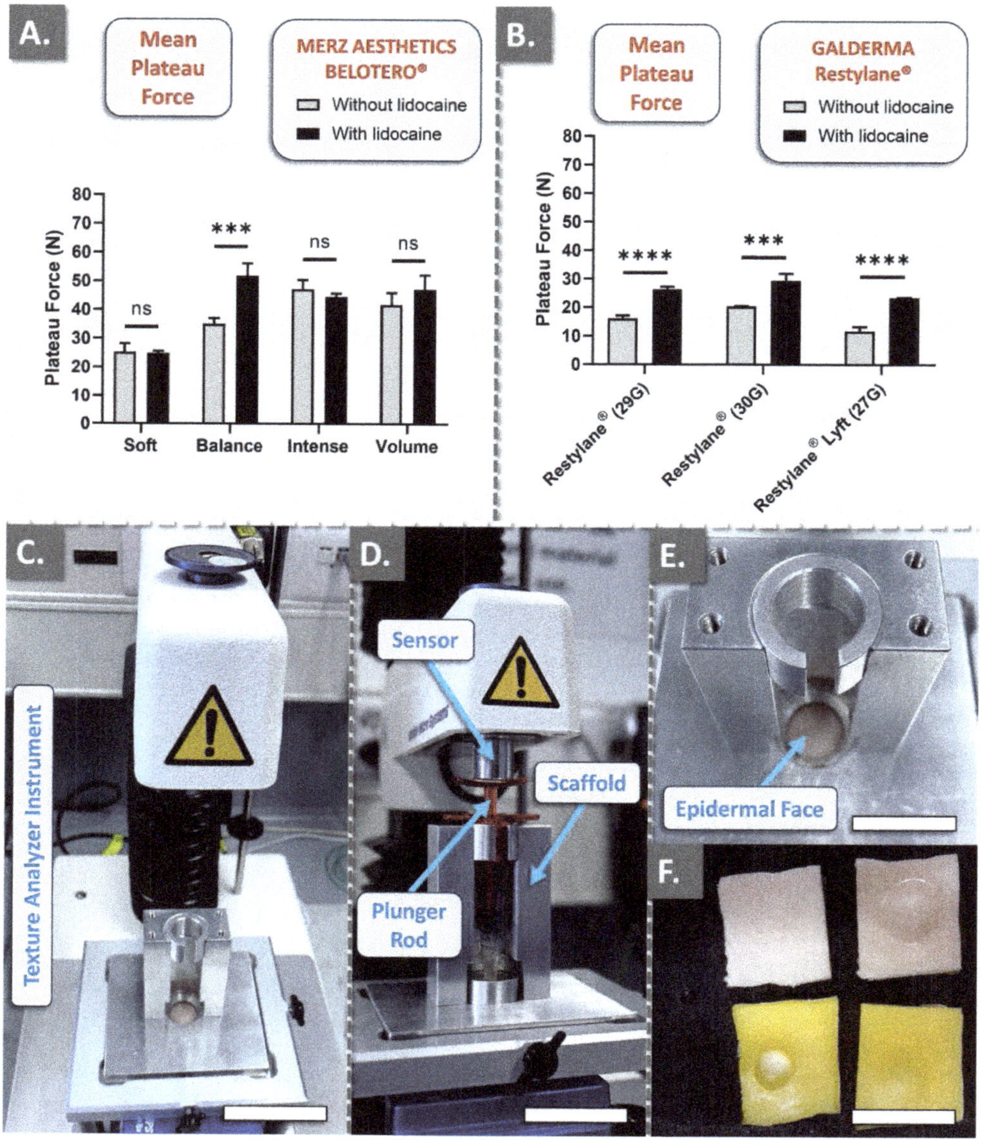

Figure 4. Results of in vitro automated injectability studies for the quantification of the influence of lidocaine on product injectability. The quantitative measurements were performed using a constant

plunger rod actuation speed of 1 mm·s^{-1}. (**A**) Comparison of mean plateau injection forces for BELOTERO® product variants (i.e., with and without lidocaine). (**B**) Comparison of mean plateau injection forces for Restylane® product variants (i.e., with and without lidocaine). Experiments were performed in triplicate, and the results were presented as average values assorted to corresponding standard deviations as error bars. Extremely significant statistical differences (i.e., *** or $0.0001 < p\text{-value} < 0.001$; **** or $p\text{-value} < 0.0001$) were found between the presented mean values. Detailed results of the statistical analyses are presented in Table S9. (**C**) Experimental setup for automated hydrogel injectability measurements. Scale bar = 5 cm. (**D**) Product syringe loaded in the automated injectability measurement setup. Scale bar = 4 cm. (**E**) Top view of the SimSkin® cutaneous equivalent (i.e., epidermal side) in the automated injectability measurement setup. Scale bar = 2 cm. (**F**) Epidermal side (top) and hypodermal side (bottom) of the SimSkin® cutaneous equivalent, before and after hydrogel product injection. Scale bar = 2 cm. ns, non-significant.

Of note, only the Balance product was found to behave specifically within the BELOTERO® brand, as all other variants were, respectively, not found to behave statistically differently from each other (Figure 4A). In detail, the Balance variant with lidocaine exhibited injection forces 1.5 times higher on average than its lidocaine-free counterpart, corroborating the manual findings of injectors PM and TB during point-by-point injection (Figure 2(A1–A3)). Importantly, the Balance variant with lidocaine was recorded as more homogeneous during manual injection than its lidocaine-free counterpart (Table S4). Consideration of the automated injection curves for the Balance variants corroborated such observations, where the variant with lidocaine presented much smoother plateau injection forces (Figure S2). Moreover, when comparing the results obtained with those of manual injections, it is interesting to note that BELOTERO® Volume (i.e., which indicated higher values for the version with lidocaine for each injector) presented similar trends at a constant speed.

Specific consideration of the Restylane® variants confirmed the systematic and significant increase in mean plateau injection forces in the lidocaine-containing products (Figure 4B). In detail, a two-fold increase was recorded for Restylane® Lyft, a 1.6-fold increase was recorded for Restylane® injected with a 29 G needle and a 1.4-fold increase was recorded for Restylane® administered with a 30 G needle (Figures 4B and S3). It is of great interest to note that there is a lack of studies investigating the interactions between lidocaine and HA, especially when considering the number of products that contain such functional excipients [41,50,51]. Nevertheless, the presented results clearly indicated that most product variants with lidocaine were more difficult to inject than their lidocaine-free counterparts (Figures 4, S2 and S3).

As regards the comparison of the injection force values obtained during manual injection and during automated injection, experimental setup-related rationale elements may be presented. Specifically, the average injection forces were notably higher across the board of investigated products when measured with the Texture Analyzer instrument. An explanation for the higher injection forces as measured in the automated setup is related to the speed of injection. Namely, the average speed for point-by-point manual injection was between 10 and 20 mm·min^{-1}. In contrast, the automated and constant injection speed was set at 60 mm·min^{-1}. For an equal counterpressure (i.e., mediated by hydrogel physical attributes, needle gauge and length, and cutaneous substrate composition), an increased injection speed will increase the plateau injection force. The corroboration of such mechanical principles with the obtained experimental datasets was confirmed for the JUVÉDERM®, TEOSYAL RHA®, and STYLAGE® brands, where automated injection forces were systematically recorded as superior to manual injection forces (Figures 5 and S4–S6).

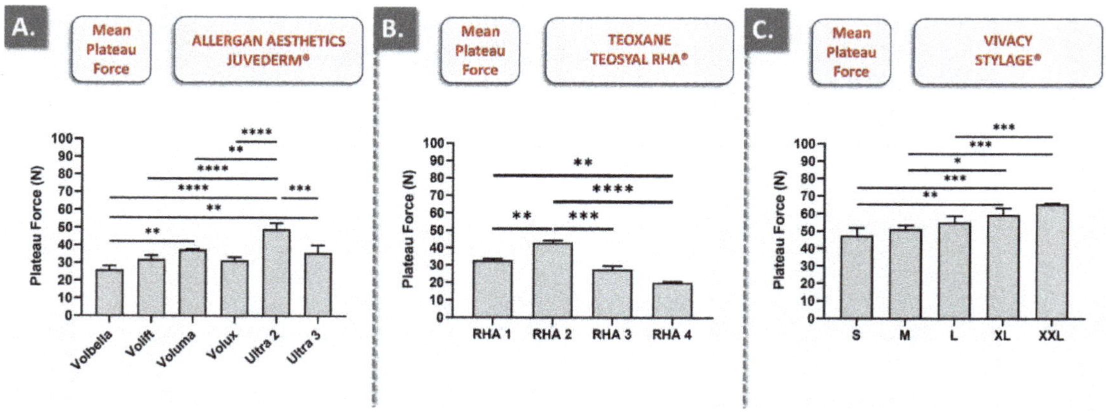

Figure 5. Results of in vitro automated injectability studies for the assessment of dermal filler intra-brand injection force variability. The quantitative measurements were performed using the SimSkin® cutaneous equivalent and a constant plunger rod actuation speed of 1 $mm \cdot s^{-1}$. (**A**) Comparison of the mean plateau injection forces required for JUVÉDERM® products. (**B**) Comparison of the mean plateau injection forces required for TEOSYAL RHA® products. (**C**) Comparison of the mean plateau injection forces required for STYLAGE® products. Experiments were performed in triplicate, and the results were presented as average values assorted to corresponding standard deviations as error bars. Statistically significant differences (i.e., * or p-value < 0.05), very significant differences (i.e., ** or 0.001 < p-value < 0.01), or extremely significant differences (i.e., *** or 0.0001 < p-value < 0.001; **** or p-value < 0.0001) were found between the presented mean values. Detailed results of the statistical analyses are presented in Table S9.

Interestingly, only TEOSYAL RHA® 1 exhibited intra-product variability comparable to that of BELOTERO® Balance without lidocaine (Figure S5). Overall, despite the presence of intra-brand statistically significant differences in automated injectability values, the relative differences were assessed as being lower than in the manual injectability measurement setup (Figures 3 and 5). This may partly be explained by the experimental standardization in the automated setup.

From a technical viewpoint, when comparing products administered with the same needle gauge (e.g., Volbella and Volift), it may be assessed that the force plateau generally rises with the increase in HA concentration (Figures 5 and S4, Table 2). This type of force plateau increase may also be observed for TEOSYAL RHA® products (Figures 5 and S5, Table 2). Additionally, the same observation may be made about STYLAGE® products, which presented the highest mean plateau injection forces (Figures 5 and S6, Table 2). Specifically, STYLAGE® products were found to be significantly tougher to inject in the automated setup as compared to TEOSYAL RHA® and JUVÉDERM® products (Figures 5 and S6). This observation, which was not made in the manual setup, could potentially be attributed to the IPN-Like® cross-linking technology or the addition of mannitol (Figure 3, Table 2). Overall, it was also assessed that plateau injection forces do not linearly correlate with the HA concentration or the needle gauge (Figures 4, 5 and S2–S6, Table 2). This indicated that the true injection force is likely conjointly influenced by HA concentration, needle gauge, product formulation, cross-linking type and degree, and the specific product manufacturing technology. Therefore, in conducting comparative product analyses encompassing all three injectors based on the values acquired via the dynamometric sensor, one encounters substantial challenges. This complexity arises due to the multitude of variable factors involved, such as the position of the thumb, the speed of injection, familiarity with the product, and the hydrogel homogeneity within the syringe.

Consideration of the obtained injectability profiles confirmed the hypothesis that significant variations in the intra-product injection force (i.e., and in the felt pressure) during product administration were possible (e.g., BELOTERO® Balance, TEOSYAL® Ultra Deep; Figures S2 and S5). The obtained data specifically demonstrated that some hydrogel systems present a higher degree of uniformity in their viscoelastic characteristics, both within a single syringe and intra-brand. An example thereof may be set forth with the comparison of the automated injection force profiles of TEOSYAL RHA® products and STYLAGE® products, the latter presenting smoother force plateaus (Figures S5 and S6). Furthermore, intra-product homogeneity was shown to be impacted in product variants containing lidocaine. In detail, non-homogeneous force plateaus were specifically identified for TEOSYAL RHA® 1, Ultra Deep, Voluma, Ultra 2, BELOTERO® Balance, BELOTERO® Volume (i.e., with and without lidocaine), BELOTERO® Intense (i.e., with and without lidocaine), and Restylane® Lyft (i.e., with and without lidocaine; Figures S2–S6).

From a formulation viewpoint, the hydrogel systems utilizing the CPM® cross-linking technology (i.e., with the exception of BELOTERO® Soft and BELOTERO® Intense) display irregular extrusion curves (Figure S2). More specifically, product variants with lidocaine showed higher plateau extrusion forces compared to their variants without lidocaine (Figure S2). For products manufactured with the NASHA® technology, the most traditional variant in the range (i.e., Restylane®) demonstrated the most homogeneous viscoelastic properties (Figure S3). In the JUVÉDERM® brand, except for the Voluma® and Ultra 2 variants, products using the VYCROSS® cross-linking technology present very stable extrusion force curves, indicating fairly homogeneous viscoelastic properties (Figure S4). Generally, excluding RHA® 1, the Preserved Network® technology provides hydrogels with excellent viscoelastic properties and high consistency within a single syringe (Figure S5). Finally, STYLAGE® hydrogels manufactured with the IPN-Like™ cross-linking technology exhibit the smoothest profiles, suggesting homogeneity in their viscoelastic properties (Figure S6). The obtained experimental results confirmed and underscored, from a clinically relevant injectability standpoint, that not all cross-linked HA-based dermal fillers are created equal. Overall, it is most likely that each cross-linking technology necessitates specific clinician experience and expertise for effective in vivo application.

2.5. Assessment of the Influence of Lidocaine Incorporation in Cross-Linked HA-Based Hydrogel Systems

As previously mentioned, the nature of the interactions between HA-based polymeric systems and lidocaine, a common dermal filler ingredient, is incompletely understood [41,50]. From a purely compositional standpoint, the presence or absence of lidocaine in a given HA-based hydrogel (i.e., linear or cross-linked polymer, lidocaine quantities usually present in dermal fillers) is insufficient to justify significant variations in system viscoelasticity attributes. A critical approach followed by well-advised dermal filler formulators consists of taking into account the product manufacturing process steps and their respective influence on the individual product components [10,36–38]. For example, it is well known that steam sterilization drastically negatively impacts the viscoelastic properties of an HA-based hydrogel system by means of HA chain breakdown. The manufacturing process therefore creates the need for pre-emptive formulation correction, typically with the selection of high molecular weight HA, which accounts for polymer breakdown during product terminal sterilization and eventually enables the reaching of appropriate endpoint molecular weight range technical specifications [52,53].

A similar approach may be used to understand the effects of lidocaine during the manufacturing process of an HA-based dermal filler. For the experimental investigation of such effects, cross-linked HA-based hydrogels with and without lidocaine were prepared and submitted to conservative steam sterilization processing. The endpoint rheological measurement results indicated that the presence of lidocaine during the sterilization step drastically negatively impacted the rheological parameters of the system (i.e., significant additional drop in storage and loss moduli in the lidocaine-containing group; Figure S7). Based on such data, the assumption may be made that the product formulation process or the manufacturing process for dermal filler variants (i.e., same product, with and without lidocaine) must be different if endpoint rheological profiles are similar. Specifically, single or multiple specific technical pre-emptive countermeasures to the negative impacts of lidocaine during sterilization must be employed for products containing lidocaine. This theory is supported by the injectability data of Restylane® variants, where lidocaine-containing products were tougher to inject (Figure 4B).

Although various experimental designs may be used to further investigate the above-mentioned theory on specific process adaptation in the case of lidocaine incorporation, further information about the retained commercial products is not available. Specifically, as each dermal filler product brand is based on different cross-linking technologies and different manufacturing processes, the technical means of lidocaine incorporation may be diverse (Table 2) [54]. Furthermore, as such details on product formulation process and manufacturing specifications are regarded as commercial trade secrets, few product manufacturers would confirm or deny the postulated elements about specific manufacturing-based lidocaine handling precautions. Furthermore, additional research is warranted for the understanding of mechanisms by which lidocaine contributes to additionally or synergistically degrade HA networks during sterilization processing. Such insights could potentially be used to develop new process enhancement options, such as the addition of thermo-protectant agents to lidocaine-containing formulas.

2.6. Qualitative Clinical Perspectives for Cross-Linked HA-Based Dermal Filler Administration: Focus on the Point-by-Point Intradermal Injection Technique

Despite the specification of defined intended clinical uses for dermal filler products, clinicians often employ off-label techniques in practice (e.g., different needles and different injection depth) based on their experience. Such off-label administrations may comprise the use of a purely volumizing agent for wrinkle filling in the epidermis or the deeper-than-average use of superficial fillers for enhanced treatment zone plasticity. Using the manual quantitative injectability evaluation setup and point-by-point injections, various off-label administration depths were investigated using the same commercial products and were compared to approved uses in terms of injectability. Injections were performed in SimSkin® substrates by injectors PM and TB.

As concerns the study of JUVÉDERM® products, the point-by-point injection of Volbella into the dermis only required low manual pressures. Conversely, the pressures required to inject Volift in the dermis (i.e., basal portion of the SimSkin® substrate) were high compared to all other HA gels intended for wrinkle filling. Voluma, administered at depths beyond its intended use, necessitated minimal exertion of pressure. Specifically, the gel's performance during injection was found to be aligned with expectations for a substance that exhibits a high degree of homogeneity in its inherent viscosity. For intradermal administration, the observations for Volux were identical to those for Voluma, namely describing a homogeneous gel requiring low pressures for injection via a 27 G needle. Excluding the Volift product, the hydrogels utilizing the VYCROSS® cross-linking technology generally demonstrated desirable viscoelastic characteristics, and their ease of injectability suggests a uniform gel composition. Regarding the Ultra 2 product, based on the HYLACROSS® cross-linking technology, exceedingly low injection pressures were recorded. Injector TB, operating at a faster pace than injector PM, recorded pressures approximately five times higher with this product. However, as the gel viscosity increased (i.e., using Ultra 3), the differences in applied pressures diminished. Interestingly, injector PM displayed more variation in the exerted pressure than injector TB (i.e., who does not commonly use HYLACROSS® and VYCROSS® gels).

As concerns the study of Restylane® products, the product variant without lidocaine required a very low pressure for intradermal injection with both needle gauge sizes (i.e., 29 G, 30 G). Restylane® without lidocaine is a non-cohesive gel, following the Sundaram-Gavard-Molliard classification [49]. For this product, the 30 G needle is not the manufacturer's recommended needle. However, the latter was used during the market introduction of the NASHA® gel technology in the late 1990s. In detail, injector PM has maintained the use of this fine needle, which facilitates detailed and precise clinical work. Restylane® Lyft without lidocaine was injected via a 27 G needle and displayed exceptionally low pressures, closely mirroring those observed with the standard Restylane® variant. For Restylane® Lyft with lidocaine, the injection pressures were notably low, albeit slightly superior to those of the lidocaine-free variant, and identical between injectors. Generally, despite being non-cohesive, hydrogels employing the NASHA® cross-linking technology demanded minimal injection pressure for administration into the dermis. However, the addition of lidocaine to the original formulation leads to a minor yet tangible alteration in the properties of the system.

As concerns the study of BELOTERO® products, only the least concentrated and least cross-linked hydrogel (i.e., BELOTERO® Soft) seemed homogeneous in its intrinsic viscosity, as assessed by manual injectability evaluation. Notably, the addition of lidocaine during the manufacture of the Balance product substantially increases the system viscoelasticity, necessitating significantly higher injection pressures. This aspect is particularly noticeable when injecting BELOTERO® Balance via a 30 G needle. Conversely, BELOTERO® Intense, which is designed for treating deep wrinkles (i.e., therefore, theoretically more viscoelastic), necessitated lower injection pressures for the lidocaine-containing variant. Finally, while BELOTERO® Volume is a volumizing agent and therefore expected to display high viscoelasticity, it was easily injected via a 30 G needle. Specifically, the required pressure was systematically low (i.e., for both injectors), regardless of whether or not lidocaine was present.

As concerns the study of TEOSYAL RHA® products utilizing the Preserved Network® cross-linking technology, all investigated products exhibited remarkably low viscoelasticity and easy injectability. Such properties enable effective administration (i.e., including the gel designed for volumetric augmentation) by applying minimal pressure on the syringe plunger rod.

Concerning the study of STYLAGE® products, STYLAGE® S and STYLAGE® M required the highest injection pressure among all the investigated fine wrinkle treatment hydrogels. Nevertheless, it demonstrated excellent intrinsic viscoelastic homogeneity. STYLAGE® XL and XXL are volumizing products and are not recommended for intradermal injection, yet they displayed acceptable pressure levels. Generally, all of the investigated hydrogels utilizing the IPN-Like® cross-linking technology displayed excellent, if not superior, intrinsic viscoelastic homogeneity.

Overall, factors such as the speed of injection, clinician familiarity with the product, and the area of the thumb used to apply pressure on the syringe plunger rod hilt all significantly impact the injectability of dermal fillers. Additionally, it was interesting to observe that during repeated injections (i.e., most often), the injector gains confidence and tends to inject more rapidly, daring to exert higher pressures than during the first injection. Importantly, the specific combination of a skilled injector and a high-quality product plays a crucial role, impacting the ease and effectiveness of the injection process.

2.7. Clinical Considerations, Performance Implications, and Perspectives on Product Injectability Attributes

Generally, the experimental results presented herein have confirmed the primary hypothesis of the study as concerns the possible intra-product variability or inhomogeneity in their injectability attributes (Figures S2–S6). Specifically, parallels were made between products with inhomogeneous force injection profiles and the available clinician feedback on injection force discrepancies. This aspect was interpreted to directly support the observations and regular communications of practicing co-author PM on product injectability, for example, in the case of BELOTERO® Balance (i.e., Patrick Micheels, unpublished communications with manufacturers, 2000–2023). Specifically, numerous clinician feedback on injection force inhomogeneity upon administration of BELOTERO® Balance were regularly notified, and such elements were experimentally confirmed by the data presented herein (Figure S2).

The concrete clinical perspectives and significance of the presented data may be approached from a patient safety and intervention quality standpoint. Specifically, in the case of a hydrogel with low homogeneity in its injectability attributes, some irregular indentation-like or jerking behavior of the plunger rod may be felt by the injector during administration. In the worst case, this can potentially lead to variations in the injected gel quantity, where an excess product amount or an inappropriately shallow product injection may be detrimental. Specifically, such administration-related defects may bear tangible consequences for the patient, such as Tyndall effects or the apparition of persistent nodules or lines if too much product is injected superficially [55,56]. Therefore, and importantly, the direct control by the clinician over the administration process of a dermal filler is dependent (i.e., among others) upon product injectability attributes, which may be linked to the safety and effectiveness of the intervention. Overall, as regulatory and reputational damages may be incurred rapidly (i.e., product manufacturers and physicians) in case of in-use safety or efficacy problematics, it is deemed critical for all stakeholders to consider product injectability with an important level of scrutiny.

The injectability data presented herein show that highly significant inter-injector differences may exist in terms of quantitative forces to apply on product syringes. Specifically, this aspect may depend on the experience of the injector with the product, the injection speed, the administration site, or the area of the hand used to inject. Notwithstanding, it may be assessed and set forward that different injection force levels do not necessarily impact the clinical performance of a given homogeneous product. Therein, the most critical factor consists of the in-use level of control of the physician over the dermal filler syringe, which may be correlated with the quality level of the obtained results. Therefore, the mean or peak quantitative aspects of product injectability and the inter-injector variability as concerns injection forces are most probably of lesser importance for clinical performance than the qualitative aspects of product injectability (i.e., intra-product homogeneity, plateau smoothness of injection curves). Such considerations may be of practical use as development perspectives for product manufacturers in particular, which may be urged to fine-tune novel formulas based on quality-driven and clinical-oriented needs.

2.8. Study Limitations

The main technical limitation of the present study consisted of the inclusion of only three injectors for the manual product injectability measurements. This specification was linked to the high purchasing costs of the commercial dermal fillers, which were each procured in multiple units. For mitigation of the limited number of injectors, the level of qualification and experience of the retained clinicians with injectable dermal fillers was set high. It should be noted that differences in clinician habits and experience with specific product brands were noted, yet reaching the same levels of clinical experience and ability with all products for all injectors is not tangibly achievable. Such aspects are confirmed by clinical practices in aesthetic medicine, where patient expectation levels are high. Therefore, practitioners generally choose a limited number of products or brands that they master and do not routinely diversify across the board of available commercial products.

A second technical limitation of the present study consisted of the use of different needle gauge sizes for the various injectability experiments. Specifically, strict benchmarking of the 28 retained products in terms of comparative quantitative injectability would have required the use of a single needle reference. However, the choice was made to mainly use the respective needles provided with the dermal fillers in order to conform to manufacturer specifications and to enhance the translational relevance of the obtained datasets.

2.9. Future Research Perspectives Based on the Study

Specific future perspectives to the present study comprise in-depth technical assessments of selected products or selected product brands in order to better understand which parameters and specifications influence the quality of the clinical administration process and the quality of the overall patient experience. Therefore, standardized assessments of larger product samples (e.g., evaluating different product lots) would provide enhanced injectability dataset robustness. Furthermore, a specific technical investigation into the formulation-based options for lidocaine incorporation is warranted in order to potentially further optimize specific parameters of product manufacturing. Finally, the datasets presented herein may be used as complements for the clinical education of specialized practitioners. Specifically, the latter classically rely on their own experience with commercial products, yet the majority of the available technical information accompanying a product is manufacturer-provided, which does not allow for the exclusion of various forms of bias.

An additional future perspective to the presented work could also include in vivo injectability pressure measurements during patient treatments. This could potentially facilitate comparative analyses between injections into synthetic cutaneous equivalents and actual facial injections, taking into account the variability in dermal thickness across different treatment areas. Such studies would provide deeper insights into the behavior and properties of these products under real-world application conditions, which could potentially further optimize their use in routine clinical practice.

3. Conclusions

The present study provided comprehensive injectability assessments and multi-level comparisons of various dermal filler products, each presenting specificities designed to address different clinical needs. Experimental injectability results revealed important variability between and within product brands, with a strong influence of lidocaine, HA contents, and needle gauge size. Specifically, it was shown that the plateau injection force for cross-linked HA-based dermal fillers varies significantly between brands and even within brands. Notably, this force was not found to correlate linearly with the HA concentration or needle gauge. Critical appraisals of the investigated products were centered on respective experimental injectability quality levels. Intra-product inhomogeneity in terms of exerted pressure during automatic injection was observed notably in systems using NASHA® and CPM® technologies. Conversely, hydrogels using the IPN-Like® cross-linking technology exhibited good homogeneity in their injection profile. Gels containing lidocaine generally displayed conserved or higher injection force trends compared to their counterparts without lidocaine. Generally, it was confirmed that each cross-linked HA-based dermal filler brand and individual product requires specific expertise for optimal injection, mainly due to differing viscoelastic characteristics and specific injectability attributes. The main conclusion of the study was that some products present qualitatively better injectability attributes than others (i.e., smoother injection curves) and that in the hands of different injectors, the same product may behave very differently in terms of injectability. Overall, the presented work underscored the central importance for injectors to work with HA-based dermal filler products with which they are experienced, comfortable, and skilled, eventually aiming toward enhancing the clinical results and overall patient experience.

4. Materials and Methods

4.1. Materials Used in the Study

Physiological saline solution (NaCl 0.9%) was purchased from Bichsel (Unterseen, Switzerland). Lidocaine was purchased from Sigma-Aldrich (Buchs, Switzerland). A total of 28 different cross-linked HA-based commercial dermal fillers from the JUVÉDERM®, Restylane®, BELOTERO®, TEOSYAL RHA®, and STYLAGE® brands were purchased from the respective product manufacturers. The various needles used in the study were taken directly from each corresponding product packaging and comprised 30 G × ½″ needles (0.30 × 13 mm; TSK Laboratories, Tochigi-Ken, Japan), 29 G × ½″ needles (0.33 × 12 mm; Terumo, Tokyo, Japan), and 27 G × ½″ needles (0.40 × 13 mm; TSK Laboratories, Tochigi-Ken, Japan). For establishing ex vivo injectability measurement conditions, human skin tissue from resected abdominoplasty waste materials was used. For establishing standardized in vitro injectability measurement conditions, SimSkin® synthetic skin was purchased from Wallcur (San Diego, CA, USA). SimSkin® synthetic skin consists of an epidermis,

dermis, and subcutaneous layer. Its total thickness is 0.6 mm (i.e., 0.3 mm for the epidermis, 0.2 mm for the dermis, and 0.1 mm for the subcutaneous layer. The exact polymeric composition of the three layers of the SimSkin® product is not provided by the manufacturer.

4.2. Comparative Manual Injectability Studies in Ex Vivo Human Skin and in SimSkin® Cutaneous Equivalents

In order to first validate the SimSkin® cutaneous equivalent model for further experimental setup standardization, comparative manual injectability studies were performed in vitro and ex vivo. Therefore, quantitative injectability measurements were performed by two specialized and experienced clinicians using a dynamometric sensor (FlexiForce® Quickstart Board, Tekscan, Boston, MA, USA) connected to myDAQ for data acquisition (National Instruments, Austin, TX, USA). The force injection parameters of a commercially available dermal filler product (i.e., TEOSYAL RHA® 2) were determined using the original syringes and needles used for clinical product administration. The TEOSYAL RHA® 2 product (i.e., based on the "Preserved Network®" cross-linking technology; Teoxane, Geneva, Switzerland) was injected into ex vivo human skin samples and in the SimSkin® scaffold. The wrinkle-filling hydrogel (i.e., TEOSYAL RHA® 2) was injected point-by-point into the superficial to medium dermis in a retro-tracing manner. For each injection, the needle was introduced tangentially to the skin plane at an angle $< 10°$. The mean forces of injection and peak forces of injection were automatically recorded.

4.3. Comparative Product Benchmarking: Multi-Injector Manual Injectability Study in Cutaneous Equivalents

A standardized comparative injectability study was performed for the 28 included commercial hydrogel products. The hydrogels were injected into SimSkin® scaffolds by three specialized and experienced clinicians in clinical conditions (i.e., products were injected as if the recipient was a live patient). All products were administered using the original syringes and needles used for clinical product administration. Injectability measurements were performed using a dynamometric sensor (FlexiForce® Quickstart Board; Tekscan, Boston, MA, USA) connected to myDAQ for data acquisition (National Instruments, Austin, TX, USA). The force injection parameters of the commercially available dermal filler products were determined and compared. Injectability results were compared between the various products and between the three injectors.

4.4. Comparative Automated Injectability Study in Cutaneous Equivalents at Constant Injection Speed

In order to enhance the setup standardization and the experimental result granularity levels, automated product injectability measurements were performed at constant speed on a Texture Analyzer TA.XT. Plus instrument (Tracomme, Schlieren, Switzerland). The 28 considered hydrogel products were sequentially injected into SimSkin® scaffolds at a constant plunger rod actuation speed of 1 $mm{\cdot}s^{-1}$ at ambient temperature (i.e., 25 °C). The relatively fast injection speed of 1 $mm{\cdot}s^{-1}$ was retained from preliminary experiments in order to best discriminate products in terms of hydrogel system intra-syringe homogeneity. The force injection profiles of the various products were determined using the respective original syringes and needles used for clinical product administration.

4.5. Comparative Analysis of Different Manual Hydrogel Injection Techniques

In order to complement the automated injectability study results, manual injections were performed by two specialized and experienced clinicians using various injection techniques. Specifically, quantitative injectability studies were performed using a dynamometric sensor (FlexiForce® Quickstart Board; Tekscan, Boston, MA, USA) connected to myDAQ for data acquisition (National Instruments, Austin, TX, USA) and SimSkin® scaffolds. The 28 retained commercial hydrogel products were injected using retro-tracing injections in the dermis and hypodermis and using the bolus technique in the hypodermis. Qualitative assessments of hydrogel product injectability were recorded for each product, injector, and injection technique, focusing on potential variations in the required pressure to inject the product.

4.6. Experimental Assessment of the Impact of Lidocaine on Hydrogel System Attributes during Product Sterilization

In order to experimentally assess the impact of lidocaine on hydrogel rheological attributes within the sterilization process, formulations with lidocaine and without lidocaine were prepared. A BDDE-cross-linked HA-based hydrogel (University of Geneva, Geneva, Switzerland) served as an experimental base hydrogel. Lidocaine was incorporated at 2 mg/mL in a fraction of the hydrogel base to approximate the quantity generally present in commercial dermal filler products. Both hydrogel groups were conditioned in 6R clear glass vials and were submitted to steam sterilization (Systec, Sysmex, Kobe, Japan) at 121 °C for 12 min, using rapid ramp heating and cooling protocols. Pre-sterilization and post-sterilization rheological attributes (i.e., storage modulus G' and loss modulus G'') were determined in oscillatory rheology on a Haake Mars rheometer (Thermo Fisher Scientific, Waltham, MA, USA). A Peltier cone plate characterized by a C35 2°/Ti measuring geometry was mounted on the instrument. The measurements were performed in triplicate at 22 °C on 450 μL of sample with a constant oscillatory frequency of 1 Hz. Shear stress was set at 3 N/m^2 in order to remain in the linear viscoelastic region. A sample hood was used during the measurements to minimize sample evaporation.

4.7. Statistical Analysis and Data Presentation

Data were reported as mean values accompanied by the corresponding standard deviations as error bars. For the statistical comparison of values from multi-group quantitative datasets, a one-way ANOVA or a two-way ANOVA test was performed and was followed by a post hoc Tukey's multiple comparison test. A p-value < 0.05 was retained as a general base for statistical significance determination. Detailed levels of statistical significance can be found in the Results section and in the Supplementary Tables. The statistical calculations and/or data presentation were performed using Microsoft Excel (Microsoft Corporation, Redmond, WA, USA), Microsoft PowerPoint, and GraphPad Prism v. 8.0.2 (GraphPad Software, San Diego, CA, USA).

Supplementary Materials: The following supporting information can be downloaded at https://www.mdpi.com/article/10.3390/gels10020101/s1. Figure S1: Illustration of different syringe handling dispositions for dermal filler injection; Figure S2: Force injection profiles of the BELOTERO® products; Figure S3: Force injection profiles of the Restylane® products; Figure S4: Force injection profiles of the JUVÉDERM® products; Figure S5: Force injection profiles of the TEOSYAL RHA® products; Figure S6: Force injection profiles of the STYLAGE® products; Figure S7: Results of experimental sterilization studies; Table S1: Technical benchmarking data on the investigated products; Table S2: Complementary statistical analysis results; Table S3: Complementary statistical analysis results; Table S4: Qualitative and quantitative injectability records for BELOTERO® products; Table S5: Qualitative and quantitative injectability records for Restylane® products; Table S6: Qualitative and quantitative injectability records for STYLAGE® products; Table S7: Qualitative and quantitative injectability records for TEOSYAL RHA® products; Table S8: Qualitative and quantitative injectability records for JUVÉDERM® products; Table S9: Complementary statistical analysis results; Table S10: Complementary statistical analysis results.

Author Contributions: Conceptualization, P.M., A.P., T.B., E.A., A.L. and O.J.; methodology, P.M., A.P., T.B., D.P., P.Q., E.A., A.L. and O.J.; software, P.M., A.P., A.L. and O.J.; validation, P.M., A.P., T.B., D.P., P.Q., Y.K., E.A., A.L. and O.J.; formal analysis, P.M., A.P., T.B., E.A., A.L. and O.J.; investigation, P.M., A.P., T.B., D.P., P.Q. and A.L.; resources, P.M., T.B., E.A. and A.L.; data curation, P.M., A.P. and A.L.; writing—original draft preparation, P.M., A.P. and A.L.; writing—review and editing, P.M., A.P., T.B., D.P., P.Q., Y.K., E.A., A.L. and O.J.; visualization, P.M., A.P., T.B., P.Q. and A.L.; supervision, P.M., T.B., P.Q., E.A. and O.J.; project administration, P.M., A.P. and E.A.; funding acquisition, P.M. All authors have read and agreed to the published version of the manuscript.

Funding: This study was not supported by specific grants or by any institutional programs.

Institutional Review Board Statement: The biological starting materials (i.e., ex vivo human skin) used for the present study were procured according to the guidelines of the Declaration of Helsinki [57]. The procurement and use of patient biological materials followed the regulations of the Biobank of the CHUV Department of Musculoskeletal Medicine (Lausanne University Hospital, Lausanne, Switzerland). Materials from adult patients having undergone abdominoplasty surgery were included in the study under the Ethics Committee Protocol N°264/12, approved by the Vaud Cantonal Ethics Committee.

Informed Consent Statement: Informed consent (i.e., formalized in a general informed consent agreement) was obtained from all patients or from their legal representatives at the time of treatment for unrestricted use of the gathered and anonymized patient data or anonymized biological materials.

Data Availability Statement: The data presented in this study are openly available in the article.

Acknowledgments: We would like to thank the laboratory TEOXANE (Geneva, Switzerland) for their assistance in demonstrating the function of the FlexiForce® dynamometric sensor, with special thanks to A. Jacquemin, D. Nuzzo, F. Bourdon, and J. Renou. We would like to thank the laboratory Wallcur® (San Diego, CA, USA) for providing a sample of SimSkin® for preliminary testing purposes. We would like to thank the various laboratories (i.e., ALLERGAN AESTHETICS, GALDERMA, MERZ AESTHETICS, TEOXANE, and VIVACY) for answering author queries and providing test syringes and product samples [58].

Conflicts of Interest: Author A.L. was employed by TEC-PHARMA SA (Bercher, Switzerland) and by LAM Biotechnologies SA (Epalinges, Switzerland) during the course of this study. The remaining authors declare no conflicts of interest for this study.

Abbreviations

BDDE	1,4-butanediol diglycidyl ether
CE	European mark of conformity
Da	Daltons
FDA	US Food and Drug Administration
G'	storage modulus
G''	loss modulus
HA	hyaluronic acid
ISO	International Standards Organization
MD	medical device
MDa	megaDaltons
min	minutes
MW	molecular weight
N	Newtons
NA	non-applicable
ns	non-significant
Pa	Pascals
Pa·s	Pascal seconds
ROS	reactive oxygen species
USA	United States of America

References

1. Vazirnia, A.; Braz, A.; Fabi, S.G. Nonsurgical jawline rejuvenation using injectable fillers. *J. Cosmet. Dermatol.* **2020**, *19*, 1940–1947. [CrossRef]
2. Bacos, J.T.; Dayan, S.H. Superficial dermal fillers with hyaluronic acid. *Facial Plast. Surg.* **2019**, *35*, 219–223. [CrossRef]
3. Fallacara, A.; Manfredini, S.; Durini, E.; Vertuani, S. Hyaluronic acid fillers in soft tissue regeneration. *Facial Plast. Surg.* **2017**, *33*, 87–96. [CrossRef]
4. American Society of Plastic Surgeons. Plastic Surgery Statistics Report. Available online: https://www.plasticsurgery.org/documents/News/Statistics/2020/plastic-surgery-statistics-full-report-2020.pdf (accessed on 17 October 2023).
5. Fagien, S.; Bertucci, V.; von Grote, E.; Mashburn, J.H. Rheologic and physicochemical properties used to differentiate injectable hyaluronic acid filler products. *Plastic Reconstr. Surg.* **2019**, *143*, 707e–720e. [CrossRef]
6. Fundarò, S.P.; Salti, G.; Malgapo, D.M.H.; Innocenti, S. The rheology and physicochemical characteristics of hyaluronic acid fillers: Their clinical implications. *Int. J. Mol. Sci.* **2022**, *23*, 10518. [CrossRef]

7. Choi, M.S. Basic rheology of dermal filler. *Arch. Plast. Surg.* **2020**, *47*, 301–304. [CrossRef]
8. Molliard, S.G.; Bétemps, J.B.; Hadjab, B.; Topchian, D.; Micheels, P.; Salomon, D. Key rheological properties of hyaluronic acid fillers: From tissue integration to product degradation. *Plast. Aesthet. Res.* **2018**, *5*, 17. [CrossRef]
9. Rosamilia, G.; Hamade, H.; Freytag, D.L.; Frank, K.; Green, J.B.; Devineni, A.; Gavril, D.L.; Hernandez, C.A.; Pavicic, T.; Cotofana, S. Soft tissue distribution pattern of facial soft tissue fillers with different viscoelastic properties. *J. Cosmet. Dermatol.* **2020**, *19*, 312–320. [CrossRef] [PubMed]
10. Sundaram, H.; Cassuto, D. Biophysical characteristics of hyaluronic acid soft-tissue fillers and their relevance to aesthetic applications. *Plast. Reconstruct. Surg.* **2013**, *132*, 5S–21S. [CrossRef] [PubMed]
11. Steenen, S.A.; Bauland, C.G.; van der Lei, B.; Su, N.; van Engelen, M.D.G.; Anandbahadoer-Sitaldin, R.D.R.R.A.L.; Koeiman, W.; Jawidan, T.; Hamraz, Y.; Lange, J. Head-to-head comparison of 4 hyaluronic acid dermal fillers for lip augmentation: A multicenter randomized, quadruple-blind, controlled clinical trial. *J. Am. Acad. Dermatol.* **2023**, *88*, 932–935. [CrossRef] [PubMed]
12. Safran, T.; Swift, A.; Cotofana, S.; Nikolis, A. Evaluating safety in hyaluronic acid lip injections. *Expert Opin. Drug Safety* **2021**, *20*, 1473–1486. [CrossRef]
13. Trévidic, P.; Kaufman-Janette, J.; Weinkle, S.; Wu, R.; Dhillon, B.; Antunes, S.; Macé, E.; Maffert, P. Injection guidelines for treating midface volume deficiency with hyaluronic acid fillers: The ATP approach (Anatomy, Techniques, Products). *Aesthet. Surg. J.* **2022**, *42*, 920–934. [CrossRef]
14. Liu, X.; Gao, Y.; Ma, J.; Li, J. The efficacy and safety of hyaluronic acid injection in tear trough deformity: A systematic review and meta-analysis. *Aesthet. Plast. Surg.* 2023, *in press*. [CrossRef]
15. Wongprasert, P.; Dreiss, C.A.; Murray, G. Evaluating hyaluronic acid dermal fillers: A critique of current characterization methods. *Dermatol. Ther.* **2022**, *35*, e15453. [CrossRef]
16. da Costa, A.; Biccigo, D.G.Z.; de Souza Weimann, E.T.; Mercadante, L.M.; Oliveira, P.R.G.; Prebianchi, S.B.; Abdalla, B.M.Z. Durability of three different types of hyaluronic acid fillers in skin: Are there differences among biphasic, monophasic monodensified, and monophasic polydensified products? *Aesthet. Surg. J.* **2017**, *37*, 573–581. [CrossRef]
17. Li, J.; Guan, S.; Su, J.; Liang, J.; Cui, L.; Zhang, K. The development of hyaluronic acids used for skin tissue regeneration. *Current Drug Deliv.* **2021**, *18*, 836–846. [CrossRef] [PubMed]
18. Tran, C.; Carraux, P.; Micheels, P.; Kaya, G.; Salomon, D. In vivo bio-integration of three hyaluronic acid fillers in human skin: A histological study. *Dermatology* **2014**, *228*, 47–54. [CrossRef] [PubMed]
19. Žádníková, P.; Šínová, R.; Pavlík, V.; Šimek, M.; Šafránková, B.; Hermannová, M.; Nešporová, K.; Velebný, V. The degradation of hyaluronan in the skin. *Biomolecules* **2022**, *12*, 251. [CrossRef] [PubMed]
20. Santer, V.; Molliard, S.G.; Micheels, P.; Río-Sancho, S.D.; Quinodoz, P.; Kalia, Y.N.; Salomon, D. Hyaluronic acid after subcutaneous injection-An objective assessment. *Dermatol. Surg.* **2019**, *45*, 108–116. [CrossRef] [PubMed]
21. Arlette, J.P.; Trotter, M.J. Anatomic location of hyaluronic acid filler material injected into nasolabial fold: A histologic study. *Dermatol. Surg.* **2008**, *34*, S56–S62. [CrossRef] [PubMed]
22. Micheels, P.; Goodman, L. Injection depth in intradermal therapy: Update and correction of published data. *J. Drugs Dermatol.* **2018**, *17*, 88–96.
23. Olenius, M. The first clinical study using a new biodegradable implant for the treatment of lips, wrinkles, and folds. *Aesthet. Plast. Surg.* **1998**, *22*, 97–101. [CrossRef]
24. Micheels, P.; Sarazin, D.; Besse, S.; Sundaram, H.; Flynn, T.C. A blanching technique for intradermal injection of the hyaluronic acid Belotero®. *Plast. Reconstr. Surg.* **2013**, *132*, 69S–76S. [CrossRef]
25. Micheels, P.; Besse, S.; Flynn, T.C.; Sarazin, D.; Elbaz, Y. Superficial dermal injection of hyaluronic acid soft tissue fillers: Comparative ultrasound study. *Dermatol. Surg.* **2012**, *38*, 1162–1169. [CrossRef]
26. Della Volpe, C.; Andrac, L.; Casanova, D.; Legré, R.; Magalon, G. Skin diversity: Histological study of 140 skin residues, adapted to plastic surgery. *Ann. Chir. Plast. Esthet.* **2012**, *5*, 423–449. [CrossRef]
27. Kaya, G.; Saurat, J.-H. Dermatoporosis: A chronic cutaneous insufficiency/fragility syndrome. Clinicopathological features, mechanisms, prevention and potential treatments. *Dermatology* **2007**, *215*, 284–294. [CrossRef]
28. Tsukahara, K.; Tamatsu, Y.; Sugawara, Y.; Shimada, K. Relationship between the depth of facial wrinkles and the density of the retinacula cutis. *Arch. Dermatol.* **2012**, *148*, 39–46. [CrossRef]
29. Kaya, G. Dermatoporose: Un syndrome emergent. *Rev. Med. Suisse* **2008**, *155*, 1078–1082.
30. Kim, J. Effects of injection depth and volume of stabilized hyaluronic acid in human dermis on skin texture, hydration, and thickness. *Arch. Aesthetic Plast. Surg.* **2014**, *20*, 97–103. [CrossRef]
31. Allergan Aesthetics, Allergan (Allergan, Annecy, France). Instructions for Use of the Juvéderm® Gel Range (Volbella®, Volift®, Voluma®, Juvéderm® Ultra 2, 3). 2022, *unpublished work*. Available online: https://media.allergan.com/actavis/actavis/media/general/Juvederm-voluma-IFU.pdf (accessed on 23 January 2024).
32. Q-Med AB, Galderma SA (Galderma, Zug, Switzerland). Instructions for Use of the Restylane® Gel Range (Restylane®, Restylane® Lyft). 2017, *unpublished work*. Available online: https://www.restylane.com/ca/sites/default/files/2018-03/Restylane%20LYFT%20Lidocaine.pdf (accessed on 23 January 2024).
33. Merz Aesthetics, Anteis SA (Anteis, Plan-les-Ouates, Switzerland). Instructions for Use of the Belotero® Gel Range (Belotero® Soft, Balance, Intense, Volume). 2015, *unpublished work*. Available online: https://www.merz.ch/wp-content/uploads/2016/03/BELOTERO_Volume_Lidocaine.pdf (accessed on 23 January 2024).

34. Teoxane SA (Teoxane, Geneva, Switzerland). Instructions for Use of the RHA® Gel Range (RHA® 1, 2, 3, 4). 2015, *unpublished work*. Available online: https://irp-cdn.multiscreensite.com/e193bd60/files/uploaded/Teosyal-Rha-Dr.-brochure-en-1.pdf (accessed on 25 January 2024).
35. Laboratoires VIVACY (Vivacy, Paris, France). Instructions for Use of the Stylage® Gel Range (Stylage® S, M, L, XL, XXL). 2024, *unpublished work*. Available online: https://vivacy.com/fr/produits/medecine-esthetique/ (accessed on 23 January 2024).
36. Faivre, J.; Pigweh, A.I.; Iehl, J.; Maffert, P.; Goekjian, P.; Bourdon, F. Crosslinking hyaluronic acid soft-tissue fillers: Current status and perspectives from an industrial point of view. *Exp. Rev. Med. Dev.* **2021**, *18*, 1175–1187. [CrossRef] [PubMed]
37. Pluda, S.; Salvagnini, C.; Fontana, A.; Marchetti, A.; Di Lucia, A.; Galesso, D.; Guarise, C. Investigation of crosslinking parameters and characterization of hyaluronic acid dermal fillers: From design to product performances. *Gels* **2023**, *9*, 733. [CrossRef] [PubMed]
38. Fidalgo, J.; Deglesne, P.A.; Arroyo, R.; Sepúlveda, L.; Ranneva, E.; Deprez, P. Detection of a new reaction by-product in BDDE cross-linked autoclaved hyaluronic acid hydrogels by LC-MS analysis. *Med. Dev.* **2018**, *11*, 367–376. [CrossRef] [PubMed]
39. Wu, G.T.; Kam, J.; Bloom, J.D. Hyaluronic acid basics and rheology. *Facial Plast. Surg. Clin. N. Am.* **2022**, *30*, 301–308. [CrossRef] [PubMed]
40. Micheels, P.; Obamba, M. Rheological properties of several hyaluronic acid-based gels: A comparative study. *J. Drugs Dermatol.* **2018**, *17*, 602–608.
41. Micheels, P.; Besse, S.; Sarazin, D.; Obamba, M. Hyaluronic acid gel based on CPM® technology with and without lidocaine: Is there a difference? *J. Cosmet. Dermatol.* **2018**, *18*, 36–44. [CrossRef] [PubMed]
42. Santoro, S.; Russo, L.; Argenzio, V.; Borzacchiello, A. Rheological properties of cross-linked hyaluronic acid dermal fillers. *J. Appl. Biomater. Biomech.* **2011**, *9*, 127–136. [CrossRef]
43. De la Guardia, C.; Virno, A.; Musumeci, M.; Bernardin, A.; Silberberg, M.B. Rheologic and physicochemical characteristics of hyaluronic acid fillers: Overview and relationship to product performance. *Facial Plast. Surg.* **2022**, *38*, 116–123. [CrossRef]
44. Conrozier, T.; Mathieu, P.; Rinaudo, M. Mannitol preserves the viscoelastic properties of hyaluronic acid in an in vitro model of oxidative stress. *Rheumatol. Ther.* **2014**, *1*, 45–54. [CrossRef]
45. Rinaudo, M.; Lardy, B.; Grange, L.; Conrozier, T. Effect of mannitol on hyaluronic acid stability in two in vitro models of oxidative stress. *Polymers* **2014**, *6*, 1948–1957. [CrossRef]
46. Available online: https://www.postersessiononline.eu/173580348_eu/congresos/WBC2020/aula/-WBC2020-LATE_4410_WBC2020.pdf (accessed on 13 November 2023).
47. Faivre, J.; Gallet, M.; Tremblais, E.; Trévidic, P.; Bourdon, F. Advanced concepts in rheology for the evaluation of hyaluronic acid-based soft tissue fillers. *Dermatol. Surg.* **2021**, *47*, e159–e167. [CrossRef] [PubMed]
48. De Maio, M. MD Codes™: A methodological approach to facial aesthetic treatment with injectable hyaluronic acid fillers. *Aesthet. Plast. Surg.* **2021**, *45*, 690–709. [CrossRef]
49. Sundaram, H.; Rohrich, R.J.; Liew, S.; Sattler, G.; Talarico, S.; Trévidic, P.; Gavard Molliard, S. Cohesivity of hyaluronic acid fillers: Development and clinical implications of a novel assay, pilot validation with a five-point grading scale, and evaluation of six U.S. Food and Drug Administration-approved fillers. *Plast. Reconstr. Surg.* **2015**, *136*, 678–686. [CrossRef]
50. Mrestani, Y.; Hammitzch, M.; Neubert, R.H.H. Investigation of the interaction between lidocaine and the components of hyaluronic acid using frontal analysis continuous capillary electrophoresis. *Chromatographia* **2009**, *69*, 1321–1324. [CrossRef]
51. Hintze, V.; Schnabelrauch, M.; Rother, S. Chemical modification of hyaluronan and their biomedical applications. *Front. Chem.* **2022**, *10*, 830671. [CrossRef]
52. Haridas, N.; Rosemary, M.J. Effect of steam sterilization and biocompatibility studies of hyaluronic acid hydrogel for viscosupplementation. *Polymer Degrad. Stab.* **2019**, *163*, 220–227. [CrossRef]
53. Chen, J.; Peng, C.; Nie, J.; Kennedy, J.F.; Ma, G. Lyophilization as a novel approach for preparation of water resistant HA fiber membranes by crosslinked with EDC. *Carbohydrate Polym.* **2014**, *102*, 8–11. [CrossRef] [PubMed]
54. Huerta-Ángeles, G.; Nešporová, K.; Ambrožová, G.; Kubala, L.; Velebný, V. An effective translation: The development of hyaluronan-based medical products from the physicochemical, and preclinical aspects. *Front. Bioeng. Biotechnol.* **2018**, *6*, 62. [CrossRef] [PubMed]
55. King, M. Management of Tyndall effect. *J. Clin. Aesthet. Dermatol.* **2016**, *9*, E6–E8.
56. Urdiales-Gálvez, F.; Delgado, N.E.; Figueiredo, V.; Lajo-Plaza, J.V.; Mira, M.; Moreno, A.; Ortíz-Martí, F.; Del Rio-Reyes, R.; Romero-Álvarez, N.; Del Cueto, S.R.; et al. Treatment of soft tissue filler complications: Expert consensus recommendations. *Aesthetic Plast. Surg.* **2018**, *42*, 498–510. [CrossRef]
57. World Medical Association. World Medical Association Declaration of Helsinki: Ethical principles for medical research involving human subjects. *JAMA* **2013**, *310*, 2191–2194. [CrossRef] [PubMed]
58. Micheels, P.; (Private Medical Practice, Chêne-Bougeries, Switzerland); Porcello, A.; (University of Geneva, Geneva, Switzerland). Characterization of Commercial Dermal Fillers for Expert Review by Dr. Patrick Micheels. 2023, *unpublished work*.

gels

MDPI

Article

Comprehensive Evaluation of Injectability Attributes in OxiFree™ Dermal Fillers: MaiLi® Product Variants and Clinical Case Reports

Patrick Micheels [1,*], Alexandre Porcello [2], Thierry Bezzola [3], Daniel Perrenoud [4], Marie-Odile Christen [5], Lee Ann Applegate [6,7,8] and Alexis Laurent [6,9,10,*]

1 Private Medical Practice, CH-1224 Chêne-Bougeries, Switzerland
2 Development Department, Abcello Sàrl, CH-1432 Belmont-sur-Yverdon, Switzerland; alexandre.porcello@abcello.com
3 Private Medical Practice, CH-1204 Geneva, Switzerland; tbezzola@gmail.com
4 Private Medical Practice, CH-1006 Lausanne, Switzerland; drperrenoud@gmail.com
5 Private Office, F-75116 Paris, France; christen.marieodile3@gmail.com
6 Regenerative Therapy Unit, Lausanne University Hospital, University of Lausanne, CH-1066 Epalinges, Switzerland; lee.laurent-applegate@chuv.ch
7 Center for Applied Biotechnology and Molecular Medicine, University of Zurich, CH-8057 Zurich, Switzerland
8 Oxford OSCAR Suzhou Center, Oxford University, Suzhou 215123, China
9 Manufacturing Department, TEC-PHARMA SA, CH-1038 Bercher, Switzerland
10 Manufacturing Department, LAM Biotechnologies SA, CH-1066 Epalinges, Switzerland
* Correspondence: patrickscab@bluewin.ch (P.M.); alexis.laurent@pharmabercher.ch (A.L.); Tel.: +41-22-347-11-13 (P.M.); +41-21-887-80-38 (A.L.)

Abstract: Dermal filler injectability is a critical factor for commercial product adoption by medical aesthetic professionals and for successful clinical administration. We have previously reported (in vitro and ex vivo) cross-linked hyaluronic acid (HA)-based dermal filler benchmarking in terms of manual and automated injectability requirements. To further enhance the function-oriented product characterization workflows and the clinical relevance of dermal filler injectability assessments, the aim of this study was to perform in vivo evaluations. Therefore, several variants of the MaiLi® product range (OxiFree™ technology) were characterized in vitro and in vivo in terms of injectability attributes, with a focus on hydrogel system homogeneity and ease of injection. Firstly, standardized in vitro assays were performed in SimSkin® cutaneous equivalents, with variations of the clinical injector, injection site, and injection technique. Then, automated injections in SimSkin® cutaneous equivalents were comparatively performed in a texture analysis setup to obtain fine-granulometry injection force profile results. Finally, five female participants were recruited for the in vivo arm of the study (case reports), with variations of the clinical injector, injection site, and injection technique. Generally, the obtained quantitative force values and injection force profiles were critically appraised from a translational viewpoint, based on discussions around the OxiFree™ manufacturing technology and on in-use specialized clinician feedback. Overall, the present study outlined a notable level of homogeneity across the MaiLi® product range in terms of injectability attributes, as well as consistently high ease of administration by medical aesthetic clinicians.

Keywords: aesthetic medicine; case reports; cross-linked hyaluronic acid; dermal fillers; hydrogel systems; injectability; medical device; needle gauge; OxiFree™ technology; technical benchmarking

1. Introduction

Dermal filler product injectability attributes are linked to the quality, safety, and efficacy outcomes of specific medical aesthetic interventions [1,2]. Furthermore, the clinical results of facial dermal filler applications are dependent upon the skill and experience level of the practicing physician. Therein, the level of control over the filler injection process is critical to ensure that the correct and desired amount of hydrogel is placed at the appropriate anatomical site and depth [1,3,4]. Importantly, the use of inhomogeneous dermal filler products (i.e., in terms of injectability attributes) may therefore result in excess in vivo product dispensing or in inappropriately shallow gel placement, potentially leading to adverse effects and clinical failure [5–9]. Of note, close consideration of dermal filler product injectability is of central importance for widely used cross-linked hyaluronic acid (HA)-based hydrogel systems. Specifically, their formulation parameters and biophysical attributes may be linked to the overall quality of the provided medical aesthetic care [10–16].

Notably, previous experimental work by the authors reported some in-use disparities in the injectability attributes of various commercial cross-linked HA-based dermal fillers, within standardized in vitro and ex vivo product benchmarking setups [1]. Specifically, the parallel characterization of 28 dermal fillers (i.e., BELOTERO®, JUVÉDERM®, VIVACY®, RESTYLANE®, and TEOSYAL® brands) revealed that inter-product and intra-product qualitative and quantitative variations existed in terms of injectability [1]. Therein, much of the reported intra-brand variability (i.e., as regards injectability attributes) was linked to the specific composition (i.e., HA content, lidocaine presence) or the manufacturing process of individual product variants [1]. Overall, the available experimental data did not enable the identification of a product manufacturer achieving a high level of brand-wide homogeneity and consistency in terms of in vitro and ex vivo hydrogel system injectability [1].

Contrasting with such findings, preliminary in-use clinical feedback by the practicing co-authors reported no tangible discrepancies or differences in the in vivo injection characteristics of several MaiLi® product variants (i.e., MaiLi® Precise, Define, Volume, Extreme; Oxi-Free™ manufacturing technology) [17]. Based on such subjective reports of qualitatively and quantitatively enhanced injectability attributes and building on previous scientific works, prospective investigations of MaiLi® dermal fillers in terms of in-use behaviour were warranted [1,18–21]. Specifically, a key design interest of the new study was to compare in vitro and in vivo MaiLi® product injectability (i.e., during clinical facial dermal filling) to maximize the translational and clinical relevance of the obtained datasets.

Therefore, the first aim of this study was to perform in vitro qualitative and quantitative injectability attribute assessments on a range of four OxiFree™-based cross-linked HA dermal fillers (i.e., MaiLi® Precise, Define, Volume, and Extreme variants). Therein, standardized in vitro manual injectability evaluations with multi-injector assessments were performed, followed by automated product extrusion in SimSkin® cutaneous equivalents. The second aim of this study was to transpose the manual product injectability assessments in vivo, under real-world in-use clinical conditions. Therefore, dynamometric evaluations of MaiLi® dermal filler injection forces were performed by two specialized clinicians during patient treatment and were analysed in the form of five clinical case reports.

Of note, the primary hypothesis of the study was that MaiLi® dermal fillers present intra-product and brand-wide homogeneity as regards their injectability attributes. Specifically, it was hypothesized that very smooth injection force plateaus could be obtained when injecting MaiLi® products, with low quantitative differences between the product variants. The secondary hypothesis of the study was that MaiLi® dermal fillers present an enhanced quality level of injectability attributes as compared to similar medical devices. Specifically, it was hypothesized that the MaiLi® product brand could display better felt behaviours (i.e., by the physician) during administration and smoother injection force plateaus than alternative dermal fillers. Overall, the originality/novelty and significance of the study resided in the use of advanced and multiparametric quantitative setups (i.e., in vitro and in vivo) for experimental commercial dermal filler product injectability attribute characterization, enabling detailed technical investigation and providing enhanced translational relevance. Notably, the dual use of standardized in vitro setups for product injectability characterization and the in vivo measurement of product injection forces was uniquely brought forth herein for optimal demonstration of the key features of the OxiFree™ technology under real-world conditions. Such an approach is of high relevance and informative value for clinical practitioners, supplementing manufacturer-provided data. Namely, the reported results significantly illustrated the importance of selecting high-quality aesthetic devices and care centres with track records of clinical expertise to obtain the highest level of medical aesthetic performance.

2. Results and Discussion

2.1. Technical Benchmarking of MaiLi® Dermal Filler Product Attributes

The retained investigational products consisted of four CE-marked dermal filler variants (i.e., Precise, Define, Volume, Extreme) of the MaiLi® brand (Sinclair Pharma Ltd., London, UK), exclusively based on the OxiFree™ HA cross-linking technology [22,23]. Prior to the experimental investigations, brief technical documentary reviews were performed based on the available manufacturer-provided elements for preliminary general product attribute benchmarking (Table 1).

Table 1. Comparative overview of the technical specifics of the MaiLi® dermal filler product variants included in the study. The data were compiled from manufacturer-provided sources. G, gauge; HA, hyaluronic acid; N/A, non-applicable.

Product Variant	Clinical Uses [1]	Needle Size (G × Length) [2]	HA Concentration (mg/mL)	Cross-Linked HA (Y/N)	Lidocaine Presence (Y/N) [3]	HA Cross-Linking Technology
Precise	Fine lines	30 G × ½″	15.0 mg/mL	Yes	Yes	OxiFree™
Define	Medium folds; lip volume	30 G × ½″	18.0 mg/mL	Yes	Yes	OxiFree™
Volume	Volumizer	27 G × ½″	21.0 mg/mL	Yes	Yes	OxiFree™
Extreme	Cheeks; temples; jaw line; chin volumizer	27 G × ½″	24.0 mg/mL	Yes	Yes	OxiFree™
NaCl Control	N/A	30 G × ½″	0.0 mg/mL	N/A	No	N/A

[1] Product clinical uses, as specified by the manufacturer [22]. [2] The original needles, as supplied with the MaiLi® product syringes, are manufactured by TSK, Tochigi-Ken, Japan. [3] Lidocaine contents are consistent across the whole MaiLi® dermal filler product range, with 0.3% m/v lidocaine.

Of note, all of the considered MaiLi® dermal filler variants are industrially manufactured using a specifically designed hydrogel production process (Table 1). Namely, enhanced hydrogel system resilience and distinctive projection capacities were previously reported for OxiFree™-based products compared to alternative commercial dermal fillers [19,20]. Specifically, such attributes are derived through an optimized manufacturing process, which may be broken down into four main phases, as follows [23]:

(a) Hydration of high molecular weight, non-animal origin, pharmaceutical grade HA;
(b) Sparing BDDE-based HA chemical cross-linking under a protective atmosphere for viscoelastic hydrogel formation;
(c) Extraction of reactive oxygen molecules prior to hydrogel product terminal sterilization;
(d) Hydrogel product terminal sterilization in oxidant-deprived conditions.

In addition, noteworthy clinical advantages were previously set forth around this dermal filler technology, comprising the sparing use of hydrogel for patient treatment, based on the enhanced functional parameters conferred by specific viscoelasticity attributes [18–21]. From a formulation viewpoint, it should be noted that manufacturers are legally required to declare the concentration of HA in their dermal filler products (Table 1). However, they are not required to disclose the molecular weight of the HA raw material nor the degree of cross-linking of the system in the finished product. Therefore, within the same technology or filler product range, manufacturers can modulate the degrees of polymer cross-linking or use multiple HA molecular weight ranges [1]. Specifically, to facilitate product injection, it is possible to incorporate an additional phase of linear HA within a cross-linked polymer system. Thus, such formulation- and processing-based elements allow for finished product technical optimization in order to achieve the desired injectability attributes.

2.2. Rheological Characterization of the MaiLi® Dermal Filler Product Range

For the specific comparative documentation of the viscoelasticity attributes of the investigated hydrogel systems, the rheological behaviours of the four MaiLi® product variants were experimentally determined. Specifically, complex viscosity η* values (i.e., in oscillatory rheology) were determined for each dermal filler variant as a function of the applied oscillation frequency (Figure 1).

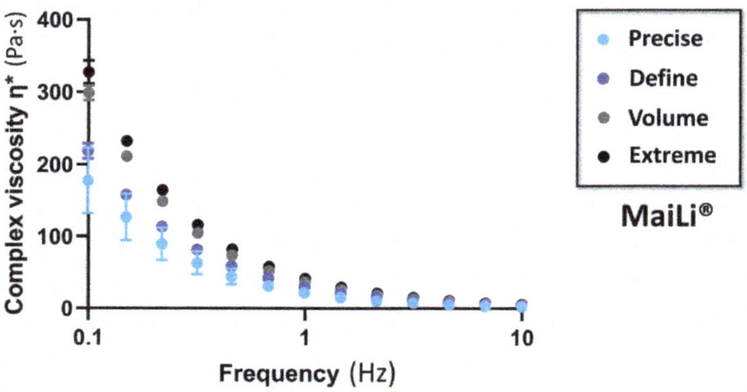

Figure 1. Rheological characterization results of the four MaiLi® dermal filler variants. Complex viscosity data are plotted as means with standard deviations as error bars. Hz, Herz; Pa·s, Pascal seconds.

Of note, complex viscosity (η*) is a measure of a material's mechanical resistance to deformation under shear stress. It is obtained by calculating the quotient of the maximum stress amplitude and maximum strain rate amplitude [2]. This rheological attribute is classically used to describe the behaviour of HA-based fillers in terms of "thickness" or "resistance to flow" during hydrogel injection [2,24–26]. Namely, viscosity appears to better quantitatively approximate HA-based dermal filler injectability when it is interpreted as the process of extrusion from the syringe/needle and for immediate tissue integration [10].

Of further note, a material with a low η* value is easy to deform, and conversely, a high η* value is characteristic of a material which is more difficult to deform. By extension, a low η* value suggests that the considered dermal filler is easier to inject, whereas a high η* value means that the hydrogel product is more challenging to inject, with a higher shear thinning point and yield stress values [6,10,11]. Interestingly, the experimental rheological data gathered in this study enabled the classification of the MaiLi® product variants based on decreasing mean complex viscosity η* values (i.e., Extreme > Volume > Define > Precise; Figure 1). Such results confirmed the technical validity of the specified indications/intended product uses for the MaiLi® product variants (i.e., ranging from powerful filling and facial sculpting to skin-finishing for smoothing wrinkles; Figure 1, Table 1).

Of final note, while hydrogel system rheological properties may be a useful technical proxy for finished product injectability during preliminary screening assays, it is important to take formulation and packaging/accessory specifics into consideration for product injectability characterization. For example, MaiLi® Precise and MaiLi® Define are both meant to be injected through a 30 G × ½″ needle, yet the HA concentration of the Precise variant is 3.0 mg/mL lower than that of the Define variant (i.e., 15.0 mg/mL against 18.0 mg/mL, respectively; Table 1). Therefore, differences in mean and peak injection forces may be expected for these two product variants based on their differential formulation-based specifics (Table 1). Overall, only experimental setups incorporating the final product packaging and the specified administration system(s) may be considered pertinent for sound injectability attribute characterization.

2.3. Manual Injectability Assessments of the MaiLi® Product Range in SimSkin® Cutaneous Equivalents

The identified relevance of performing manual injectability evaluations of dermal fillers is based on the low informative value of product extrusion forces (i.e., automatic protocols, extrusion in atmospheric air), generally used as functional characteristics or as features by product manufacturers [1]. Specifically, it was previously noted that the manual injection forces of cross-linked HA-based dermal fillers were systematically inferior to the values obtained in automated laboratory settings [1].

For the needs of the present study, each MaiLi® product variant was injected in vitro at the indicated anatomic depth (i.e., intradermal or hypodermal injections in SimSkin® substrates, Figure S1) by three practicing physicians. Among the three operators, only co-author PM had clinical experience with the use of MaiLi® products, whereas co-authors TB and DP were initially not familiar with this medical device brand. Additionally, based on off-label (i.e., yet widespread) clinical practice of hypodermal injections with wrinkle-filling products, the latter were injected in the artificial hypodermis component of the SimSkin® substrate (Figure S1).

Overall, the three operators each performed four injection regimens for each MaiLi® product variant (i.e., intradermal point-by-point and retro-tracing injections, hypodermal bolus and retro-tracing injections, Figure S1) [1]. For the technical needs of this study, the plunger rods from the MaiLi® dermal filler variants were replaced by those of TEOSYAL RHA® 2 dermal fillers (TEOXANE, Geneva, Switzerland) comprising a hilt-mounted dynamometric sensor [1]. All injections were performed in triplicate by each operator and for each MaiLi® product variant, where the mean forces of injection were recorded (Tables 2–5).

Table 2. Quantitative data on the manual injection forces required to inject the MaiLi® Precise dermal filler in SimSkin® cutaneous equivalents. Experiments were performed by three qualified operators using various injection techniques and in vitro administration sites. DP, Daniel Perrenoud; N, Newtons; PM, Patrick Micheels; TB, Thierry Bezzola.

Injection Parameters		Injection Force in the Dermis (N)		Injection Force in the Hypodermis (N) [1]		Mean Values/Injector (N)
		Point-by-Point	Retro-Tracing	Bolus	Retro-Tracing	
Injectors	PM	0.46	0.36	0.54	0.73	0.52 ± 0.16
	TB	2.01	1.60	0.83	1.65	1.52 ± 0.50
	DP	0.33	0.52	0.48	0.53	0.46 ± 0.09
Mean Values/Injection Site (N)		0.93 ± 0.93	0.83 ± 0.67	0.62 ± 0.19	0.97 ± 0.60	Global Mean Value (N) 0.84 ± 0.58

[1] Hypodermal injections are not part of the specified indications for use of this MaiLi® product variant (Table 1).

Table 3. Quantitative data on the manual injection forces required to inject the MaiLi® Define dermal filler in SimSkin® cutaneous equivalents. Experiments were performed by three qualified operators using various injection techniques and in vitro administration sites. DP, Daniel Perrenoud; N, Newtons; PM, Patrick Micheels; TB, Thierry Bezzola.

Injection Parameters		Injection Force in the Dermis (N)		Injection Force in the Hypodermis (N) [1]		Mean Values/Injector (N)
		Point-by-Point	Retro-Tracing	Bolus	Retro-Tracing	
Injectors	PM	0.82	1.04	1.28	1.77	1.23 ± 0.41
	TB	1.15	1.13	1.42	1.25	1.24 ± 0.13
	DP	0.77	0.91	0.87	0.75	0.83 ± 0.08
Mean Values/Injection Site (N)		0.91 ± 0.21	1.02 ± 0.11	1.19 ± 0.29	1.26 ± 0.51	Global Mean Value (N) 1.10 ± 0.30

[1] Hypodermal injections are not part of the specified indications for use of this MaiLi® product variant (Table 1).

Table 4. Quantitative data on the manual injection forces required to inject the MaiLi® Volume dermal filler in SimSkin® cutaneous equivalents. Experiments were performed by three qualified operators using various injection techniques and in vitro administration sites. DP, Daniel Perrenoud; N, Newtons; PM, Patrick Micheels; TB, Thierry Bezzola.

Injection Parameters		Injection Force in the Dermis (N) [1]		Injection Force in the Hypodermis (N)		Mean Values/Injector (N)
		Point-by-Point	Retro-Tracing	Bolus	Retro-Tracing	
Injectors	PM	1.84	1.75	1.91	2.28	1.95 ± 0.23
	TB	0.81	0.79	0.91	0.85	0.84 ± 0.05
	DP	0.51	0.45	0.51	0.44	0.48 ± 0.04
Mean Values/Injection Site (N)		1.05 ± 0.70	1.00 ± 0.67	1.11 ± 0.72	1.19 ± 0.97	Global Mean Value (N) 1.09 ± 0.66

[1] Intradermal injections are not part of the specified indications for use of this MaiLi® product variant (Table 1).

Table 5. Quantitative data on the manual injection forces required to inject the MaiLi® Extreme dermal filler in SimSkin® cutaneous equivalents. Experiments were performed by three qualified operators using various injection techniques and in vitro administration sites. DP, Daniel Perrenoud; N, Newtons; PM, Patrick Micheels; TB, Thierry Bezzola.

Injection Parameters		Injection Force in the Dermis (N) [1]		Injection Force in the Hypodermis (N)		Mean Values/Injector (N)
		Point-by-Point	Retro-Tracing	Bolus	Retro-Tracing	
Injectors	PM	0.58	0.35	0.60	0.54	0.52 ± 0.13
	TB	0.96	1.02	1.32	1.35	1.16 ± 0.20
	DP	0.46	0.41	0.51	0.70	0.52 ± 0.13
Mean Values/Injection Site (N)		0.67 ± 0.26	0.59 ± 0.37	0.81 ± 0.44	0.86 ± 0.43	Global Mean Value (N) 0.73 ± 0.35

[1] Intradermal injections are not part of the specified indications for use of this MaiLi® product variant (Table 1).

For the MaiLi® Precise variant, the corresponding in vitro injection force profiles were presented in Figure S2 (Table 2). Overall, while the injection force curves were found to be consistent and similar in profile across all three injectors, the recorded mean values were systematically higher for injector TB (Table 2, Figure S2). It is noteworthy that injector DP administered the product rapidly in all experimental settings as compared to injectors PM and TB. The experimental results obtained for the MaiLi® Define variant were on average higher in value than for the MaiLi® Precise variant and were found to be less dispersed between injectors (Tables 2 and 3).For the MaiLi® Define variant, the corresponding in vitro injection force profiles were presented in Figure S3 (Table 3). For this hydrogel, injector DP administered the product rapidly again, with systematically and surprisingly lower mean injection forces as compared to injectors PM and TB (Table 3). For the MaiLi® Volume variant, high inter-injector variability was noted for the measured injection forces, yet low inter-injection site variability was noted for each injector (Table 4).

For the MaiLi® Volume variant, the corresponding in vitro injection force profiles were presented in Figure S4 (Table 4). It should be noted that the MaiLi® Volume product is not indicated for intradermal injection, as it is primarily destined for hypodermal or close-to-the-bone placement (Table 1). Notably, it was observed that hypodermal bolus injections by injector PM were comparatively less homogeneous and required more force, which was attributed to the use of the P1–P2 thumb joint for the injections. Additionally, these relatively high force values were correlated with the lower clinical experience level of injector PM with volumizing agents (Figure S4). Here again, injector DP administered the product relatively rapidly and with low overall mean injection forces (Table 4, Figure S4). Finally, as concerns the MaiLi® Extreme variant, the experimental injection force values were found to be closer between injectors PM and DP, with the values of injector TB being systematically recorded as higher (Table 5).

For the MaiLi® Extreme variant, the corresponding in vitro injection force profiles were presented in Figure S5 (Table 5). Here again, it was noted that the considered hydrogel product is not indicated for hypodermal injection (Table 1). In this case, it was set forth that the injection forces exerted by injector TB could possibly be superior to those of injector PM because of the use of the thenar eminence for plunger rod actuation by the latter (Table 5). Furthermore, the injection force curves were found to be notably more variable on average for the MaiLi® Extreme product variant, as compared to the other three MaiLi® product variants (Figures S2–S5).

Generally, the recorded inter-injector variability as regards in vitro product injection forces was attributed to varying levels of experience with the MaiLi® product range, varying speeds of injection, and the use of different thumb portions by the injectors (Tables 2–5, Figures S2–S5). Specifically, it was previously reported that the use of the P1-P2 thumb joint or the thenar eminence generally resulted in superior mean injection forces as compared to the use of the thumb pulp [1].

As regards the off-label injection of the various MaiLi® product variants, such approaches were experimentally performed based on real-world clinical practice and in view of methodological continuity with previous reports by the authors [1,17]. Generally, it was noted that the recorded injection forces were relatively low as compared to manufacturer-provided MaiLi® injectability data [22]. Finally, the recorded injection force profiles were assessed to be qualitatively consistent across the considered MaiLi® product range and for all three clinical injectors (Figures S2–S5). Importantly, such elements suggested that hydrogel systems based on the OxiFree™ technology possess high levels of intra-product homogeneity, contrasting with some declinations of the BELOTERO® or TEOSYAL® product ranges [1].

2.4. Automated In Vitro MaiLi® Product Injectability Assessments: Comparative Injection Force Curves for Standardized Dermal Filler Product Benchmarking

In order to gain further insights into the injectability attributes of the considered MaiLi® product variants (i.e., and to enhance the qualitative and quantitative levels of product characterization), in vitro automated injections were performed. Therein, repeated texture analysis measurements enabled us to visualize and quantify the plateau forces of product injection in SimSkin® cutaneous equivalents at two specified and constant injection speeds (Figures 2 and 3, Table 6).

Table 6. Quantitative data of plateau injection force [1] determination in automated in vitro injectability assays at a constant plunger rod actuation speed of 0.2 $mm{\cdot}s^{-1}$ and 1 $mm{\cdot}s^{-1}$ for the four MaiLi® product variants. No statistical difference (i.e., p-value > 0.05) was found as regards the inter-variant mean plateau force of injection at a given speed. Extremely significant statistical differences (i.e., p-value < 0.0001) were found between all values when comparing the respective results at different injection speeds. N, Newtons.

Product Variant	Low Injection Speed (0.2 $mm{\cdot}s^{-1}$)			High Injection Speed (1 $mm{\cdot}s^{-1}$)		
	Plateau Minimum Force (N)	Plateau Maximum Force (N)	Plateau Mean Force (N)	Plateau Minimum Force (N)	Plateau Maximum Force (N)	Plateau Mean Force (N)
MaiLi® Precise	10.2 ± 0.3	11.8 ± 0.2	11.2 ± 0.5	32.7 ± 0.4	37.4 ± 0.6	35.8 ± 1.3
MaiLi® Define	10.0 ± 0.4	11.3 ± 0.3	10.1 ± 1.1	34.8 ± 1.9	40.5 ± 0.6	36.4 ± 2.7
MaiLi® Volume	9.2 ± 0.5	10.8 ± 0.2	10.6 ± 0.6	33.6 ± 1.3	36.4 ± 2.3	34.9 ± 1.9
MaiLi® Extreme	9.8 ± 0.4	10.8 ± 0.4	10.0 ± 0.9	35.0 ± 1.1	39.5 ± 1.2	36.8 ± 1.1
NaCl Control	3.8 ± 0.1	4.3 ± 0.2	4.0 ± 0.2	3.9 ± 0.1	4.2 ± 0.1	4.0 ± 0.1

[1] Plateau force values were determined between 40% and 95% of the plunger travel distance during a full hydrogel unit extrusion cycle.

Two highly interesting aspects were noted about the automated injection force profiles of the considered MaiLi® dermal filler variants. Firstly, from a qualitative standpoint, the very high degree of force plateau smoothness (i.e., as observed across all groups) was assessed as specifically informative regarding intra-product system homogeneity in terms of viscoelasticity and general biophysical attributes (Figures 2 and 3) [1]. Additionally, the high reproducibility of the analyses (i.e., superposed injection force curves for the experimental replicates and across all groups) denoted inter-syringe and brand-wide system homogeneity, despite the respective formulation-related specificities (Table 1, Figures 2 and 3).

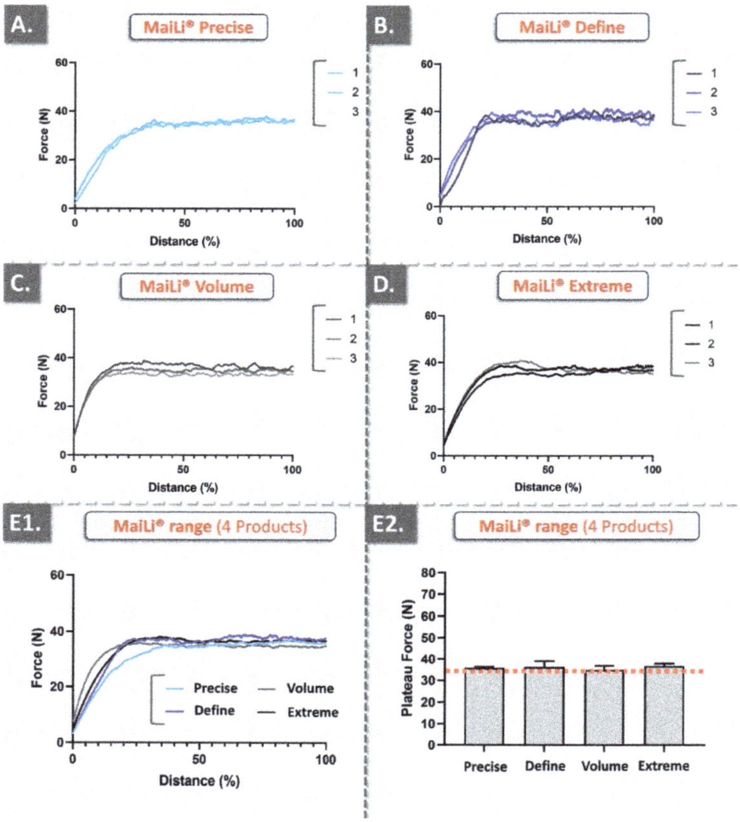

Figure 2. Experimental results of automated injectability evaluations at a constant plunger rod actuation speed of 1 mm·s^{-1} with MaiLi® product in vitro injection in SimSkin® cutaneous substrates. (**A**) Injection force curves of the MaiLi® Precise product. (**B**) Injection force curves of the MaiLi® Define product. (**C**) Injection force curves of the MaiLi® Volume product. (**D**) Injection force curves of the MaiLi® Extreme product. (**E1**) Combined mean injection force curves of the four MaiLi® product variants. (**E2**) Comparison of the mean injection force plateaus of the four MaiLi® product variants. The overall mean plateau injection force (i.e., dotted red line) was found to be 35 N (Table 6). N, Newtons.

Importantly, the gathered in vitro automated injection force data were found to clearly stand out as compared to the profiles which were previously gathered for BELOTERO® or TEOSYAL® product variants [1]. Therein, some BELOTERO® (e.g., lidocaine-free Balance variant) or TEOSYAL® (e.g., RHA® 1, Ultra Deep®) groups were found to present inter-variant variability and intra-syringe inhomogeneity (i.e., in terms of injectability) [1]. Of highest importance, the obtention of superposing injection force curves in an automated setup for all the variants of a product brand (i.e., MaiLi® Precise, Define, Volume, and Extreme) was interpreted as extremely technically interesting, contrasting with previous commercial product screenings (Figures 2 and 3) [1]. Furthermore, from a quantitative standpoint, it was noted that the mean plateau injection forces were highly consistent between the experimental groups (Figures 2(E2) and 3(E2), Table 6).

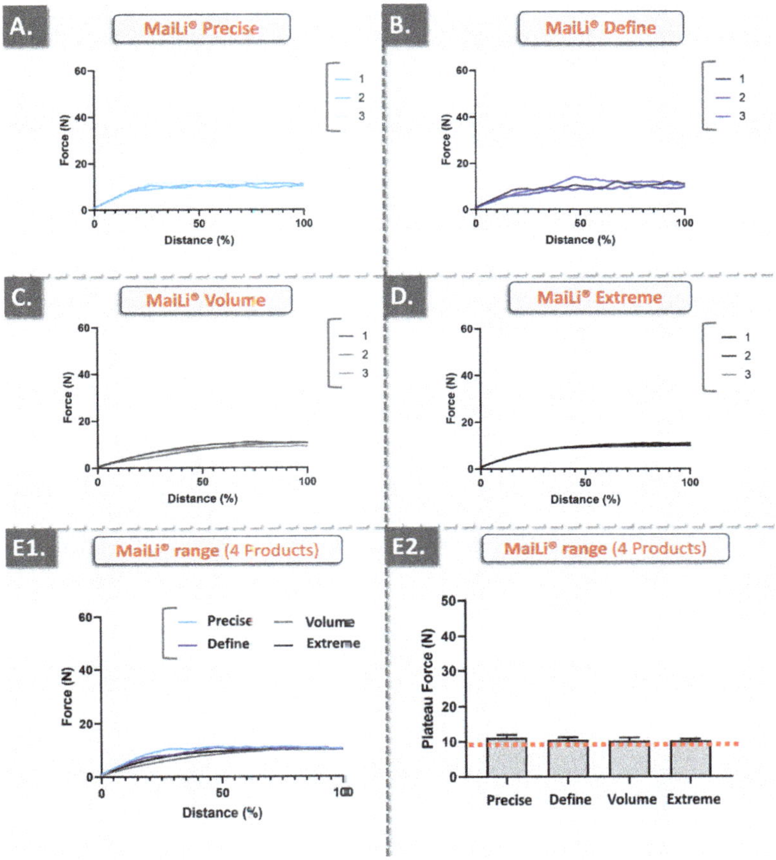

Figure 3. Experimental results of automated injectability evaluations at a constant plunger rod actuation speed of 0.2 mm·s^{-1} with MaiLi® product in vitro injection in SimSkin® cutaneous substrates. (**A**) Injection force curves of the MaiLi® Precise product. (**B**) Injection force curves of the MaiLi® Define product. (**C**) Injection force curves of the MaiLi® Volume product. (**D**) Injection force curves of the MaiLi® Extreme product. (**E1**) Combined mean injection force curves of the four MaiLi® product variants. (**E2**) Comparison of the mean injection force plateaus of the four MaiLi® product variants. The overall mean plateau injection force (i.e., dotted red line) was found to be 10 N (Table 6). N, Newtons.

Such closely distributed experimental data underscored the tight control over MaiLi® product biophysical attributes during the hydrogel manufacturing phases (Table 6). Of note, as regards the quantitative levels of injection force for the MaiLi® product variants, the automated in vitro setup yielded significantly higher values than the manual in vitro setups (Tables 2–6). In detail, the relatively rapid and constant injection speed of 1 mm·s^{-1} was used herein for optimal data comparability with previous reports, yet such speeds are not common in clinical practice (i.e., even for bolus injections) [1]. Therefore, the slower injection speed of 0.2 mm·s^{-1} (i.e., closer to that of clinical practice) was additionally used herein and confirmed that the hydrogel injection force is positively correlated to the injection speed in an automated setup (Table 6).

Specifically, the obtained automated injectability results were found to be systematically higher in value than those obtained in the various iterations of the manual setup (i.e., 10–35 N versus 0.50–2.00 N, respectively; Tables 2–6). Importantly, the quantitative difference between the automated and manual injectability setups, in terms of injection force requirements, may be attributed to the differing injection geometries. Specifically, the automated setup comprised the perpendicular insertion of the needle in the SimSkin® cutaneous equivalent, whereas the manual injections were performed using needle angles of 10–19° with the skin plane (i.e., the most obtuse angles being used for hypodermal bolus injections).

Overall, the key clinical interest of obtaining consistent injection forces across an entire product range is to optimize the in-use experience of the injector with a given dermal filler technology [1]. In detail, while various MaiLi® product variants exist and are finely tuned in terms of biophysical attributes (i.e., for differential clinical indications), conserved injection forces may be used for each variant (Tables 2–6). The resulting direct advantage for the clinician is that all MaiLi® product variants should behave the same in terms of injectability in defined settings (Figures 2 and 3). Thus, a high level of consistency of in-use product behaviour may be anticipated, thereby conferring optimal control over the administration process to the injector. As previously mentioned, this aspect is to be considered as a critical component (i.e., for optimizing patient experience) of the provided medical aesthetic care quality [1].

2.5. Clinical Case Reports on In Vivo MaiLi® Dermal Filler Injectability during Patient Treatments

In order to finally provide data with the highest level of clinical relevance, in vivo quantitative product injectability assessments were performed on five female participants. In detail, consenting participants were included in the study and were treated with MaiLi® product variants by injector PM or by injectors PM and TB in the cases where the participant accepted treatment by two different injectors (i.e., one injector per face side). The anonymized in vivo study parameters are summarized in Table 7.

Table 7. Summary of the anonymized demographic and treatment-related data for the in vivo portion of the study. The five included female participants were all treated with MaiLi® dermal filler variants during routine maintenance visits. G, Gauge.

Participant N° (Age)	Fitzpatrick/Glogau Class	Treated Areas (Injection Techniques)	MaiLi® Product Variant	Needle Size (G) [1]	Injection Depth
1 (64)	II/III	Galbella (Point-by-point); Nasolabial folds (Antero-tracing, Retro-tracing); Marionet (Point-by-point)	Precise	30 G × ½″	Mid-reticular dermis
2 (54)	II/III	Glabella (Point-by-point); Nasolabial folds (Point-by-point); Marionet (Retro-tracing)	Define	30 G × ½″	Mid-reticular dermis
3 (75)	III/III	Cheeks (Bolus); Nasolabial folds (Retro-tracing, Point-by-point); Marionet (Retro-tracing, Point-by-point)	Volume	27 G × ½″	Fat; deep dermis/hypodermis; deep dermis/hypodermis
4 (64)	IV (hispanic)/III	Cheeks (Bolus)	Extreme	27 G × ½″	Close-to-the-bone
5 (53)	III/III	Cheeks (Bolus)	Extreme	27 G × ½″	Fat

[1] All MaiLi® dermal filler injections were performed with needles and no cannulas were used in the study.

The retained MaiLi® dermal filler injection techniques corresponded to those routinely used in the clinical practice of injector PM for facial wrinkle filling and volumetric correction (Table 7). Importantly, it should be noted that the in vivo portion of the study only focussed on the filler administration-related injectability measurements (i.e., quantitative force values) and did not focus on efficacy-related outcomes. The various MaiLi® products were injected using either the thenar eminence, the pulp, or the P1–P2 joint of the thumb (Figure 4).

It is of note that the FlexiForce® dynamometric sensor and the attached cable may have slightly modified the position of the product syringe in the hand of the injector, as compared to routine clinical practice (Figure 4B). Generally, the recorded in vivo manual injection forces were found to be inferior in value to the data gathered in vitro in the automated setup, in adequation with the in vitro manual data (Tables 2–6). Furthermore, quantitative differences were observed between the injectors (i.e., injector PM was well experienced with MaiLi® products, injector TB was a new user) and between the areas of the thumb used for the injections (i.e., higher forces for the P1–P2 joint compared to the thumb pulp; Figure 4).

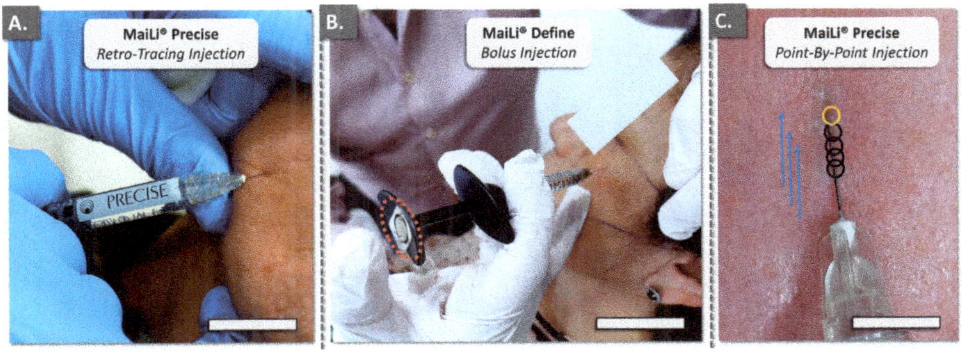

Figure 4. Photographic records of the clinical case reports for in vivo MaiLi® dermal filler product injectability assessment. (**A**) Injection force was applied with the thenar eminence of the thumb. Scale bar = 20 mm. (**B**) Injection force was applied with the thumb pulp. The dynamometric sensor is visible on the plunger rod hilt (i.e., dotted red outline). Scale bar = 30 mm. (**C**) Illustration of the point-by-point injection technique methodology. Black circles illustrate injections which were already realised, where the bevel of the needle was introduced in the centre of the circle. The area of the next injection is outlined in orange. The directionality and direction of progression are depicted by blue arrows. Additional methodological details are presented in Figure S6. Scale bar = 6 mm.

2.5.1. First Case Report: In Vivo MaiLi® Precise Injectability Assessments for Participant N°1

Participant N°1 was treated for general wrinkle filling of the face, using the MaiLi® Precise dermal filler (Table 7). The recorded forces for point-by-point injection were found to be close between injectors PM and TB (Figure 5, Table S1).

The recorded forces were found to be systematically higher in this case when the P1-P2 joint of the thumb was used (Figure 5). Of note, injector PM performed antero-grade injections using the thumb pulp, where the recorded forces were slightly inferior to those recorded during point-by-point injections (Table 7 and Table S1). As this type of injection is less common for facial wrinkle filling, it is possible that slower injection speeds were used, contributing to the lower recorded injection forces (Table S1). As concerns the retro-tracing injections, higher values were recorded for injector TB, which may be attributed to superior injection speeds as compared to injector PM (Table S1).

2.5.2. Second Case Report: In Vivo MaiLi® Define Injectability Assessments for Participant N°2

Participant N°2 was treated for general wrinkle filling of the face, using the MaiLi® Define dermal filler (Table 7). Point-by-point and retro-tracing injections were performed by injector PM, yielding low quantitative force values (Figure 6, Table S2).

Here again, the peak injection forces were observed to be high as compared to the mean injection forces (Figure 6). Conversely, the mean injection forces for point-by-point injections were recorded as relatively low (e.g., 1.45 N to 1.51 N; Table S2). Such values were linked to careful and slow injection in the glabella zone, which comports increased risks of adverse intravascular administration as compared to other zones of the face (Table S2). Of note, the antero-tracing and retro-tracing injections required relatively high injection forces as compared to the point-by-point injections. Here again, the higher values recorded for injector TB most probably resulted from higher injection speeds (Table S2).

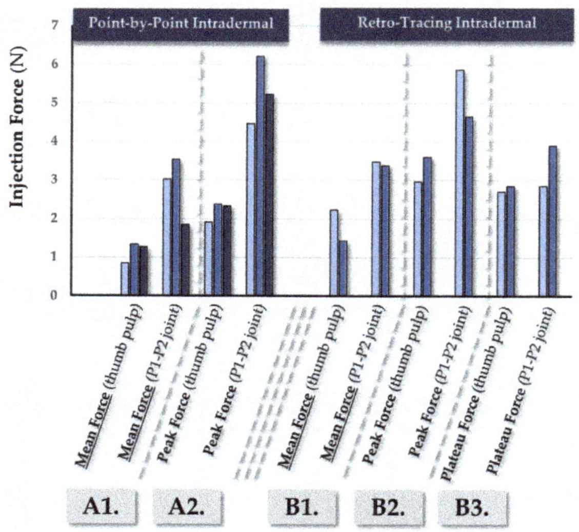

Figure 5. Quantitative results of in vivo dermal filler product injectability evaluation in Participant N°1, using the MaiLi® Precise dermal filler. Mean forces (A1) and peak forces (A2) were plotted for intradermal point-by-point injections. Mean forces (B1), peak forces (B2), and plateau forces (B3) were plotted for intradermal retro-tracing injections. Selected injection force data and injection force profiles are presented in Figures S7–S9 and in Table S1. N, Newtons.

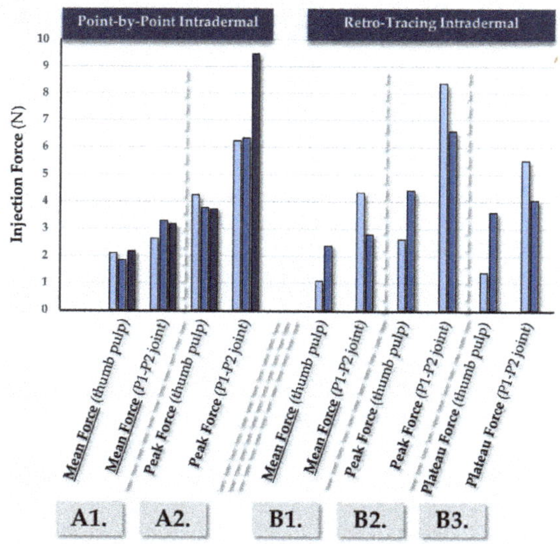

Figure 6. Quantitative results of in vivo dermal filler product injectability evaluation in Participant N°2, using the MaiLi® Define dermal filler. Mean forces (A1) and peak forces (A2) were plotted for intradermal point-by-point injections. Mean forces (B1), peak forces (B2), and plateau forces (B3) were plotted for intradermal retro-tracing injections. Selected injection force data and injection force profiles are presented in Figures S7 and S9 and in Table S2. N, Newtons.

2.5.3. Third Case Report: In Vivo MaiLi® Volume Injectability Assessments for Participant N°3

Participant N°3 was treated for deep wrinkle filling of the face, using the MaiLi® Volume dermal filler (Table 7). The recorded forces for point-by-point injections were found to be close between injectors PM and TB (Figure 7, Table S3).

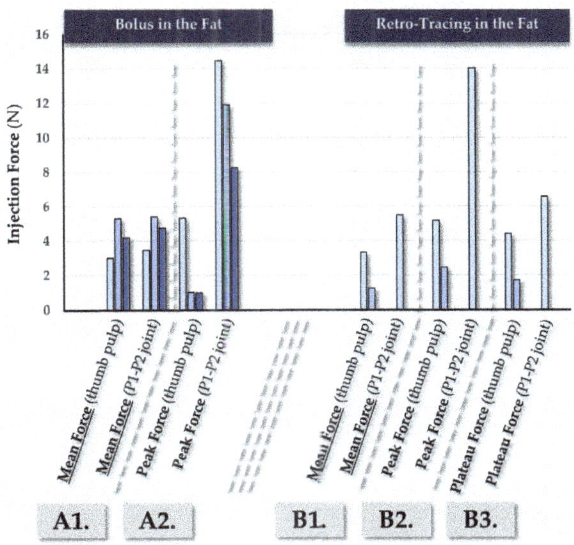

Figure 7. Quantitative results of in vivo dermal filler product injectability evaluation in Participant N°3, using the MaiLi® Volume dermal filler. Mean forces (A1) and peak forces (A2) were plotted for bolus injections in the fat. Mean forces (B1), peak forces (B2), and plateau forces (B3) were plotted for retro-tracing injections in the fat. Selected injection force data and injection force profiles are presented in Figure S10 and in Table S3. N, Newtons.

In this case, various injection depths were used (Table 7). The recorded injection forces were found to be significantly higher than those observed in previous participants, due to the fact that the injections were performed in the fat (Figure 7, Table S3). Confirming previous observations, the peak forces which were exerted using the P1-P2 thumb joint were again found to be higher in value than those exerted using the thumb pulp (Figure 7). It is noted that off-label injection in the deep reticular dermis or at the dermal-hypodermal junction is sometimes performed, yet this is only technically possible in zones of thick skin such as the nasogenian folds.

2.5.4. Fourth Case Report: In Vivo MaiLi® Extreme Injectability Assessments for Participant N°4

Participant N°4 was treated by injector PM for volumizing of the face, using the MaiLi® Extreme dermal filler (Table 7, Figure 8).

In this case, the MaiLi® Extreme dermal filler was injected close-to-the-bone, for the obtention of maximal volumizing effects (Table 7). Highly significant differences were noted in terms of the recorded injection force between injection modalities, where the use of the P1-P2 thumb joint produced excessive injection forces (Table S4). Similarly to the first three clinical cases, high recorded peak injection forces were noted (Figures 5–8).

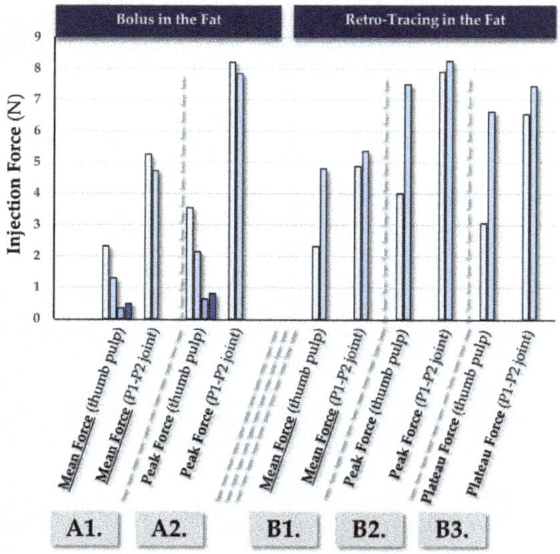

Figure 8. Quantitative results of in vivo dermal filler product injectability evaluation in Participant N°4, using the MaiLi® Extreme dermal filler. Mean forces (A1) and peak forces (A2) were plotted for bolus injections in the fat. Mean forces (B1), peak forces (B2), and plateau forces (B3) were plotted for retro-tracing injections in the fat. Selected injection force data and injection force profiles are presented in Figure S10 and in Table S4. N, Newtons.

2.5.5. Fifth Case Report: In Vivo MaiLi® Extreme Injectability Assessments for Participant N°5

Participant N°5 was treated for volumizing of the face, using the MaiLi® Extreme dermal filler (Table 7, Figure 9).

In this case, the MaiLi® Extreme dermal filler was injected close-to-the-bone by injector PM, using the thumb pulp and a bolus regimen (Figure 9). Therein, varying injection speeds were used and a positive trend between the injection speed and injection force was evidenced (Table S5). Specifically, the use of rapid injection regimens resulted in peak or plateau injection forces which were approximately double in value as compared to slower injections (Table S5). Furthermore, significantly higher injection forces were recorded with P1-P2 joint injections as compared to thumb pulp injections, which was consistent with the previous findings of the study.

2.6. Clinical Considerations on In Vitro and In Vivo MaiLi® Dermal Filler Injectability

Global consideration of the presented experimental results has enabled us to confirm the primary and secondary hypotheses of the study. Namely, it was specifically shown that MaiLi® product variants present intra-product and inter-product homogeneity as regards their injectability attributes (Figures 2 and 3, Table 6). Furthermore, it was set forth that MaiLi® dermal fillers present enhanced injectability attributes (i.e., smooth and consistent automated injection force plateaus) as compared to similar medical devices [1]. Overall, the gathered data were interpreted to concur with the available clinical feedback on the seamless injectability of MaiLi® dermal fillers, contrasting with that of alternative brands (e.g., specific BELOTERO® or TEOSYAL® variants) [1,17].

The significance of the present study resided in the novelty of reporting in vivo dermal filler in-use injectability assessments, as well as the documentation of the distinctive injectability behaviour of MaiLi® products (Figures 2–4). In detail, the presented automated injection force profiles were interpreted to qualitatively stand out from those of five commercial competitor brands (i.e., BELOTERO®, JUVÉDERM®, VIVACY®, RESTYLANE®, and TEOSYAL®), as assessed by the same operators in the same experimental setups [1]. Therein, no other product manufacturer was found to attain the quality level of the MaiLi® product brand in terms of plateau injection force smoothness and inter-product injectability consistency (Figure 3, Table 6) [1].

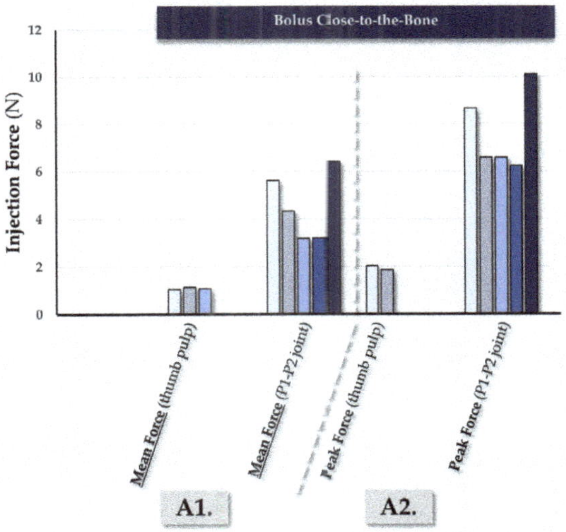

Figure 9. Quantitative results of in vivo dermal filler product injectability evaluation in Participant N°5, using the MaiLi® Extreme dermal filler. Mean forces (A1) and peak forces (A2) were plotted for close-to-the-bone bolus injections. Selected injection force data are presented in Table S5. N, Newtons.

Importantly, hydrogel product injectability attributes may be studied in diverse experimental setups and are often leveraged by manufacturers as technical features (i.e., to optimize clinician and patient experience of care) [27–30]. Notwithstanding, tangible correlations may be made between the quality level of dermal filler product injectability and the safety or efficacy outcomes of the treatment [1]. In the case of a homogeneous hydrogel system presenting smooth injection force plateaus (e.g., MaiLi® variants), the clinician may expect no or low levels of jerking/indentation-like behaviour of the plunger rod during filler administration (Figures 3 and S2–S5). Thereby, the risk of clinically injecting excess hydrogel amounts or of shallow product placement (i.e., potential nodule creation or Tyndall effect) is diminished [31,32].

From a quantitative standpoint, the comparative analysis of MaiLi® hydrogel product manual injection force profiles between SimSkin® cutaneous equivalents and human participants showed similarities for all four product variants (Figures S2–S5 and S7–S10). Furthermore, qualitative aspects of hydrogel product behaviour upon injection in both conditions were largely conserved, confirming the high translational relevance of the in vitro SimSkin® model (Figures S7–S10). Notwithstanding, several force values were recorded as being higher in vivo, especially for peak values (Figures 5–9, Tables S1–S5). Such results were considered to be mainly linked to the differences in composition and topography between the artificial SimSkin® constructs and human skin [33–37]. Specifically, while differential layers are present in SimSkin® substrates, it is probable that the synthetic material does not exactly match the diversity and nuances of human skin structures (i.e., epidermis, dermis, and hypodermis) and their biomechanical attributes [33,35].

Of further note, the presented in vivo data of MaiLi® product injectability also confirmed the clear trend of lower injection forces as compared to the automated in vitro injectability setup (Table 6, Figures 5–9). Specifically, these results were interpreted to be mainly linked to the relatively low manual injection speeds used in clinical settings (e.g., 0.05–0.1 mm·s^{-1}) as compared to the investigated automated injection speeds (i.e., 0.2–1.0 mm·s^{-1}, Table 6). Furthermore, significant inter-injector quantitative differences were recorded as regards MaiLi® product manual injectability (Tables 2–5). Confirming previous assessments, the present experiments outlined the significant influence of the injection speed, syringe handling modalities (i.e., injection with different portions of the thumb), administration site, and physician experience with a given product on the effective injection force [1]. Finally, it was shown that the depth of injection largely dictated the practically required injection forces of MaiLi® dermal fillers (Tables 2–5). Specifically, in vivo superficial reticular dermis injections generally required more force than deeper injections, in all probability due to the higher fibre density [33]. However, the experimental values gathered for in vivo MaiLi® injections in the fat were notably higher than those obtained in vitro for hypodermal injections (Figures S2–S5, Tables S1–S5).

Overall, the reported homogeneity of MaiLi® products ensured consistent force levels during injection, addressing previous concerns about in-use pressure variations with alternative filler products [1]. Such attributes may lead to smoother, more uniform aesthetic results, potentially reducing complications like the Tyndall effect or unevenness. Notwithstanding, while the study's in vitro analyses and the comparisons with previous observations of 28 HA gels offered valuable insights, an important gap between laboratory conditions and clinical practices was recognized [1]. Notably, most manufacturer-provided data, including product extrusion force curves, are usually obtained under conditions which are not fully representative of actual treatment scenarios. Namely, injections are often performed in the air rather than in skin or skin analogues. Therefore, the use of relevant in vitro and in vivo experiments may provide enhanced clinical predictability of the behaviour of dermal filler products [1]. Furthermore, expanding research through multicentre studies involving diverse patient populations and multiple skin types would help to validate the presented laboratory findings in a broader clinical context.

2.7. Hydrogel System Biophysical Attributes and Injection Technical Specifications Are Key Modulators of Product Injectability and Performance

In theory, the extrusion force required to push the dermal filler through the needle is mainly impacted by the viscosity of the system. Specifically, higher-viscosity fillers will require more force to extrude, while lower-viscosity fillers will require less force. During the first part of the injection process (i.e., elastic regime), the slope of the injection force curve is proportional to the G' storage modulus of the system [38]. Then, beyond a yield point, the gel begins to flow and the viscous regime dominates. Depending on the homogeneity of the hydrogel system, the injection of the filler is smooth and occurs at a steady rate [38]. Generally, a close relationship exists between the injection force profiles and clinical outcomes [1]. If the clinician suddenly pauses and stops applying a steady force at any point during the injection, the entire process must be restarted to reach the viscous regime once more. Thus, starting and stopping can lead to an uneven distribution of the product, resulting in an undesirable and uneven outcome. The same issues can be observed in cases of non-homogeneity during the viscous regime [1,38].

From a clinical standpoint, the relationship between injection speed, geometry, and injection force is complex. This relationship is highly influenced by the anatomical site of injection and the specific characteristics of the injector's technique. As previously demonstrated, reducing the applied and constant plunger rod speed leads to lower extrusion forces during automated in vitro injections [1]. This was confirmed herein and suggested that there is a nuanced interplay between the physical properties of the injected material, the technique used, and the anatomical site of injection (Table 6). Additionally, in comparison with automated setups, clinicians pause (i.e., while applying a steady force) during the injection process, to reposition the needle. In more detail, the mathematical formula that describes the depth of HA gel injections is based on the sinus of the needle penetration angle and the length of the needle (i.e., as implanted in the skin), which further illustrates the precision required in performing these procedures [5]. This precision level, coupled with the variability in skin thickness across different anatomical sites, underscores the challenges in standardizing injection techniques.

Given the diversity in injector sensitivity and technique (e.g., choice of needle or cannula, point-by-point, retro-tracing, or bolus methods), there is a significant variation in how each injector applies force. This variability is compounded by the inherent differences in skin anatomy, where the thickness of the epidermis, dermis, and hypodermis varies significantly across different body regions [5,8]. Moreover, ultrasound and MRI examinations align with histological data in revealing the layered complexity of the skin, challenging the notion of a uniform approach to injections [5,8]. The fact that some specialists identify only two layers within the dermis, likely due to limitations in imaging sensitivity, further complicates the endeavour to standardize injection techniques.

Overall, it appears clearly that variations in dermal filler injection speed, geometry, and applied force play a critical role in clinical practice, affecting the efficacy and outcomes of HA-based hydrogel injections. Notably, the individualized nature of the retained injection techniques, influenced by an injector's aesthetic perspective and delicacy in gesture, makes it difficult to uniformize practices for treating wrinkles or volume loss. This variability, alongside the anatomical and histological diversity of the skin, calls for further research, including multicentric studies involving injectors from various medical specializations. Such studies would potentially provide deeper insights into the optimal applications of specific injection techniques, accommodating the broad range of human anatomy and injector preferences.

2.8. Influence of Tailored Clinical Protocols on Dermal Filler Injection Force Profiles

Generally, the impact of the retained injection technique on the resulting injection force profile is significant, as highlighted by the variation in resistance encountered when injecting into the dermis as compared to the hypodermis (Tables S1–S5). The dermis, composed of elastin and collagen fibres, offers more resistance to HA gel flow than the less dense hypodermic fat tissue [1,5]. Techniques such as bolus injection require higher pressures for a larger gel volume, contrasting with the more delicate point-by-point technique used primarily in the dermis [1]. This variability is further complicated by factors such as the injector's visibility during the procedure and the speed of injection, which affect the extrusion force profile, especially in simulated skin environments [1]. Therefore, addressing the challenge of standardizing injection protocols is complex due to the inherent diversity in clinical injector sensitivity and preferences. Therein, some practitioners favour sharp needles, while others prefer cannulas, and similarly, there is high diversity in the adoption of point-by-point versus retro-tracing or bolus techniques across different layers of the skin. Importantly, this individualized or tailored approach to dermal filler injections reflects a broader resistance to clinical protocol standardization that could potentially homogenize aesthetic outcomes, erasing the unique characteristics that define our diverse global identities. The concern over a "one-size-fits-all" approach extends to cultural and aesthetic diversity, underscoring the importance of maintaining individuality in aesthetic treatments.

Notwithstanding, the potential for standardizing dermal filler injection forces through technology, such as the electronic "gun"-like TeosyalPen®, is acknowledged. This device offers a more uniform injection pressure, which some patients may find less painful, yet it may not accommodate all injection techniques. This consideration highlights a potential avenue for reducing variability in injection outcomes without compromising the essential personalized approach in aesthetic medicine. Overall, while the different injection techniques and the anatomical site of injection significantly influence the force profiles of HA-based hydrogel injections, the practical feasibility of quantitatively standardizing injection protocols is limited by the diversity of practitioner techniques and the importance of maintaining individualized treatment approaches. Specifically, future strategies may include technological advancements that offer consistent injection pressures while still allowing for the nuanced, personalized techniques that reflect the unique identities and preferences of both practitioners and patients.

2.9. Study Limitations

The main technical limitations of the present study comprised the limited number of injectors and participants in the study. Furthermore, the number of in vivo MaiLi® dermal filler injections was limited by the clinical treatment modalities, which did not enable the gathering of large datasets. As a result, the statistical comparison of the quantitative in vivo injectability data was not possible. As regards the low number of included patients, the study faced the challenge of finding individuals willing to be injected by two different professionals using different HA gels. Furthermore, the patients were made aware of the potential feeling of the FlexiForce® sensor cable, which was not part of their usual treatment and could have occasioned minor discomfort. Moreover, the study's methodology, particularly the exclusive use of sharp needles over cannulas, may not fully align with modern aesthetic practices, potentially biasing outcomes. The small sample size of five patients, due to the challenge of obtaining consent without financial incentives in Switzerland, severely limited the study's generalizability. Furthermore, the non-multicentric nature of the study and the financial constraints restricted the scope and applicability of the findings to a broader context. Despite the clinical credibility added by the diverse backgrounds of the injectors, such elements introduced variability that could have influenced the results. Overall, the outlined limitations (i.e., methodological choices, small cohort, singular geographic focus, and financial restrictions) significantly impacted the findings' generalizability. Therefore, future research should aim for larger, multicentric studies incorporating diverse injection techniques, HA gels, and injector specializations to enhance validity and applicability across the aesthetic medicine field.

As regards the low number of clinical injectors, the study included qualified and experienced practitioners as a counterbalancing measure, albeit two of them were new users of the MaiLi® product range. Therefore, the discrepancy in specific clinical experience with the MaiLi® product range could potentially have introduced some bias in the quantitative injectability measurements. Notwithstanding, all the injectors involved in this study had over a decade of experience in various injection techniques, administering treatments to approximately 10 patients daily in their respective practices. This extensive experience encompassed the different injection methods discussed in this paper, mirroring the expertise shared in our previous work examining 28 gels [1]. In that study, as in the current one, not all injectors utilized every gel, but they were all proficient in the four injection techniques evaluated both in vitro (i.e., using SimSkin®) and in vivo for injectors PM and TB. Specifically, PM has been using the entire range of the previously described 28 gels and the MaiLi® brand since their introduction on the Swiss and European markets, even before FDA approval. Thus, while injectors TB and DP were not initially familiar with the MaiLi® brand, it is important to highlight that all injectors possessed significant expertise in the required injection techniques. This foundational skillset ensured a level of consistency and proficiency across all treatments, mitigating potential biases related to unfamiliarity with specific products. Moreover, the diverse experiences and practices among the included injectors enriched the study's findings, providing a comprehensive understanding of MaiLi® product performance under real-world conditions. The injectors' long-standing experience and adaptability to different HA gel technologies, including the adoption of new products like MaiLi®, further diminished the likelihood of technique bias or skewed measurement/outcome evaluations.

Another limitation of the study consisted of the basic experimental biophysical attribute characterization for the considered hydrogels (i.e., rheology), as the applied methodologies constitute common debate elements [38–40]. Specifically, additional investigation into chemical and structural aspects of the cross-linked polymeric network of MaiLi® products could potentially have enabled some form of elucidation of their optimal injectability attributes (i.e., as demonstrated herein). Specifically, it is well known that chemical modifications of the HA polymer network and the addition of functional excipients bear critical impacts on hydrogel system behaviour and efficacy [41–45]. Notwithstanding, it is also recognized that the manufacturing process parameters themselves (e.g., homogenization, sterilization) and their fine-tuning bear significant impacts on finished product attributes, thus endpoint analytical controls would probably prove insufficient to holistically assess any given manufacturing technology (e.g., OxiFree™) [1,46,47].

Finally, few comparisons of the reported original injectability data with literature sources were performed herein, as the study was mainly designed as a continuation of previous work by the authors [1]. Specifically, while similar trends and influencing factors were found to be conserved in the present study (i.e., same operators and experimental setups), the obtained datasets confirmed that MaiLi® dermal fillers clearly stand out in terms of in-use injectability behaviour [1]. Of note, the domain of cross-linked HA-based dermal filler injectability (i.e., especially in vivo) is poorly represented in the literature (which focuses primarily on system rheology), which constituted a main rationale element for the design of the present study [48–50].

2.10. Future Perspectives

Specific future perspectives to the present study comprise the use of larger injector and participant pools in order to further augment the robustness of the obtained datasets. Specifically, larger and multicentric investigations by injectors of diverse clinical experience levels would add to the translational relevance of the study, with in vivo data gathered under real-world conditions. Secondly, the use of micro-canulas instead of the provided needles would be of high interest for MaiLi® dermal filler product injectability assessments, as such administration devices are often used in modern clinical practice. Thirdly, it would be of high interest to prospectively investigate if the reported qualitatively enhanced injectability levels of MaiLi® dermal fillers correlate with an enhanced care experience by the patients and/or the practitioners.

Furthermore, future research should focus on quantifying the optimum force profile required for even and smooth dermal filler extrusion. This research would not only deepen the understanding of HA gel properties but also ameliorate clinical practices by guiding physicians in achieving better, more consistent aesthetic results. Finally, future directions of research could comprise the use of advanced imaging techniques like ultrasound and MRI to observe the behaviour of HA gels in vivo, providing non-invasive and detailed insights into how these gels perform within the skin and hypodermis over time. Addressing these points could tangibly bridge the gap between bench research and bedside application, enhancing the utility of HA gels in aesthetic medicine.

3. Conclusions

The present study provided comprehensive injectability datasets on four widely used MaiLi® dermal filler variants. The study was designed to build on previous reports of cross-linked HA dermal filler injectability attribute assessments, with enhanced translational and clinical relevance. While quantitative differences were observed between the automated and manual injectability setups (i.e., in vitro and in vivo), highly consistent qualitative aspects of filler injectability were set forth for the MaiLi® product variants. Specifically, it was shown that very smooth injection force plateaus could be obtained for all four MaiLi® variants and that the respective mean plateau forces were not statistically different.

Overall, the experimental results confirmed a high degree of hydrogel system homogeneity in terms of viscoelasticity and injectability attributes, which is most notable for a whole range of products with varying quantitative compositions. The observational case reports for in vivo product injectability evaluation underscored the importance of injection speed and the area of the thumb used for the administration. The main conclusion of the study was that MaiLi® products present highly conserved qualitative and quantitative injectability attributes across multiple variants, which is distinctive in the field. Overall, the presented work underscored the central importance of thoroughly and systematically examining product attributes from a clinically relevant quality viewpoint, in order to rationally select the best available options for medical aesthetic care.

4. Materials and Methods

4.1. Reagents and Consumables Used in the Study

Physiological saline solution (NaCl 0.9%) was purchased from Bichsel (Unterseen, Switzerland). A total of four different cross-linked HA-based commercial dermal fillers from the MaiLi® brand were purchased from the product manufacturer (Sinclair Pharma Ltd., London, UK). The investigated MaiLi® Precise lots were N°211322-1, N°222456-2, N°211322-2, and N°213561-1. The investigated MaiLi® Define lots were N°202775-1, N°223419-2, and N°212916-1. The investigated MaiLi® Volume lots were N°202775-2, N°211228-2, and N°211803-2. The investigated MaiLi® Extreme lots were N°202843-1, N°222773-1, N°211678-2, and N°213561-2. The various needles used in the study were taken directly from each corresponding product packaging and comprised 30 G × ½″ needles (0.30 × 13 mm; TSK Laboratories, Tochigi-Ken, Japan) and 27 G × ½″ needles (0.40 × 13 mm; TSK Laboratories, Tochigi-Ken, Japan). For the technical needs of the study, TEOSYAL RHA® 2 products were purchased from TEOXANE (Geneva, Switzerland). For establishing standardized in vitro dermal filler injectability assessment conditions, synthetic SimSkin® cutaneous equivalents were purchased from Wallcur (San Diego, CA, USA).

4.2. MaiLi® Dermal Filler Product Rheological Characterization

The basic rheological behaviours of the four investigated MaiLi® product variants were determined in oscillatory rheology. Measurements were performed on an HR 10 rheometer (TA Instruments, Guyancourt, France) equipped with a Peltier plate–plate measuring geometry. Each measurement was performed in triplicate on 700 µL of undiluted hydrogel sample. The complex viscosity (η^*) values of the samples were determined as a function of the applied oscillatory frequency. To this end, a frequency sweep was performed from 0.1 Hz to 10 Hz at a fixed temperature of 25 °C. The constant shear stress was set at 2 $N{\cdot}m^{-2}$ in all experiments.

4.3. MaiLi® Dermal Filler Comparative Manual Injectability Studies in SimSkin® Cutaneous Equivalents

For the in vitro comparative assessments of MaiLi® dermal filler product injectability attributes, a SimSkin® cutaneous equivalent model was used. SimSkin® units consist of polymeric epidermis, dermis, and subcutaneous layers. The total substrate thickness is of 0.6 cm (i.e., 0.3 cm for the epidermis, 0.2 cm for the dermis, and 0.1 cm for the subcutaneous layer) [1]. Quantitative injectability measurements were performed by three experienced clinicians, using a dynamometric sensor (FlexiForce® Quickstart Board, Tekscan, Norwood, MA, USA) connected to myDAQ for data acquisition (National Instruments, Austin, TX, USA). For the assays, the plunger rods of the four MaiLi® product variants were replaced with those of TEOSYAL RHA® 2 products, which were equipped with the hilt-mounted sensor, as previously described [1].

The injection force parameters of the four MaiLi® dermal fillers were all determined using the original syringes and needles provided by the manufacturer. The mean forces of injection and peak forces of injection were automatically recorded. For each injection, the needle was introduced tangentially to the skin plane, at an angle of <10° for intradermal injections and of <19° for hypodermal injections. For point-by-point injections, only the bevel of the needle was introduced into the skin and small quantities of hydrogel were placed at regular intervals. For retro-tracing injections, the whole needle body was introduced into the skin and was pulled out simultaneously with the hydrogel injection.

For each MaiLi® product variant, four types of hydrogel product administration modalities (i.e., both indicated uses and off-label uses) were retained and were performed in triplicate by each injector. Namely, intradermal point-by-point and retro-tracing injections were performed, followed by hypodermal bolus and retro-tracing injections in SimSkin®

substrates. Based on the routine clinical practices of each injector, the thumb pulp or the P1-P2 thumb joint was used to manually and sequentially inject the investigated products. Specifically, the methods of injection closely mirrored those routinely used in the clinical practices of the co-authors for facial wrinkle filling or for volumetric supplementation. Of note, only injector PM was clinically experienced with the MaiLi® product range, whereas injectors TB and DP had not previously used the investigated products [17].

4.4. Comparative Automated Injectability Studies of MaiLi® Dermal Fillers in SimSkin® Cutaneous Equivalents at Constant Injection Speeds

To complement the manual injectability experiments, automated injectability assessments were performed in a dedicated in vitro texture analysis setup. The injection force profiles of the four commercially available MaiLi® product variants were determined in triplicate using the original syringes and needles supplied by the manufacturer. For the assays, the plunger rods of the four MaiLi® product variants were replaced with those of TEOSYAL RHA® 2 products, as previously described [1]. The hydrogel products were injected into SimSkin® cutaneous equivalents by a Texture Analyzer TA.XT. Plus instrument (Tracomme, Schlieren, Switzerland). The needle body was inserted perpendicularly to the skin plane at the appropriate depth within the synthetic scaffold for hydrogel injection. Constant plunger rod actuation speeds of 0.2 $mm \cdot s^{-1}$ and 1 $mm \cdot s^{-1}$ were used at ambient temperature (i.e., 25 °C). The higher injection speed value of 1 $mm \cdot s^{-1}$ was used for preliminary experiments in order to best discriminate the products in terms of hydrogel system intra-syringe homogeneity. The lower injection speed value of 0.2 $mm \cdot s^{-1}$ was then used to better approximate clinical manual dermal filler product injection conditions. The analysis of the injection force profiles enabled us to calculate the mean plateau injection forces for each MaiLi® product variant.

4.5. Clinical Case Reports on the In Vivo Injectability Attributes of MaiLi® Dermal Fillers

For the obtention of in vivo injectability data for MaiLi® dermal fillers, a series of five case reports was gathered from the routine clinical practice of co-author PM. Therefore, five female participants were included in this study, following oral and written information and a fixed consideration period of fifteen days. This study was compliant with the declaration of Helsinki and the applicable Swiss laws on copyright [51,52]. Four female participants were of Caucasian ethnicity and one female participant was of Hispanic ethnicity. The mean participant age was 62 years. All participants explicitly consented to the gathering of in vivo injection force data using the FlexiForce® dynamometric sensor during their routine clinical treatment. In parallel and subsequently to the in vivo injections, the same MaiLi® dermal filler administration regimens were carried out with SimSkin® cutaneous equivalents. Generally, the product administration methodology employed for the in vivo part of the presented study was consistent with that of previous reports by the authors [5].

4.5.1. In Vivo Intradermal Point-by-Point Injection Methodology

For intradermal point-by-point injections, the needle bevel (i.e., approximately 1 mm in length) was introduced under the skin at an angle of <10°, reaching the medium dermis. Plunger rod actuation then allowed to dispense a small quantity of hydrogel in order to slightly raise the skin surface into an artificial papule. The needle bevel was then introduced under the skin again, following the wrinkle line and adjacent to the first papule, before the injection process was repeated. This sequence was performed iteratively until the whole wrinkle was assessed as being filled.

4.5.2. In Vivo Intradermal Retro-Tracing Injection Methodology

For intradermal retro-tracing injections, the whole needle body was introduced under the skin at an angle of <10°, reaching the medium dermis. Plunger rod actuation was then performed simultaneously with the slow backwards pulling out of the needle from the skin. This sequence was performed iteratively until the whole wrinkle was assessed as being filled.

4.5.3. In Vivo Intradermal Antero-Tracing Injection Methodology

This type of injection was initiated by placing the needle bevel in the same position as for the point-by-point injections. Then, the whole needle body was progressively introduced further into the skin (i.e., in a trajectory parallel to the skin surface), where simultaneous and constant plunger rod actuation enabled to gently dispense the hydrogel. This sequence was performed iteratively until the whole wrinkle was assessed as being filled.

4.5.4. In Vivo Hypodermal Bolus Injection Methodology

For hypodermal bolus injections, the needle bevel was placed in the hypodermis, at a perpendicular angle with the skin plane (i.e., to reach the distal close-to-the-bone layer) or at an angle of <19° with the skin plane (i.e., to reach the fat layer). Continuous plunger rod actuation then enabled to deposit a relatively large amount of hydrogel at a constant depth.

4.5.5. In Vivo Hypodermal Retro-Tracing Injection Methodology

This type of injection was performed in the same way as the intradermal retro-tracing injection but using an angle of <19° between the needle and the skin plane.

4.6. Statistical Analysis and Data Presentation

Data were reported as mean values accompanied by the corresponding standard deviations as error bars, wherever applicable. For the statistical comparison of values from multi-group quantitative datasets, a one-way ANOVA or a two-way ANOVA test was performed and was followed by a post hoc Tukey's multiple comparison test. A p-value < 0.05 was retained as a general base for statistical significance determination. Detailed levels of statistical significance can be found in the Section 2. The statistical calculations and/or data presentation were performed using Microsoft Excel (Microsoft Corporation, Redmond, WA, USA), Microsoft PowerPoint, and GraphPad Prism v. 8.0.2 (GraphPad Software, San Diego, CA, USA).

Supplementary Materials: The following supporting information can be downloaded at https://www.mdpi.com/article/10.3390/gels10040276/s1, Figure S1: Illustrations of the SimSkin® cutaneous equivalent and its uses; Figure S2: In vitro injection force profiles for the MaiLi® Precise dermal filler; Figure S3: In vitro injection force profiles for the MaiLi® Define dermal filler; Figure S4: In vitro injection force profiles for the MaiLi® Volume dermal filler; Figure S5: In vitro injection force profiles for the MaiLi® Extreme dermal filler; Figure S6: Illustration of dermal filler injection techniques; Figure S7: In vivo dermal filler injection force profiles; Figure S8: In vitro and in vivo dermal filler injection force profiles; Figure S9: In vitro and in vivo dermal filler injection force profiles; Figure S10: In vitro and in vivo dermal filler injection force profiles; Table S1: In vivo injectability data for the MaiLi® Precise dermal filler; Table S2: In vivo injectability data for the MaiLi® Define dermal filler; Table S3: In vivo injectability data for the MaiLi® Volume dermal filler; Table S4: In vivo injectability data for the MaiLi® Extreme dermal filler; Table S5: In vivo injectability data for the MaiLi® Extreme dermal filler.

Author Contributions: Conceptualization, P.M., A.P., T.B. and A.L.; methodology, P.M., A.P., T.B., D.P., M.-O.C. and A.L.; software, P.M., A.P. and A.L.; validation, P.M., A.P., T.B., D.P., M.-O.C., L.A.A. and A.L.; formal analysis, P.M., A.P., T.B., D.P. and A.L.; investigation, P.M., A.P., T.B., D.P. and A.L.; resources, P.M., T.B. and A.L.; data curation, P.M., A.P. and A.L.; writing—original draft preparation, P.M., A.P. and A.L.; writing—review and editing, P.M., A.P., T.B., D.P., M.-O.C., L.A.A. and A.L.; visualization, P.M., A.P., T.B. and A.L.; supervision P.M., T.B. and A.L.; project administration, P.M., A.P. and A.L.; funding acquisition, P.M. All authors have read and agreed to the published version of the manuscript.

Funding: This study received no external funding and was not supported by any specific grants or institutional programs.

Institutional Review Board Statement: This study was conducted according to the guidelines of the Declaration of Helsinki [51]. The in vivo section of this study was conducted in Switzerland under the responsibility of registered medical professionals. Application with the local institutional review board was waived, as the study parameters (i.e., small case report series) placed it out of the scope of the Swiss Federal Act on Research involving Human Beings (Human Research Act, HRA, 2011) [53].

Informed Consent Statement: Informed consent (i.e., formalized in a general informed consent agreement) was obtained from all patients at the time of treatment for unrestricted use of the gathered and anonymized patient data or anonymized biological materials.

Data Availability Statement: The data presented in this study are openly available within the article files.

Acknowledgments: We would like to sincerely thank the five participants to this study. We would like to thank the Wallcur laboratory for the provision of SimSkin® cutaneous equivalents. Artwork templates were partly generated with www.biorender.com, accessed on 10 February 2024.

Conflicts of Interest: Author A.F. was employed by Abcello Sàrl (Belmont-sur-Yverdon, Switzerland) during this study. Author M.-O.C. was a consultant for Sinclair Pharma Ltd. during this study. Author A.L. was employed by TEC-PHARMA SA (Bercher, Switzerland) and by LAM Biotechnologies SA (Epalinges, Switzerland) during this study. The remaining authors declare no conflicts of interest for this study.

Abbreviation List

BDDE	1,4-butanediol diglycidyl ether
CE	European mark of conformity
DP	Daniel Perrenoud
η^*	complex viscosity
G	gauge
HA	hyaluronic acid
Hz	Hertz
MRI	magnetic resonance imaging
N	Newtons
N/A	non-applicable
Pa·s	Pascal seconds
PM	Patrick Micheels
TB	Thierry Bezzola
UK	United Kingdom
USA	United States of America

References

1. Micheels, P.; Porcello, A.; Bezzola, T.; Perrenoud, D.; Quinodoz, P.; Kalia, Y.; Allémann, E.; Laurent, A.; Jordan, O. Clinical perspectives on the injectability of cross-linked hyaluronic acid dermal fillers: A standardized methodology for commercial product benchmarking with inter-injector assessments. *Gels* **2024**, *10*, 101. [CrossRef] [PubMed]
2. Fundarò, S.P.; Salti, G.; Malgapo, D.M.H.; Innocenti, S. The rheology and physicochemical characteristics of hyaluronic acid fillers: Their clinical implications. *Int. J. Mol. Sci.* **2022**, *23*, 10518. [CrossRef] [PubMed]
3. Ginter, A.; Lee, T.; Woodward, J. How much does filler apparatus influence ease of injection (and hence, potential safety)? *Ophthal. Plast. Reconstr. Surg.* **2023**, *39*, 76–80. [CrossRef] [PubMed]
4. Rosamilia, G.; Hamade, H.; Freytag, D.L.; Frank, K.; Green, J.B.; Devineni, A.; Gavril, D.L.; Hernandez, C.A.; Pavicic, T.; Cotofana, S. Soft tissue distribution pattern of facial soft tissue fillers with different viscoelastic properties. *J. Cosmet. Dermat.* **2020**, *19*, 312–320. [CrossRef]
5. Micheels, P.; Goodmann, L. Injection depth in intradermal therapy: Update and correction of published data. *J. Drugs Dermatol.* **2018**, *17*, 611–618. [PubMed]
6. Sundaram, H.; Cassuto, D. Biophysical characteristics of hyaluronic acid soft-tissue fillers and their relevance to aesthetic applications. *Plast. Reconstr. Surg.* **2013**, *132*, 5S–21S. [CrossRef] [PubMed]
7. Liu, X.; Gao, Y.; Ma, J.; Li, J. The efficacy and safety of hyaluronic acid injection in tear trough deformity: A systematic review and meta-analysis. *Aesthet. Plast. Surg.* **2024**, *48*, 478–490. [CrossRef] [PubMed]
8. Kim, J. Effects of injection depth and volume of stabilized hyaluronic acid in human dermis on skin texture, hydration, and thickness. *Arch. Aesthet. Plast. Surg.* **2014**, *20*, 97–103. [CrossRef]
9. Urdiales-Gálvez, F.; Delgado, N.E.; Figueiredo, V.; Lajo-Plaza, J.V.; Mira, M.; Moreno, A.; Ortíz-Martí, F.; Del Rio-Reyes, R.; Romero-Álvarez, N.; Del Cueto, S.R.; et al. Treatment of soft tissue filler complications: Expert consensus recommendations. *Aesthet. Plast. Surg.* **2018**, *42*, 498–510. [CrossRef]

10. Gavard Molliard, S.; Bon Bétemps, J.B.; Hadjab, B.; Topchian, D.; Micheels, P.; Salomon, D. Key rheological properties of hyaluronic acid fillers: From tissue integration to product degradation. *Plast. Aesthet. Res.* **2018**, *5*, 17. [CrossRef]
11. Gavard Molliard, S.; Albert, S.; Mondon, K. Key importance of compression properties in the biophysical characteristics of hyaluronic acid soft-tissues fillers. *J. Mech. Behav. Biomed. Mater.* **2016**, *61*, 290–298. [CrossRef] [PubMed]
12. Faivre, J.; Pigweh, A.I.; Iehl, J.; Maffert, P.; Goekjian, P.; Bourdon, F. Crosslinking hyaluronic acid soft-tissue fillers: Current status and perspectives from an industrial point of view. *Exp. Rev. Med. Dev.* **2021**, *18*, 1175–1187. [CrossRef] [PubMed]
13. De Boulle, K.; Glogau, R.; Kono, T.; Nathan, M.; Tezel, A.; Roca-Martinez, J.X.; Paliwal, S.; Stroumpoulis, D. A review of the metabolism of 1,4-butanediol diglycidyl ether-crosslinked hyaluronic acid dermal fillers. *Dermatol. Surg.* **2013**, *39*, 1758–1766. [CrossRef] [PubMed]
14. Žádníková, P.; Šínová, R.; Pavlík, V.; Šimek, M.; Šafránková, B.; Hermannová, M.; Nešporová, K.; Velebný, V. The degradation of hyaluronan in the skin. *Biomolecules* **2022**, *12*, 251. [CrossRef] [PubMed]
15. Flégeau, K.; Jing, J.; Vantou, C.; Brusini, R.; Bourdon, F.; Faivre, J. Strengthening the key features of volumizing fillers: Projection capacity and long-term persistence. *Pharmaceutics* **2023**, *15*, 2585. [CrossRef] [PubMed]
16. Ho, D.; Jagdeo, J. Biological properties of a new volumizing hyaluronic acid filler: A systematic review. *J. Drugs Dermatol.* **2015**, *14*, 50–54. [PubMed]
17. Micheels, P. (Private Medical Practice, Chêne-Bougeries, Switzerland). Personal Communications on the Clinical Injectability of MaiLi® Dermal Fillers, 2023. (Unpublished work).
18. Micheels, P.; Poiraud, M. A preliminary report on the safety and efficacy of a novel MaiLi HA filler for facial correction of wrinkles, folds, volume and lips. *J. Clin. Cosmet. Dermatol.* **2023**, *7*, 1–8. [CrossRef]
19. Bon Bétemps, J.B.; Marchetti, F.; Lim, T.; Hadjab, B.; Micheels, P.; Salomon, D.; Gavard Molliard, S. Projection capacity assessment of hyaluronic acid fillers. *Plast. Aesthet. Res.* **2018**, *5*, 19. [CrossRef]
20. Gavard Molliard, S.; Bon Bétemps, J.; Hadjab, B.; Ghazal, A.; Badi, M.; Cerrano, M. Stabilized composition of 26 mg/mL of high molecular weight HA for subcutaneous injection to improve skin quality. *Plast. Aesthet. Res.* **2022**, *9*, 52. [CrossRef]
21. Bon Bétemps, J.; Gavard Molliard, S.; Hadjab, B.; Badi, M.; Ghazal, A.; Cerrano, M. Study of biological markers in skin quality treatment by subcutaneous injection of a stabilized composition of 26 mg/mL of high molecular weight HA. *Plast. Aesthet. Res.* **2022**, *9*, 59. [CrossRef]
22. Sinclair Pharma (Sinclair Pharma Ltd., London, UK). Instructions for Use of the MaiLi® Gel Range. 2023. Available online: https://maili.com/physicians/maili-an-expert-s-guide/ (accessed on 7 March 2024).
23. Tauzin, B.V. Method for Preparing an Injectable Cross-Linked Hydrogel, Hydrogel Obtained; and Use of the Obtained Hydrogel. WO2016207496A1, 29 December 2016; WIPO: Geneva, Switzerland, 2016.
24. Billon, R.; Hersant, B.; Meningaud, J.P. Hyaluronic acid rheology: Basics and clinical applications in facial rejuvenation. *Ann. Chir. Plast. Esthet.* **2017**, *62*, 261–267. [CrossRef] [PubMed]
25. La Gatta, A.; De Rosa, M.; Frezza, M.A.; Catalano, C.; Meloni, M.; Schiraldi, C. Biophysical and biological characterization of a new line of hyaluronan-based dermal fillers: A scientific rationale to specific clinical indications. *Mater. Sci. Eng. C* **2016**, *68*, 565–572. [CrossRef] [PubMed]
26. Fagien, S.; Bertucci, V.; von Grote, E.; Mashburn, J.H. Rheologic and physicochemical properties used to differentiate injectable hyaluronic acid filler products. *Plast. Reconst. Surg.* **2019**, *143*, 707e–720e. [CrossRef] [PubMed]
27. De la Guardia, C.; Virno, A.; Musumeci, M.; Bernardin, A.; Silberberg, M.B. Rheologic and physicochemical characteristics of hyaluronic acid fillers: Overview and relationship to product performance. *Facial Plast. Surg.* **2022**, *38*, 116–123. [CrossRef]
28. Pluda, S.; Salvagnini, C.; Fontana, A.; Marchetti, A.; Di Lucia, A.; Galesso, D.; Guarise, C. Investigation of crosslinking parameters and characterization of hyaluronic acid dermal fillers: From design to product performances. *Gels* **2023**, *9*, 733. [CrossRef] [PubMed]
29. Robinson, T.E.; Hughes, E.A.B.; Eisenstein, N.M.; Grover, L.M.; Cox, S.C. The quantification of injectability by mechanical testing. *J. Vis. Exp.* **2020**, *159*, e61417. [CrossRef]
30. Feng, X.; Wu, K.W.; Balajee, V.; Leissa, J.; Ashraf, M.; Xu, X. Understanding syringeability and injectability of high molecular weight PEO solution through time-dependent force-distance profiles. *Int. J. Pharm.* **2023**, *631*, 122486. [CrossRef] [PubMed]
31. King, M. Management of Tyndall effect. *J. Clin. Aesthet. Dermatol.* **2016**, *9*, E6–E8. [PubMed]
32. Colon, J.; Mirkin, S.; Hardigan, P.; Elias, M.J.; Jacobs, R.J. Adverse events reported from hyaluronic acid dermal filler injections to the facial region: A systematic review and meta-analysis. *Cureus* **2023**, *15*, e38286. [CrossRef]
33. Della Volpe, C.; Andrac, L.; Casanova, D.; Legré, R.; Magalon, G. Skin diversity: Histological study of 140 skin residues, adapted to plastic surgery. *Ann. Chir. Plast. Esthet.* **2012**, *57*, 423–449. [CrossRef]
34. Yousef, H.; Alhajj, M.; Sharma, S. Anatomy, skin (integument), epidermis. In *StatPearls*; StatPearls Publishing: St. Petersburg, FL, USA, 2022.
35. Bonifant, H.; Holloway, S. A review of the effects of ageing on skin integrity and wound healing. *Br. J. Commun. Nurs.* **2019**, *24*, S28–S33. [CrossRef]
36. Davydov, D.A.; Budylin, G.S.; Baev, A.V.; Vaypan, D.V.; Seredenina, E.M.; Matskeplishvili, S.T.; Evlashin, S.A.; Kamalov, A.A.; Shirshin, E.A. Monitoring the skin structure during edema in vivo with spatially resolved diffuse reflectance spectroscopy. *J. Biomed. Opt.* **2023**, *28*, 057002. [CrossRef] [PubMed]

37. Starcher, B.; Aycock, R.L.; Hill, C.H. Multiple roles for elastic fibers in the skin. *J. Histochem. Cytochem.* **2005**, *53*, 431–443. [CrossRef]
38. Wongprasert, P.; Dreiss, C.A.; Murray, G. Evaluating hyaluronic acid dermal fillers: A critique of current characterization methods. *Dermatol. Ther.* **2022**, *35*, e15453. [CrossRef] [PubMed]
39. Dugaret, A.S.; Bertino, B.; Gauthier, B.; Gamboa, B.; Motte, M.; Rival, Y.; Piwnica, D.; Osman-Ponchet, H.; Bourdès, V.; Voegel, J.J. An innovative method to quantitate tissue integration of hyaluronic acid-based dermal fillers. *Ski. Res. Technol.* **2018**, *24*, 423–431. [CrossRef] [PubMed]
40. La Gatta, A.; Schiraldi, C.; Zaccaria, G.; Cassuto, D. Hyaluronan dermal fillers: Efforts towards a wider biophysical characterization and the correlation of the biophysical parameters to the clinical outcome. *Clin. Cosmet. Investg. Dermatol.* **2020**, *13*, 87–97. [CrossRef]
41. Kim, J.; Kim, S.; Son, D.; Shin, M. Phenol-hyaluronic acid conjugates: Correlation of oxidative crosslinking pathway and adhesiveness. *Polymers* **2021**, *13*, 3130. [CrossRef] [PubMed]
42. Schanté, C.; Zubera, G.; Herlinb, C.; Vandamme, T.F Synthesis of N-alanyl-hyaluronamide with high degree of substitution for enhanced resistance to hyaluronidase-mediated digestion. *Carbohydr. Polym.* **2011**, *86*, 747–752. [CrossRef]
43. Conrozier, T.; Mathieu, P.; Rinaudo, M. Mannitol preserves the viscoelastic properties of hyaluronic acid in an in vitro model of oxidative stress. *Rheumatol. Ther.* **2014**, *1*, 45–54. [CrossRef]
44. Conrozier, T. Is the addition of a polyol to hyaluronic acid a significant advance in the treatment of osteoarthritis? *Curr. Rheumat. Rev.* **2018**, *14*, 226–230. [CrossRef]
45. Hintze, V.; Schnabelrauch, M.; Rother, S. Chemical modification of hyaluronan and their biomedical applications. *Front. Chem.* **2022**, *10*, 830671. [CrossRef]
46. Micheels, P.; Sarazin, D.; Tran, C.; Salomon, D. Effect of different crosslinking technologies on hyaluronic acid behavior: A visual and microscopic study of seven hyaluronic acid gels. *J. Drugs Dermatol.* **2016**, *15*, 600–606. [PubMed]
47. Zhao, P.; Zhao, W.; Zhang, K.; Lin, H.; Zhang, X. Polymeric injectable fillers for cosmetology: Current status, future trends, and regulatory perspectives. *J. Appl. Polym. Sci.* **2020**, *137*, 48515. [CrossRef]
48. Guo, J.; Fang, W.; Wang, F. Injectable fillers: Current status, physicochemical properties, function mechanism, and perspectives. *RSC Adv.* **2023**, *13*, 23841–23858. [CrossRef] [PubMed]
49. Cotofana, S.; Hamade, H.; Bertucci, V.; Fagien, S.; Green, J.B.; Pavicic, T.; Nikolis, A.; Lachman, N.; Hadjab, A.; Frank, K. Change in rheologic properties of facial soft-tissue fillers across the physiologic angular frequency spectrum. *Plast. Reconstr. Surg.* **2021**, *148*, 320–331. [CrossRef]
50. Lee, W.; Oh, W.; Moon, H.J.; Koh, I.S.; Yang, E.J. Soft tissue filler properties can be altered by a small-diameter needle. *Dermatol. Surg.* **2020**, *46*, 1155–1162. [CrossRef]
51. World Medical Association. World Medical Association Declaration of Helsinki: Ethical principles for medical research involving human subjects. *JAMA* **2013**, *310*, 2191–2194. [CrossRef]
52. The Federal Assembly of the Swiss Confederation Federal Act on Copyright and Related Rights (Copyright Act, CopA). RS 231.1. 1992. Available online: https://www.fedlex.admin.ch/eli/cc/1993/1798_1798_1798/en (accessed on 8 March 2024).
53. The Federal Assembly of the Swiss Confederation Federal Act on Research involving Human Beings (Human Research Act, HRA). RS 810.30. 2011. Available online https://www.fedlex.admin.ch/eli/cc/2013/617/en (accessed on 8 March 2024).

gels

MDPI

Article

Customizable Hydrogel Coating of ECM-Based Microtissues for Improved Cell Retention and Tissue Integrity

Shani Elgin [1,†], Eric Silberman [1,2,3,†], Assaf Shapira [1,2,3] and Tal Dvir [1,2,3,4,5,*]

1 The Shmunis School of Biomedicine and Cancer Research, Faculty of Life Sciences, Tel Aviv University, Tel Aviv 6997801, Israel
2 The Sagol Center for Regenerative Biotechnology, Tel Aviv University, Tel Aviv 6997801, Israel
3 Tel Aviv University Center for Nanoscience & Nanotechnology, Tel Aviv University, Tel Aviv 6997801, Israel
4 Department of Biomedical Engineering, Faculty of Engineering, Tel Aviv University, Tel Aviv 6997801, Israel
5 Sagol School for Neuroscience, Tel Aviv University, Tel Aviv 6997801, Israel
* Correspondence: tdvir@tauex.tau.ac.il
† These authors contributed equally to this work.

Abstract: Overcoming the oxygen diffusion limit of approximately 200 μm remains one of the most significant and intractable challenges to be overcome in tissue engineering. The fabrication of hydrogel microtissues and their assembly into larger structures may provide a solution, though these constructs are not without their own drawbacks; namely, these hydrogels are rapidly degraded in vivo, and cells delivered via microtissues are quickly expelled from the area of action. Here, we report the development of an easily customized protocol for creating a protective, biocompatible hydrogel barrier around microtissues. We show that calcium carbonate nanoparticles embedded within an ECM-based microtissue diffuse outwards and, when then exposed to a solution of alginate, can be used to generate a coated layer around the tissue. We further show that this technique can be fine-tuned by adjusting numerous parameters, granting us full control over the thickness of the hydrogel coating layer. The microtissues' protective hydrogel functioned as hypothesized in both in vitro and in vivo testing by preventing the cells inside the tissue from escaping and protecting the microdroplets against external degradation. This technology may provide microtissues with customized properties for use as sources of regenerative therapies.

Keywords: microfluidics; ECM-based hydrogel; nanoparticles; microtissues; hydrogels

1. Introduction

Perhaps the most significant challenge that must be addressed in the field of tissue engineering is overcoming the oxygen diffusion limit of approximately 200 μm. In native tissues, a dense and complex vascular network fully penetrates the entire structure [1], such that cells are rarely found more than 100 μm away from the nearest capillary [2]. During gestation, the vasculature grows jointly within the developing tissues to ensure that the oxygen needs of all cells continue to be met, even as the organism grows. However, without the benefits of natural embryonic development, tissue engineering must find alternative approaches to overcome the diffusion limitation of oxygen and develop tissues thicker than a few hundred microns.

Many strategies have been proposed to deliver oxygen to cells within thick tissues. Strategies include using endothelial cells and allowing them to naturally form blood vessels [3,4], producing tissues via advanced fabrication techniques (such as lithography and 3D printing) that contain hollow lumens through which cell culture medium can be perfused [5], and designing oxygen-producing scaffolds that generate oxygen from within [6,7]. Another strategy that has gained attention is encapsulating cells within micrometric hydrogel particles, or "microgels".

Microgels provide several distinct benefits over traditional hydrogels [8–10]. First, microgels synergistically combine multiple mechanisms that promote oxygen diffusion. One method by which microgels mitigate hypoxia is by conglomerating to form macroscopic structures with large interstitial volumes that fill the void between particles. Because blood, cell growth medium, or serum can easily fill this entire volume, the diffusion distance required of oxygen is limited to the radius of a single particle [8,11]. At the same time, the large surface-area-to-volume ratio of spheres maximizes the diffusion of oxygen into the individual particles [11]. Besides the benefits that relate to mass transport, microgels provide several other advantages. For example, individual particles from within a macrostructure can be separated with minimal force. As a result, hydrogels composed of microdroplets almost always demonstrate shear thinning behavior, which makes them suitable for delivery via minimally invasive procedures [10]. Additionally, microgels encapsulating different cell types can be assembled in a modular fashion to easily create complex tissues [12]. This strategy can be further enhanced by combining microgels with a suitable technology (such as 3D printing) to provide total control over the placement of cells within the tissue [13–15].

Despite their obvious benefits, microgels also suffer from a distinct drawback. Namely, the same factors that promote nutrient diffusion also lead to rapid degradation [16]. In fact, microgels suffer from both the outward movement of encapsulated cells away from the implantation site, termed "cell escape" [17], and the inward flux of macrophages and their associated foreign body response. Multiple techniques have been employed to overcome these challenges, including centering cells within microparticles to retard their escape [18], manipulating the size of the microparticles [19], and developing core–shell microparticles, often based on synthetic polymers that cells cannot easily penetrate [20].

Our lab has previously developed a microfluidic system for encapsulating cells within a micrometric hydrogel derived from native extracellular matrix (ECM) [9]. There, we developed a protocol for the microfluidic system that led to precisely controlled cell encapsulation and high cell viability and functionality. The hydrogel was selected for its biological relevance [21] and the microfluidic process for its high repeatability and throughput.

Here, we sought to enhance the microgels' stability by generating a protective shell around the microparticles. To that end, we encapsulated cells and nanoparticles of calcium carbonate ($^{np}CaCO_3$) within the ECM-based hydrogel and used the aforementioned microfluidic system to generate microtissues containing $^{np}CaCO_3$. When these microtissues were immersed in a solution of sodium alginate, the $^{np}CaCO_3$ diffused out of the ECM-based hydrogel and interacted with the alginate molecules in the immediate vicinity of the tissue to form an alginate-based hydrogel capsule around it. Importantly, the formation of this alginate shell was highly tunable, and it could be controlled by altering several different parameters individually or in parallel. We demonstrated the ability to control the thickness of the alginate hydrogel layer by controlling multiple parameters, including the type of alginate used, its concentration, the concentration of the $^{np}CaCO_3$, and the time allowed for the reaction. Microtissues protected in this way were then assessed both in vitro and in vivo. While the alginate coating did not in any way impair cell viability, it did retard cell escape and protect the microgel from cell-based and enzymatic cleavage. This technology was ultimately demonstrated in vivo, and we showed that the presence of the alginate coating led to enhanced cell retention and tissue integrity at the injection site.

2. Results and Discussion

In order to formulate a protective coating for microtissues formed from an ECM-based hydrogel, we decided to work with alginate. Alginate is well-documented as a biomaterial owing to its low cost and overall biocompatibility [22]. Additionally, the crosslinking of alginate is orthogonal to that of the ECM-based hydrogel, which allows for each hydrogel to be crosslinked independently of the others [23].

An alginate coating around the ECM-based microgels was generated using an internal gelation strategy [24]. In this method, non-soluble calcium is preloaded in the form of calcium carbonate ($CaCO_3$), which is insoluble at physiological pH. However, because

$CaCO_3$ readily dissolves in acidic environments, lowering the pH leads to the dissolution of the salt and the subsequent crosslinking of the alginate. While this general strategy is well known, several considerations were taken into account when adapting this procedure for use as a coating technique for microtissues.

First, in order to localize the crosslinking to the surface of the microtissues, $CaCO_3$ was loaded into the microgels, rather than the alginate solution. This forced us to find a solution that would minimize the volume occupied by the $CaCO_3$, so as not to crowd out the cells. Additionally, it was important that the presence of the insoluble crystals not change the rheological properties of the hydrogel and thus disrupt the microfluidic process. Second, because this technique must be relevant for cell-laden microtissues, we sought to minimize the microgels' exposure to the acidic alginate environment. For these reasons, we chose to work with nanoparticles of $CaCO_3$ ($^{np}CaCO_3$), which, owing to their extremely high surface-area-to-volume ratio, dissolve rapidly upon exposure to a low pH, minimizing the time required for the reaction to occur [25,26].

We therefore devised a two-step process for the creation of ECM-based microtissues with an alginate coating (Figure 1). First, cells were encapsulated within an ECM-based hydrogel, and then $^{np}CaCO_3$ were suspended within the gel. This tissue precursor was then run through a custom-made microfluidic system to generate microgels containing $^{np}CaCO_3$. The microfluidic system generates a water-in-oil emulsion in which the hydrogel is the aqueous component. By separately modulating the flow rates of the hydrogel and the surrounding oil phase, it is possible to control the size of the collected droplets [8]. After being collected, the microgels were incubated at 37 °C for 15 min to allow them to thermally gel, at which point the surfactants that had been used to form the droplets were removed by washing. Second, the microgels were immersed in a slightly acidified (pH = 6) solution of sodium alginate. This step was performed on ice in order to both minimize the negative impacts of the acidic pH on the cells and to provide better control over the crosslinking kinetics of the alginate. This protocol led to a robust crosslinking, and an alginate shell was visible around the microdroplets even after repeated washings (Figure 2a).

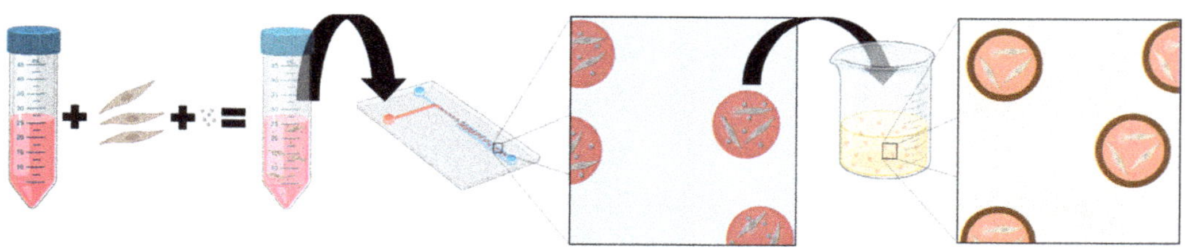

Figure 1. Overall schematic. Cells (brown) and nanoparticles of $CaCO_3$ (gray) were encapsulated within an ECM-based hydrogel (red), and a microfluidic system was used to generate microtissues. When the microdroplets were subsequently suspended in a solution of sodium alginate (yellow), the calcium ions crosslinked the alginate and generated a customizable protective shell around the microgel. Created with BioRender.com.

In order to maximize the versatility of this system, we next undertook a thorough study of the factors controlling the formation of the alginate coating (Figure 2b). Although any one factor might be enough to control the alginate hydrogel layer, by assessing the impact of multiple factors, we can more thoroughly customize the protocol and expand its range of utility. Unsurprisingly, the concentration of $^{np}CaCO_3$ incorporated in the hydrogel had a significant impact on the thickness of the alginate coating (Figure 2c). Even at concentrations as low as 0.83 mg $^{np}CaCO_3$ per mL hydrogel, an alginate coating with a thickness of 10.73 ± 0.62 µm could be observed.

One of the benefits of using alginate as a biomaterial is that alginate is not, in fact, a single molecule, but rather a class of polymers that vary in length and composition depending on the species of algae from which the material is collected [22]. It is often possible, therefore, to fine-tune a process based on alginate by simply varying the species of algae used. In order to assess the impact of alginate type on our coating protocol, we used both Protanal® LF 200 FTS and Protanal® LF 10/60 FT. Protanal® LF 200 FTS is composed of longer polymeric chains than other forms of alginate and forms significantly more viscous solutions. The reason for the increased viscosity is that the extended chains form significant interactions between themselves, and we hypothesized that these intermolecular forces would likewise act to increase the thickness of the coating when applied to a microtissue. As predicted, the use of the longer Protanal® LF 200 FTS chains led to the formation of a thicker coating (28.39 ± 6.80 μm as compared to 15.25 ± 4.29 μm) around the microgels (Figure 2d).

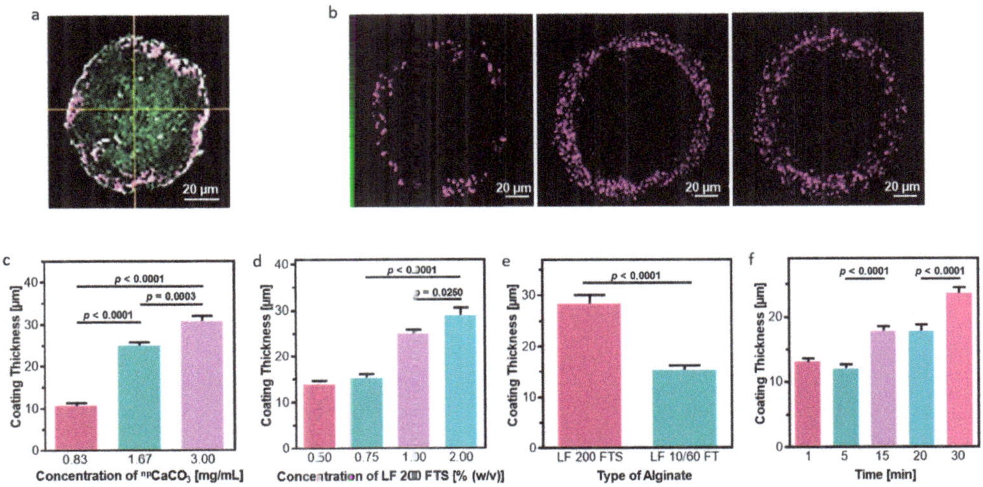

Figure 2. Controlling the Alginate Capsule. (**a**) The 3D confocal imaging revealed that the alginate shell completely surrounded the microdroplet from all sides. For visualization, FITC liposomes were incorporated in the ECM-based hydrogel and alginate was tagged with fluorescent moieties. (**b**) Representative images of alginate barriers with varying thickness and integrity. For visualization, fluorescent microparticles were incorporated in the alginate layer. (**c**) The thickness of the alginate capsule varies proportionally with the concentration of $^{np}CaCO_3$ used. (**d**) Increasing the concentration of alginate in which the microdroplets were submersed increases the thickness of the capsule. (**e**) Using longer alginate chains (Protanal® LF 200 FTS vs. Protanal® LF 10/60 FT) led to an increase in capsule thickness. (**f**) Increasing the time during which the reaction was allowed to proceed led to an increase in the thickness of the alginate layer. Scale bars = 20 μm.

Alongside the dependence on the concentration of $^{np}CaCO_3$, the coating thickness was investigated as a function of the concentration of alginate (Figure 2e). While varying the concentration of alginate below 0.75% did not affect the coating, using concentrations of 1% or 2% led to the creation of a significantly thicker shell around the microgel.

Finally, we investigated our ability to control the thickness of the alginate coating by exposing the microdroplets to acidic alginate for variable lengths of time (Figure 2f). Within one minute of exposure, a discernible coating of 13.10 ± 2.85 μm was already present around the microgel. After 30 min of exposure, the coating thickness had increased to 23.59 ± 5.61 μm. In order to preserve cell viability, the coating procedures were not conducted for time periods exceeding 30 min.

Having developed a robust protocol for customizing the thickness of the alginate layer, we next sought to determine the implications of the coating on the behavior of cells encapsulated within the microtissues. First, as it is well known that the mechanical properties of a cell's microenvironment can significantly impact the cell's functionality [27–32], the rheological characteristics of the $^{np}CaCO_3$-infused ECM-based hydrogel were investigated. In our work, the rheological properties of this hydrogel are doubly important as they can both impact the microfluidic process as well as the ability of the encapsulated cells to mature properly. As shown, the addition of the $^{np}CaCO_3$ did not lead to any significant changes in the rheological behavior of the hydrogel (Figure 3a). Next, cells were encapsulated within the hydrogels and the entire protocol—the addition of $^{np}CaCO_3$, the microfluidic generation of microtissues, and the formation of the alginate coating—was performed. As a proof-of-concept, we chose to work with fibroblasts as a model system to study cell viability, proliferation, and migration. It was observed that the coating protocol did not cause cell death immediately following the treatment, nor in the three days following (Figure 3b). One week after treatment, cell viability decreased slightly for non-coated microtissues, while microtissues that had received the alginate treatment had slightly higher cell viability. This trend was also observed in a PrestoBlue™ assay for cell metabolism (Figure 3c). We attribute the impaired cell viability and proliferation in uncoated droplets to the degradation of the droplets. Whereas the microtissues with an alginate barrier maintained their full volume over the course of the experiment, the non-coated microtissues shrank significantly. Therefore, the remaining cells become significantly denser over time, likely causing a contact-inhibition effect.

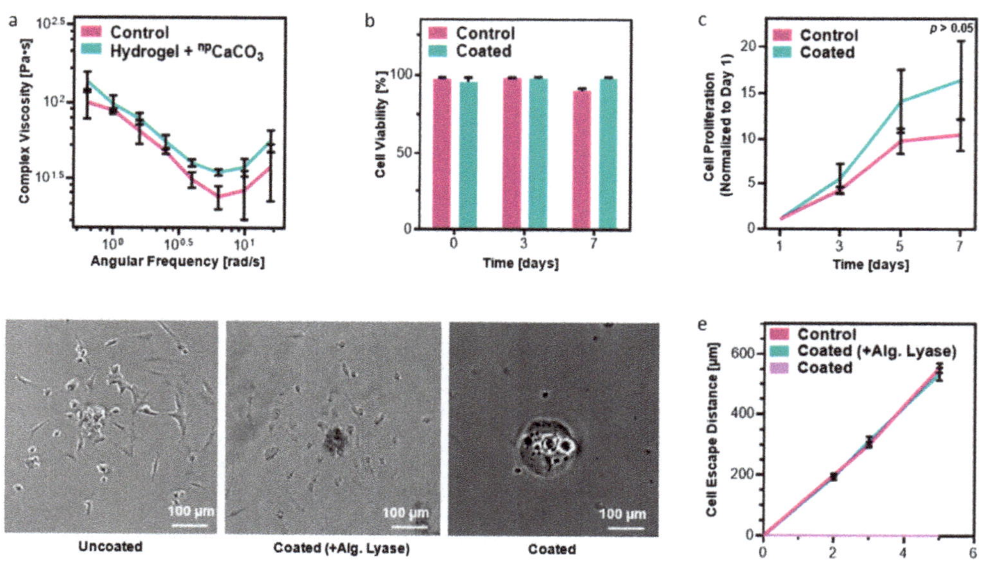

Figure 3. Functionality and biocompatibility of the alginate coating. (**a**) The addition of $^{np}CaCO_3$ caused no significant changes in the rheological properties of the ECM-based hydrogel. (**b**) The cell-coating protocol had no negative impact on cell viability, and, indeed, showed a positive increase in cell viability after a week of cultivation. (**c**) The presence of the alginate coating did not impair cell viability. In fact, because the coating acted to protect the microtissue, coated microtissues remained larger and thus were able to accommodate larger and more active cell populations. (**d**) Representative images showing cells that had escaped from the droplet and adhered to the well plate on Day 3 in both uncoated (left) and coated-then-enzymatically-degraded (center) droplets. Droplets with an intact alginate barrier showed negligible cell escape (right). Scale bars = 100 µm. (**e**) The presence of the coating almost entirely halted the ability of the cells to migrate out of and away from the microtissue. The impact of the alginate was further verified by enzymatically degrading the alginate, which led to unimpaired "cell escape".

In addition, the alginate barrier's ability to prevent "cell escape" was assessed. While cells encapsulated in non-coated microdroplets migrated 110.3 ± 2.58 µm per day, cells in protected microdroplets showed minimal migration over the course of five days. As an additional test, to ensure that the difference was due to the alginate coating and not to another factor that may have occurred during the crosslinking process, coated droplets were treated with alginate lyase. Alginate lyase is known to cleave alginate chains without harming cells [33]. After treatment with alginate lyase, cells encapsulated in droplets that had a protected layer were suddenly able to "escape" the microtissue. Furthermore, their migration proceeded at precisely the same rate (106.9 ± 3.53 µm per day) as cells that had never undergone the alginate-coating protocol (Figure 3d,e).

Having ascertained that the presence of an alginate coating not only did not harm cells, but in fact led to greater retention and viability, the behavior of the coating itself was investigated. First, microdroplets were exposed to collagenase to mimic the degradation processes that would occur in vivo, and the degradation kinetics of the ECM-based hydrogel was measured (Figure 4a). Droplets without a protective alginate layer rapidly decreased in size, with a reduction in the average droplet diameter to only 45.9 ± 1.06% of its original diameter over the course of 24 h. Interestingly, even without collagenase, unprotected droplets gradually decreased in size over the course of 48 h. This suggests that water molecules gradually pry apart the biopolymers that compose the ECM-based hydrogel even in the absence of relevant degradation enzymes. On the other hand, droplets protected by a layer of alginate showed no decrease in diameter over the course of the experiment. This demonstrates that the alginate shell acted as an effective two-way barrier, both preventing the ECM-based hydrogel's biopolymers from washing away, even if they became dislodged from the bulk, and preventing the collagenase enzymes from entering the droplet and cleaving the biopolymers, which would have resulted in smaller fragments of polymer that may be able to diffuse across the alginate barrier.

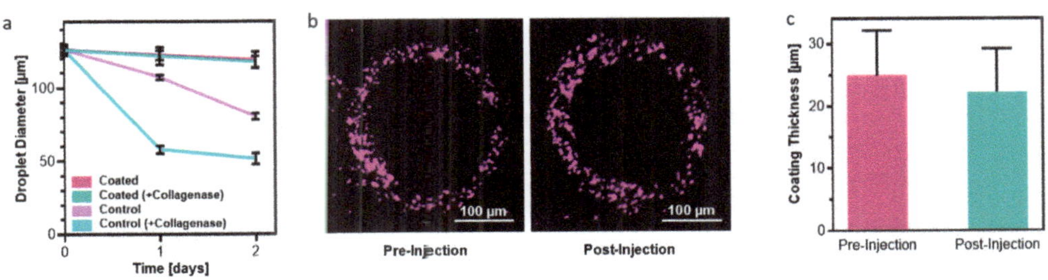

Figure 4. Integrity of alginate barrier. (**a**) The alginate barrier prevented water molecules and even collagenase enzymes from degrading the hydrogel, while unprotected droplets rapidly degraded. (**b**) Representative images showing the thickness and integrity of the alginate shell before (left) and after (right) being injected through a hypodermic needle. Scale bar = 100 µm. (**c**) Quantification of the barrier integrity showed no significant change as a result of injection.

Next, the barrier's integrity was assessed after being injected through a 25 G hypodermic needle. As explained previously, one of the advantages of generating microtissues as opposed to a single, bulk macrotissue is the inherent injectability of the droplets. Therefore, it was essential that the protective alginate barrier formed around the particles maintained its integrity during injection. To evaluate the coating integrity, microtissues were injected and the thickness of the shell was measured (Figure 4b,c). As expected, individual microgels were able to dissociate from each other and move through a needle with minimal shear stresses. As a result, the injection process did not impact the microdroplets or their coating.

Finally, as the motivation for this research was the formulation of protected microtissues to be used for regenerative medicine, microtissues were injected into the gastrocnemius of six-week-old, female C57/BL mice. The mice were sacrificed either immediately or five days post-injection, and the injection site was analyzed post mortem. In particular, immunofluorescent staining was performed to detect the presence of cell nuclei, F-actin (to stain native mouse tissue), collagen type I (the main component of the ECM hydrogel), and alginate. Additionally, the cells used in generating the microtissues were genetically modified to express red fluorescent protein (RFP) so that they could be readily viewed and distinguished from the native mouse cells. The success of the injection can be easily ascertained from the presence of cells and collagen on Day 0 (Figure 5a,b).

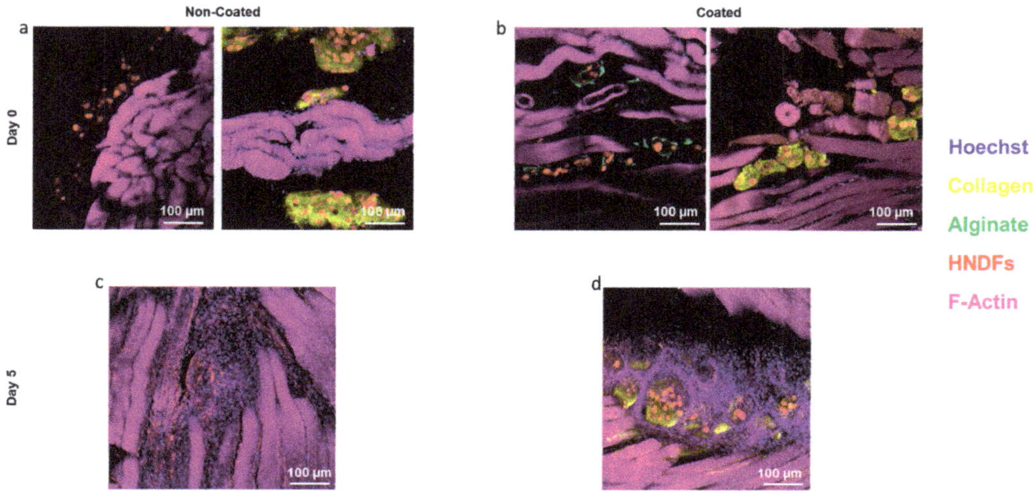

Figure 5. Injection of microtissues into in vivo murine model. (**a**) Immunofluorescent staining performed on Day 0 of the in vivo experiment confirms that the microtissues were successfully deployed within the murine muscle. (**b**) Immunofluorescent staining on Day 0 of the microtissues containing the protective alginate shell. (**c**) On Day 5, unprotected microtissues had largely been degraded and injected cells had escaped or been removed from the area. (**d**) Microtissues that contained a protective alginate coating maintained their integrity and, more importantly, maintained high cell concentrations at the site of injection. Scale bars = 100 µm.

Five days after the surgery, the foreign body response (FBR) was easily observed by the recruitment of large numbers of cells to the site of the injection. While the FBR is a natural and beneficial response to injury, one of the challenges of working with microtissues is that macrophages and other cells summoned to the area during the FBR rapidly degrade the tissue and displace the injected cells [34]. Indeed, as expected, by Day 5, none of the initial ECM-based hydrogel was visible in the unprotected droplets, and the concentration of the injected, RFP-expressing cells is noticeably decreased (Figure 5c). On the other hand, the alginate-coated microtissues were largely unaffected by the FBR (Figure 5d). The ECM-based hydrogel was still present, and the RFP-expressing cells were clearly seen within the droplets. This result strongly suggests that the alginate barrier functions as hypothesized to protect cells and enhance their retention at the site of injection, which should make the use of these microtissues more beneficial to patients.

In particular, the protected microtissues described here are envisioned as a regenerative treatment option for patients needing implanted-cell therapies that do not require extensive tissue integration. That is, for conditions such as Parkinson's disease (which may be treatable by implanting dopaminergic neurons) or diabetes (for which insulin-producing beta cells are required), the protection afforded by the alginate coating will provide a significant advantage over unprotected droplets.

3. Conclusions

We have developed a method for creating a protective barrier around microtissues based on the interaction between nanoparticles of calcium carbonate, encapsulated within the tissues, and alginate, which is externally added to them. Because the nanoparticles can be incorporated without reducing the volume available to cells and without changing the mechanical properties of the cells' microenvironment, the entire process was shown to be fully biocompatible and even demonstrated a potential benefit to the cell viability. Furthermore, the two-step process we developed naturally lends itself to customized processing, as each step along the way can be independently modified to gain full control over both the microtissues and their protective barrier. Additionally, we have demonstrated in both in vitro and in vivo models that the barrier prevented both hydrolytic and enzymatic degradation of the microtissue's ECM, which suggests that it could be utilized in tissue engineering applications that require an extended time for the cells to integrate into the host tissue. At the same time, we have demonstrated orthogonal control of the barrier's degradation, and future work may include applications in which alginate lyase is encapsulated within the microtissue during the microfluidic fabrication in order to create a controlled degradation gradient for the barrier. Because of their inherent injectability and modularity, and because of how effectively they promote oxygen transport, microtissues promise to provide novel solutions to the challenges of regenerative medicine, and our technology, by providing more versatility and control over the long-term integrity of the tissues, moves this important area of research one step closer to clinical applicability.

4. Materials and Methods

ECM-based hydrogel preparation: Omental tissues were decellularized according to previously published protocols [8]. Briefly, porcine tissues (Kibbutz Lahav, Israel) were washed with phosphate-buffered saline (PBS), then incubated in a hypotonic buffer consisting of 10 mM Tris @ pH = 8 (Sigma-Aldrich; Rehovot, Israel), 5 mM ethylenediamine tetra-acetic acid (EDTA) (Sigma-Aldrich), and 1 μM phenylmethanesulfonyl-fluoride for 1 h. The tissue was then frozen and thrice thawed in the hypotonic buffer and gradually washed with 70% (v/v) ethanol followed by 100% ethanol for a half-hour each. Lipids were extracted via three half-hour washes with 100% acetone, followed by a 24 h soak in a 60:40 volumetric ratio of a solution of hexane and acetone (exchanged thrice during the 24 h). The tissue was next washed in 100% ethanol for a half-hour, transferred to 70% ethanol, and left to soak. The next day, the tissues were washed with PBS four times then soaked overnight in a solution of 0.25% trypsin–EDTA solution (Biological Industries; Beit Haemek, Israel). The tissue was thoroughly washed with PBS and soaked in a solution of 1.5 M NaCl (exchanged three times) overnight, at which time it was washed in a solution of 50 mM Tris and 1% Triton-X100 (Sigma-Aldrich) for 1 h. The decellularized tissue was finally washed in PBS, then DDW and then frozen (−20 °C) and freeze-dried.

Following this, the dry, decellularized omental tissue was ground into a powder with a Wiley Mini-Mill (Thomas Scientific; Swedesboro, NJ, USA). The omentum powder was digested enzymatically for 96 h at RT while stirring in a solution of 1 mg/mL pepsin (Sigma-Aldrich, 4000 U/mg) in 0.1 M HCl. Once enzymatic digestion was completed, the pH was adjusted to 7.4 by titration with 5 M NaOH, and concentrated Dulbecco's modified Eagle medium (DMEM) was added until the concentration specified by the manufacturer was reached (Biological Industries) The concentration of decellularized omentum in the final solution was 1% (w/v).

Microfluidic device generation: Soft lithography techniques were used to fabricate the microfluidic devices. Negative photoresist SU-8 (3050, MicroChem Corp.; Newton, MA, USA) was first spin-coated onto a clean silicon wafer with a thickness of 300 μm (University Wafer; Boston, MA, USA) to either a thickness of 50 or 75 μm, and this was then patterned by UV exposure through a transparency photomask. Next a 10:1 ratio of PDMS prepolymer and curing agent (Sylgard 184, Dow Corning Corp.; Midland, MI, USA) was poured onto the master wafer. After one hour of curing at 65 °C, the replicas were extricated from the master wafer. Inlet holes and outlet holes were punctured via a 0.75 mm diameter biopsy punch (World Precision Instruments; Sarasota, FL, USA). Oxygen plasma (Diener Electronic GmbH & Co; KG; Ebhausen, Germany) was then used to permanently adhere the PDMS replicas to each other. Finally, the device channels were washed using Aquapel (PPG Industries; Pittsburgh, PA, USA) and immediately air-dried.

Droplet generation with $CaCO_3$ nanoparticles: A suspension of 10 mg/mL of 50 nm calcium carbonate nanoparticles ($^{np}CaCO_3$) (US Research Nanomaterials Inc.; Houston, TX, USA) was mixed with 1% (*w*/*v*) omentum hydrogel in a 1:5 ratio, yielding a final omentum concentration of 0.83% (*w*/*v*) and a final concentration of $^{np}CaCO_3$ of 0.16% (*w*/*v*). This mixture was then passed through a 70 μm nylon strainer (Corning; Airport City, Israel) and loaded into a plastic, 1 mL syringe. A mixture of 2% Pico-Surf 2 surfactant in Novec-7500 oil (Sphere Fluidics; Cambridge, UK) was also loaded into a plastic 1 mL syringe for use as the outer phase. Two fine-bore polythene tubes (Smiths Medical International Ltd.; Ashford, UK) with an outer diameter of 1.09 mm and an inner diameter of 0.38 mm were used to connect the syringes to the device inlets. Flow rates were controlled by an NE-1000 syringe pump (New Era Pump Systems; Farmingdale, NY, USA). Droplet generation was monitored using a digital microscope (Dino-lite Digital Microscope; New Taipei City, Taiwan). Flow rates of 100 μL/h (oil phase) and 50 μL/h (hydrogel phase) were used to obtain the desired droplet size of either 65–85 μm or 110–120 μm depending on the thickness of the device's channels. The system was operated at room temperature. Generated droplets were collected in an Eppendorf tube for 15 min and transferred immediately to 37 °C for 15 min to allow gelation.

After gelation, droplets were mixed with 50 μL of 20% (*v*/*v*) 1H,1H,2H,2H-perfluoro-1-octanol (Sigma-Aldrich, SKU 370533) in perfluoro-compound FC-40 (Sigma-Aldrich), which serves as an emulsion destabilizer. Finally, 600 μL of Dulbecco's modified Eagle medium (DMEM) was added to the tube to transfer the droplets to an aqueous phase.

Droplet generation with liposomes: A 1% (*w*/*v*) ECM-based omentum hydrogel was mixed in a ratio of 1:10 with FITC liposomes (provided by Dr. Dan Peer's lab, Tel Aviv University). The mixture was passed through a 70 μm nylon strainer and loaded into a plastic 1 mL syringe. The same process described above was then carried out on the hydrogel–liposome mixture.

Cell Culture: NIH/3T3 fibroblast cells (ATCC) were grown in DMEM supplemented with a 10% (*v*/*v*) fetal bovine serum (FBS) (Biological Industries), 1% (*v*/*v*) L-glutamine (L-Glu) (Biological Industries), and 1% (*v*/*v*) penicillin/streptomycin (Biological Industries). Upon reaching approximately 80% confluence, cells were passaged for expansion under sterile conditions and incubated at 37 °C in a humidified, 5% CO_2 incubator. Subculture was carried out by washing with PBS for 30 s and incubating in 0.05% trypsin/EDTA for 3 min.

Red fluorescent protein–human neonatal dermal fibroblasts (RFP-HNDFs) (Lonza; Haifa, Israel) were grown in DMEM supplemented with 10% (*v*/*v*) FBS, 1% (*v*/*v*) penicillin/streptomycin, 1% L-Glu, MEM non-essential amino acids solution (Gibco; Paisley Scotland), and 0.2% β-mercaptoethanol 50 mM (Gibco; Paisley, Scotland). At approximately 80% confluence, cells were passaged for expansion under sterile conditions and incubated at 37 °C in a humidified 5% CO_2 incubator.

Generation of microtissues with encapsulated cells: The 40 mg/mL $^{np}CaCO_3$ suspensions were mixed with 1% (*w*/*v*) ECM-based hydrogel in a 1:20 volumetric ratio, resulting in a final concentration of 0.95% (*w*/*v*) hydrogel and 0.19% (*w*/*v*) $^{np}CaCO_3$. The mixture was then passed through a 70 μm nylon strainer. Next, cells were suspended in seven times their volume of the filtered hydrogel mixture, leading to final concentrations of: 0.83% (*w*/*v*) hydrogel, 0.16% (*w*/*v*) $^{np}CaCO_3$, and 20×10^6 cells/mL (in the case of NIH/3T3 cells) or 25×10^6 cells/mL (RFP-HNDFs). This mixture was loaded into a 1 mL syringe, and droplets with $^{np}CaCO_3$ and cells were created by following the procedure described above. Prior to mixing with the hydrogel, cell viability was assessed by Trypan Blue (Biological

Industries) exclusion, and the cells were counted and centrifuged at 600 RPM for 5 min to concentrate them to the desired volume.

Creation of alginate capsule: Alginate powder (LFR 5/60, LF 10/60 FT, or LF 200 FTS) (FMC BioPolymer; Philadelphia, PA, USA) was dissolved in MES buffer (pH = 6) to reach the desired concentration. The solution was filtered through 0.22 µm filters (Millex Syringe-driven Filter Units, Merck; Rehovot, Israel) and stored at 4 °C until use. Microgels were generated as described and collected in a 15 mL Falcon tube. Following collection, the droplets were concentrated by centrifugation at 1000 RPM for 10 min, and the concentrated droplets were gently suspended and added dropwise to a 3 mL alginate solution in a 6-well tissue culture plate on ice. Immediately following the addition of the droplets to the alginate solution, the plate was gently shaken for 30 min and was then washed twice with DMEM and centrifuged at 1000 RPM for 10 min.

Synthesis of alginate with CY5 fluorescent marker: A solution of 1% LFR 5/60 was prepared in 0.1 M MES buffer and added to a round bottom flask. A 5 M excess of 10 mg/mL ethylene dicarbodiimide (EDC, ThermoFisher, Waltham, MA, USA) solution was added dropwise and allowed to react for 1 hr. The pH was then raised to 8–8.5 by titration with triethylamine, and a 5 M excess of 10 mg/mL N-hydrosuccinimide (NHS) solution was added alongside an equivalent excess of a primer amine-CY5 fluorescent molecule that was a kind gift from Prof. Roey Amir. This reaction was allowed to proceed for two hours. The reaction mixture was dialyzed (SnakeSkin™ dialysis tubing, 10 kDa MWCO, 16 m, ThermoFisher Scientific) for 4 days against a 0.1 M NaCl solution which was refreshed twice daily. After dialysis, the solution was frozen in liquid nitrogen and lyophilized.

Preparation of alginate with fluorescent microparticles: First, 20% bovine serum albumin (BSA) in DDW was mixed with 1 µm polystyrene dark blue particles (Sigma-Aldrich) in a 1:10 volumetric ratio and left at room temperature for 30 min. The particles were then mixed at a 1:10 volumetric ratio with the alginate solution, and this mixture was used for coating the ECM-based hydrogel. The thickness of the coating was measured via an upright confocal microscope (Nikon ECLIPSE NI-E; Melville, NY, USA). For each droplet that was examined, the central plane of the droplet was determined by finding the midpoint between the top and bottom of the droplet in the Z-axis. The coating thickness was recorded in at least five different locations along the central plane, and these values were averaged to generate the estimated coating thickness for each droplet. The average thickness for a given condition is the average of at least 17 droplets.

Rheology: Calcium carbonate nanoparticles were suspended at concentrations of 0.5, 1, 2, 5, 10, or 40 mg/mL, and the suspension was mixed with 1% (*w*/*v*) hydrogel in a volumetric ratio of 1:5. Then, 80 µL of the mixture was loaded into a PDMS mold (that had been previously punched with an 8 mm puncher) and incubated at 37 °C in a humidified, 5% CO_2 incubator for 1 h to allow the hydrogel to set. Next, gels were removed from the mold, exposed to a 1% alginate solution for 30 min, and washed with 0.1 M HEPES. Rheological measurements were performed using a Discovery HR-3 Hybrid Rheometer (TA Instruments; New Castle DE, USA) with 8 mm diameter parallel-plate geometry. The samples were loaded at a temperature of 37 °C, and their viscoelastic properties were measured by performing a frequency sweep between 0.01 and 10 rad/s at a constant 1% strain. At least three gels were assessed and averaged for each condition.

Live/Dead Assay: Cell viability within coated or non-coated droplets was determined by incubating cells in fluorescein diacetate (Sigma-Aldrich, 7 µg/mL) and propidium iodide (Sigma-Aldrich, 5 µg/mL) for fifteen minutes in a humidified incubator with 5% CO_2, at 37 °C. Three samples of droplets were observed with an inverted fluorescence microscope (Nikon Eclipse TI). The numbers of live and dead cells were determined by manual counting using the NIS Elements software (Version 4.13) (Nikon; Melville, NY, USA) from at least three different microscopic field images in each sample. For each time point, cells were counted in 40 droplets.

Proliferation Assay: Encapsulated cells in either coated or non-coated droplets were placed in a 24-well tissue culture plate that was pre-coated with polyHEMA (Sigma;

Rehovot, Israel). For each time point, three measurements were recorded per group. PrestoBlue™ reagent (Thermo Fisher) was added to each well at a 1:9 ratio with cell medium and incubated for 5 h. The absorbance was measured at 570 nm (600 nm serving as the reference wavelength) using a Tecan Plate Reader InfiniteM200pro. All values were normalized to Day 1.

Cell Escape: Five Eppendorf tubes of encapsulated cells were collected and merged into a pool of cellular droplets in a 15 mL Falcon tube. Later, the suspension of droplets was split into three groups: (1) coated droplets, (2) coated droplets with the addition of 1 U/mL of alginate lyase (Sigma-Aldrich), and (3) non-coated droplets. Groups 1 and 2 underwent the aforementioned coating process. All droplets then underwent three rounds of pre-plating for 30 min to remove non-encapsulated cells. Next, 10 μL from each group was placed in 6 different wells of a 24-well tissue culture plate, and 1 mL of medium was added to each well. The migration distance of cells outside the droplets was quantified based on brightfield (BF) images taken on Days 2, 3, and 5 by inverted fluorescence microscopy. For each image, the average migration distance outside the droplet was quantified by using the NIS Elements software (Nikon). Results were averaged over 50 droplets per group.

Degradation assay: NIH/3T3 fibroblasts cells were encapsulated to form microtissues. All droplets were pooled in a 15 mL Falcon tube. Half of the suspension was then coated. Both groups were then separately concentrated to a volume of 100 μL, resuspended gently, and evenly split into 4 wells (25 μL per well) of a 24-well tissue culture plate that had been previously coated with polyHEMA, and cell medium was added to reach a final volume of 500 μL per well. Next, 1 μL of 500 U/mL collagenase type II powder (Worthington; Lakewood, NJ, USA) solution was added to each well, achieving a final concentration of 1 U/mL. Results were compared against a control group without collagenase. Degradation was assessed at Days 0, 1, 2, and 3 by measuring the diameter of the droplets using the NIS Elements software (Nikon). The results presented represent the average of 50 droplets per condition.

In vivo testing: All mice were treated according to the ethical regulations of Tel Aviv University. Permission was granted by the ethics committee, protocol number 04-19-028. Six-week-old C57/BL female mice, purchased from ENVIGO (Jerusalem, Israel), were randomly divided into groups. Two mice were allotted to each group other than Day 0 for which only one mouse was used. The animals were anaesthetized subcutaneously with a ketamine (100 mg/kg)–xylazine (10 mg/kg) cocktail in saline solution. The gastrocnemius muscle of the mice was shaved in the area of the injection and either coated or non-coated droplets were injected.

Sixteen Eppendorf tubes of microtissues with encapsulated RFP-HNDF were generated and collected. The collected microtissues were pooled in a 15 mL Falcon tube and subsequently split into a coated group and a non-coated group. The coated droplets were coated with 1% alginate LF 200 FTS at pH = 6 for 30 min. The droplets (for each condition) were concentrated by centrifugation (300× g for 5 min) and resuspended in 600 μL of medium. The 600 μL suspension was divided equally into six Eppendorf tubes. Then, each tube's 100 μL suspension was loaded into a 1 mL syringe with a 25 G needle and 60 μL was injected intramuscularly into the gastrocnemius muscle of each mouse.

In vivo tissue fixation and staining: Mice were sacrificed after one hour, two days, or five days. The gastrocnemius muscles were extracted, fixed in 4% formaldehyde, and embedded in optimal cutting temperature (OCT) compound (Tissue-Tek® OCT Compound Perrigo). Using a Cryotome™ FSE (Thermo Scientific), 60 μm thick sections were prepared and affixed to X-tra® adhesive glass slides (Leica Biosystems; Wetzler, Germany). The fixed samples were washed twice in PBS to extract the OCT compound and permeabilized with 0.05% (v/v) Triton X-100 (Sigma) in a blocking solution of PBS containing 1% BSA and 10% FBS for 1 h. The samples were incubated for 90 min with primary antibody Mouse αAlginate (1:200, Sigma 5200132), washed three times, and incubated for another 90 min with FITC-Goat-αMouse (1:250, Abcam; Cambridge, UK, ab150113), Phalloidin-iFluor 647 (1:1000, Abcam, ab176759) for immunostaining of actin filaments, and Hoechst 33258 (1:20;

Sigma-Aldrich) for nuclei detection. The samples were washed three times, mounted, and visualized by upright confocal microscopy.

Statistical analysis: Statistical analysis was performed using the GraphPad Prism software (Version 8.4.2). All values are given as mean ± SEM. For experiments with only two groups, a *t*-test was performed. We applied one-tail standards for experiments in which we hypothesized that the alginate-coated droplets would tend to a particular outcome. In experiments without an obvious outcome, two-tail standards were applied. For experiments with three groups or more, one- or two-way ANOVA was used as appropriate. For multiple comparison ANOVA analyses, the Tukey post hoc test was applied unless otherwise stated. Statistical significance was set at $p < 0.05$.

Author Contributions: S.E. and E.S. contributed equally to this paper. Conceptualization, T.D. and S.E.; methodology, S.E. and A.S.; software, S.E.; validation, S.E; formal analysis, S.E.; investigation, S.E.; resources, T.D.; data curation, S.E.; writing—original draft preparation, E.S.; writing—review and editing, E.S., S.E. and T.D.; visualization, S.E. and T.D.; supervision, T.D.; project administration, T.D. and A.S.; funding acquisition, T.D. All authors have read and agreed to the published version of the manuscript.

Funding: This research was funded by ERC Consolidator Grant (101001242), Israel Science Foundation (972/21), and NSF-BSF (2020733).

Institutional Review Board Statement: The animal study protocol was approved by the Institutional Review Board (or Ethics Committee) of Tel Aviv University (protocol code 04-19-028, approval date 1 May 2019).

Data Availability Statement: The data presented in this study are available on request from the corresponding author.

Conflicts of Interest: The authors declare no conflicts of interest.

References

1. Yadid, M.; Oved, H.; Silberman, E.; Dvir, T. Bioengineering Approaches to Treat the Failing Heart: From Cell Biology to 3D Printing. *Nat. Rev. Cardiol.* **2022**, *19*, 83–99. [CrossRef] [PubMed]
2. Place, T.L.; Domann, F.E.; Case, A.J. Limitations of Oxygen Delivery to Cells in Culture: An Underappreciated Problem in Basic and Translational Research. *Free Radic. Biol. Med.* **2017**, *113*, 311–322. [CrossRef] [PubMed]
3. Orellano, I.; Thomas, A.; Herrera, A.; Brauer, E.; Wulsten, D.; Petersen, A.; Kloke, L.; Duda, G.N. Engineering Vascular Self-Assembly by Controlled 3D-Printed Cell Placement. *Adv. Funct. Mater.* **2022**, *32*, 2208325. [CrossRef]
4. Szklanny, A.A.; Machour, M.; Redenski, I.; Chochola, V.; Goldfracht, I.; Kaplan, B.; Epshtein, M.; Simaan Yameen, H.; Merdler, U.; Feinberg, A.; et al. 3D Bioprinting of Engineered Tissue Flaps with Hierarchical Vessel Networks (VesselNet) for Direct Host-To-Implant Perfusion. *Adv. Mater.* **2021**, *33*, 2102661. [CrossRef]
5. Silberman, E.; Oved, H.; Namestnikov, M.; Shapira, A.; Dvir, T. Post-Maturation Reinforcement of 3D-Printed Vascularized Cardiac Tissues. *Adv. Mater.* **2023**, *35*, 2302229. [CrossRef]
6. Chávez, M.N.; Schenck, T.L.; Hopfner, U.; Centeno-Cerdas, C.; Somlai-Schweiger, I.; Schwarz, C.; Machens, H.G.; Heikenwalder, M.; Bono, M.R.; Allende, M.L.; et al. Towards Autotrophic Tissue Engineering: Photosynthetic Gene Therapy for Regeneration. *Biomaterials* **2016**, *75*, 25–36. [CrossRef]
7. Abdullah, T.; Gauthaman, K.; Hammad, A.H.; Navare, K.J.; Alshahrie, A.A.; Bencherif, S.A.; Tamayol, A.; Memic, A. Oxygen-Releasing Antibacterial Nanofibrous Scaffolds for Tissue Engineering Applications. *Polymers* **2020**, *12*, 1233. [CrossRef]
8. Gal, I.; Edri, R.; Noor, N.; Rotenberg, M.; Namestnikov, M.; Cabilly, I.; Shapira, A.; Dvir, T. Injectable Cardiac Cell Microdroplets for Tissue Regeneration. *Small* **2020**, *16*, 1904806. [CrossRef] [PubMed]
9. Velasco, D.; Tumarkin, E.; Kumacheva, E. Microfluidic Encapsulation of Cells in Polymer Microgels. *Small* **2012**, *8*, 1633–1642. [CrossRef]
10. Xin, S.; Deo, K.A.; Dai, J.; Pandian, N.K.R.; Chimene, D.; Moebius, R.M.; Jain, A.; Han, A.; Gaharwar, A.K.; Alge, D.L. Generalizing Hydrogel Microparticles into a New Class of Bioinks for Extrusion Bioprinting. *Sci. Adv.* **2021**, *7*, 3087–3102. [CrossRef]
11. Daly, A.C.; Riley, L.; Segura, T.; Burdick, J.A. Hydrogel Microparticles for Biomedical Applications. *Nat. Rev. Mater.* **2020**, *5*, 20–43. [CrossRef] [PubMed]
12. Drury, J.L.; Mooney, D.J. Hydrogels for Tissue Engineering: Scaffold Design Variables and Applications. *Biomaterials* **2003**, *24*, 4337–4351. [CrossRef] [PubMed]
13. Zhang, P.; Abate, A.R. High-Definition Single-Cell Printing: Cell-by-Cell Fabrication of Biological Structures. *Adv. Mater.* **2020**, *32*, 2005346. [CrossRef]

14. Mahmoudi, Z.; Sedighi, M.; Jafari, A.; Naghieh, S.; Stefanek, E.; Akbari, M.; Savoji, H. In Situ 3D Bioprinting: A Promising Technique in Advanced Biofabrication Strategies. *Bioprinting* **2023**, *31*, e00260. [CrossRef]
15. Hirsch, M.; Charlet, A.; Amstad, E. 3D Printing of Strong and Tough Double Network Granular Hydrogels. *Adv. Funct. Mater.* **2021**, *31*, 2005929. [CrossRef]
16. Jiang, W.; Li, M.; Chen, Z.; Leong, K.W. Cell-Laden Microfluidic Microgels for Tissue Regeneration. *Lab Chip* **2016**, *16*, 4482–4506. [CrossRef] [PubMed]
17. Hidalgo San Jose, L.; Stephens, P.; Song, B.; Barrow, D. Microfluidic Encapsulation Supports Stem Cell Viability, Proliferation, and Neuronal Differentiation. *Tissue Eng. Part C Methods* **2018**, *24*, 158–170. [CrossRef] [PubMed]
18. Kamperman, T.; Henke, S.; Visser, C.W.; Karperien, M.; Leijten, J. Centering Single Cells in Microgels via Delayed Crosslinking Supports Long-Term 3D Culture by Preventing Cell Escape. *Small* **2017**, *13*, 1603711. [CrossRef] [PubMed]
19. Lowen, J.M.; Bond, G.C.; Griffin, K.H.; Shimamoto, N.K.; Thai, V.L.; Leach, J.K. Multisized Photoannealable Microgels Regulate Cell Spreading, Aggregation, and Macrophage Phenotype through Microporous Void Space. *Adv. Healthc. Mater.* **2023**, *12*, 2202239. [CrossRef] [PubMed]
20. Bhujbal, S.V.; De Haan, B.; Niclou, S.P.; De Vos, P. A Novel Multilayer Immunoisolating Encapsulation System Overcoming Protrusion of Cells. *Sci. Rep.* **2014**, *4*, 6856. [CrossRef]
21. Noor, N.; Shapira, A.; Edri, R.; Gal, I.; Wertheim, L.; Dvir, T. 3D Printing of Personalized Thick and Perfusable Cardiac Patches and Hearts. *Adv. Sci.* **2019**, *6*, 1900344. [CrossRef] [PubMed]
22. Lee, K.Y.; Mooney, D.J. Alginate: Properties and Biomedical Applications. *Prog. Polym. Sci.* **2012**, *37*, 106–126. [CrossRef] [PubMed]
23. Shapira, A.; Noor, N.; Oved, H.; Dvir, T. Transparent Support Media for High Resolution 3D Printing of Volumetric Cell-Containing ECM Structures. *Biomed. Mater.* **2020**, *15*, 045018. [CrossRef] [PubMed]
24. Chan, L.W.; Lee, H.Y.; Heng, P.W.S. Mechanisms of External and Internal Gelation and Their Impact on the Functions of Alginate as a Coat and Delivery System. *Carbohydr. Polym.* **2006**, *63*, 176–187. [CrossRef]
25. Paques, J.P.; Sagis, L.M.C.; van Rijn, C.J.M.; van der Linden, E. Nanospheres of Alginate Prepared through w/o Emulsification and Internal Gelation with Nanoparticles of CaCO3. *Food Hydrocoll.* **2014**, *40*, 182–188. [CrossRef]
26. Zhao, P.; Tian, Y.; You, J.; Hu, X.; Liu, Y. Recent Advances of Calcium Carbonate Nanoparticles for Biomedical Applications. *Bioengineering* **2022**, *9*, 691. [CrossRef] [PubMed]
27. Barczyk, M.; Carracedo, S.; Gullberg, D. Integrins. *Cell Tissue Res.* **2010**, *339*, 269–280. [CrossRef] [PubMed]
28. Wells, R.G. The Role of Matrix Stiffness in Regulating Cell Behavior. *Hepatology* **2008**, *47*, 1394–1400. [CrossRef]
29. Lv, H.; Li, L.; Sun, M.; Zhang, Y.; Chen, L.; Rong, Y.; Li, Y. Mechanism of Regulation of Stem Cell Differentiation by Matrix Stiffness. *Stem Cell Res. Ther.* **2015**, *6*, 103. [CrossRef]
30. Mih, J.D.; Marinkovic, A.; Liu, F.; Sharif, A.S.; Tschumperlin, D.J. Matrix Stiffness Reverses the Effect of Actomyosin Tension on Cell Proliferation. *J. Cell Sci.* **2012**, *125*, 5974–5983. [CrossRef]
31. Ehrbar, M.; Sala, A.; Lienemann, P.; Ranga, A.; Mosiewicz, K.; Bittermann, A.; Rizzi, S.C.; Weber, F.E.; Lutolf, M.P. Elucidating the Role of Matrix Stiffness in 3D Cell Migration and Remodeling. *Biophys. J.* **2011**, *100*, 284–293. [CrossRef] [PubMed]
32. Han, W.T.; Jang, T.; Chen, S.; Chong, L.S.H.; Jung, H.D.; Song, J. Improved Cell Viability for Large-Scale Biofabrication with Photo-Crosslinkable Hydrogel Systems through a Dual-Photoinitiator Approach. *Biomater. Sci.* **2020**, *8*, 450–461. [CrossRef] [PubMed]
33. Zhu, B.; Yin, H. Alginate Lyase: Review of Major Sources and Classification, Properties, Structure-Function Analysis and Applications. *Bioengineered* **2015**, *6*, 125–131. [CrossRef] [PubMed]
34. Morais, J.M.; Papadimitrakopoulos, F.; Burgess, D.J. Biomaterials/Tissue Interactions: Possible Solutions to Overcome Foreign Body Response. *AAPS J.* **2010**, *12*, 188–196. [CrossRef]

gels

MDPI

Article

Fast Wound Healing with a New Functional Hyaluronic Acid Dual Network Hydrogel

Lichun Wu [1], Yu Zhou [1], Yi Zhang [2], Jia Hu [1], Yasuhiro Ikegami [1], Shinichi Aishima [3] and Hiroyuki Ijima [1,*]

1 Department of Chemical Engineering, Faculty of Engineering, Graduate School of Engineering, Kyushu University, 744 Motooka, Nishi-ku, Fukuoka 819-0395, Japan; wu.lichun.820@s.kyushu-u.ac.jp (L.W.); zhou.yu.222@s.kyushu-u.ac.jp (Y.Z.); hu.jia.511@s.kyushu-u.ac.jp (J.H.); yikegami@chem-eng.kyushu-u.ac.jp (Y.I.)
2 Institute for Materials Chemistry and Engineering, Kyushu University, 744 Motooka, Nishi-ku, Fukuoka 819-0395, Japan; zhang_yi@ms.ifoc.kyushu-u.ac.jp
3 Department of Scientific Pathology, Graduate School of Medical Sciences, Kyushu University, 3-1-1 Maidashi, Higashi-ku, Fukuoka 812-8582, Japan; aishima.shinichi.476@m.kyushu-u.ac.jp
* Correspondence: ijima@chem-eng.kyushu-u.ac.jp; Tel./Fax: +81-92-802-2748

Abstract: As dressings for moist wound healing, hyaluronic acid hydrogels play a significant role in maintaining moisture and promoting wound healing. However, existing hydrogel dressings are inadequate in terms of slow gelation time, weak mechanical performance, and fast degradation, which increases the risk of secondary infections during treatment. Therefore, we developed a hyaluronic acid double network hydrogel (DNH). Compared to single-network hydrogels (hydrazone and Diels–Alder), DNH shows a short gelation time (25 s) and strong mechanical properties (Young's modulus = 82 kPa). These advantages enable DNH to immediately fill the irregular shape of the wound after gelation and remain intact after being squeezed. Swelling tests indicated that DNH had a suitable swelling ratio and maintained its structural integrity after swelling. We evaluated the use of DNH as a moist dressing for full-thickness wound healing in vivo. DNH-treated wounds healed faster, with enhanced blood vessel formation and macrophage polarization than gauze-treated wounds. These findings suggest that DNH not only accelerates wound healing but also improves tissue regeneration. Therefore, DNH may be a suitable moist dressing for wound healing.

Keywords: hyaluronic acid; double network hydrogels; wound healing

1. Introduction

Skin is the first line of defense in animals and protects the body from environmental damage. However, the skin is vulnerable to being damaged by injury, surgery, and disease [1]. Appropriate wound dressings are essential for both surgical wounds and daily life. Conventional gauze bandages are used for wound treatment because of their ability to absorb exudates. However, it often causes immense pain owing to the adhesion between the wound scabs and the gauze network, causing inflammation, which delays wound healing [1]. With growing concerns about wound care, it is essential to develop effective and comfortable dressings for wound healing. Hydrogel wound dressings are promising materials in wound care to keep wound moisture, absorb excess exudate, transfer oxygen, provide protection, and promote skin regeneration [2–4].

Natural polysaccharides (hyaluronic acid (HA) [5,6], chitosan [4,7], and alginate [8]) and proteins (e.g., gelatin [2,7], silk fibroin [6], and albumin [9]) have been widely reported as wound dressing materials owing to their biocompatibility, biodegradability, and physical tunability. HA was used to prepare the hydrogels because of its unique chemical structure and biofunction in wound healing. HA is a linear glycosaminoglycan with a large number of modified groups, such as carboxyl and hydroxyl groups, which facilitate the development of functional hydrogels by HA chemical modification. During wound healing, HA, an important component of the extracellular matrix (ECM), can modulate inflammation [10], promote angiogenesis, and accelerate collagen deposition [5,11]. In addition, HA hydrogels show excellent properties in absorbing excess wound exudate and keeping the wound moist because of their high hydrophilicity [12]. However, most HA hydrogels exhibit fast degradation and poor mechanical properties [13]. To solve this problem, researchers have developed functional HA hydrogels with the required mechanical properties and good healing efficiency [14,15].

The double network hydrogel (DNH) is an inter-crosslinking network consisting of two interpenetrated networks [16]. Compared to single-network HA hydrogels, two-covalent networks have proven to be promising materials with tunable mechanical properties [17,18]. HA derivatives provide additional options for the preparation of hydrogel [5,11]. Recently, various DNH based on HA derivatives have been used to develop hydrogels with strong mechanical properties, such as a combination of hydrazone crosslinking and Diels–Alder crosslinking (methyl-gifted HA) [17], hydrazone crosslinking (aldehyde-added HA), and photo-crosslinking [18]. Efficient exudate absorption and structural stability are critical for effective wound healing.

Based on the above considerations, we fabricated DNH composed of HA derivatives. In this double-crosslinking system, the hydrazone hydrogel (HYH) was crosslinked via hydrazone bonds using oxidized HA and adipic dihydrazide (ADH), which provided fast gelation to DNH [11,19,20]. HA with a furan group was crosslinked with maleimide groups to form a Diels–Alder hydrogel (DAH) to enhance its mechanical properties. To verify this, the structural, mechanical, and swelling properties of three types of hydrogels (HYH, DAH, and DNH) were investigated. Finally, DNH was applied to a wound to assess its ability to promote full-thickness wound healing based on the wound closure rate and histological observations. We expected DNH to maintain wound moisture, absorb exudates, and accelerate wound healing.

2. Results and Discussion

2.1. Synthesis and Characterization of DNH

To design a DNH with suitable gelation time and sufficient mechanical properties, different ratios of raw materials were investigated, as shown in Table S1. Based on the gelation time of the single-network hydrogel, ADH was added sooner or later after the other components to uniformly form DNH.

The formation of a hydrazone bond was confirmed by an FTIR spectrum (peaks at -1549 cm^{-1} and 1633 cm^{-1}), and the newly formed aldehyde was confirmed by ^{1}H-NMR (peaks at 5.01 ppm, 5.11 ppm, and 5.21 ppm from OAH, as in Figure 1A,B, confirming the successful formation of HYH. The optimized formulation, HYH4 (with a volume ratio of OHA to ADH of 4:1), was obtained using an ADH concentration of 8 mg/mL owing to the shortest gelation time (10.00 $\pm$ 1.73 s) and the highest Young's modulus (28.87 $\pm$ 1.92 kPa), as shown in Table S2 and Figure 1C. HYH4 exhibited a transparent appearance, which is beneficial for wound observation during the healing process. Finally, HYH4 was used to prepare the DNH.

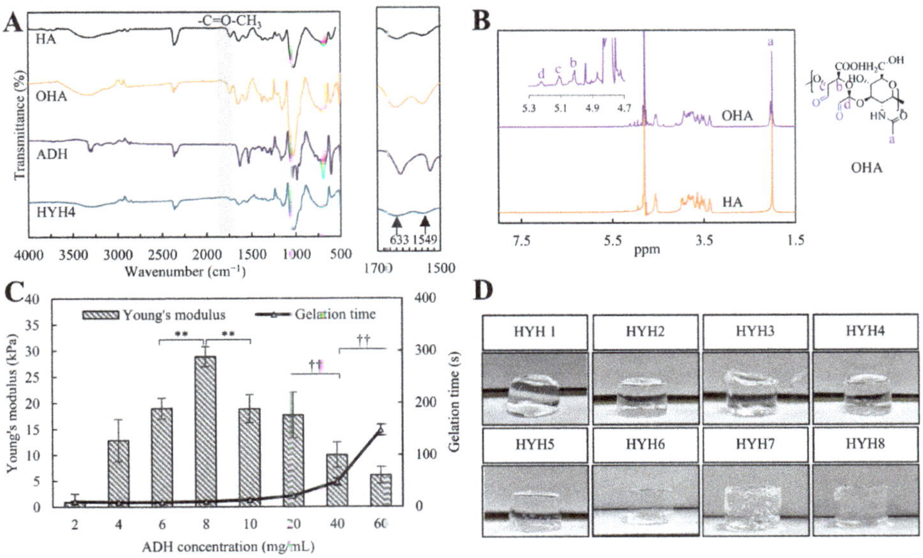

Figure 1. Properties of HYH. (**A**) ATR-FTIR spectra of HA, OHA, ADH, and HYH. (**B**) ^{1}H-NMR of HA and OHA. (**C**) Young's modulus and gelation time of HYH ($n = 3$, the value indicates mean ± S.D.). (**D**) Appearance of HYH with increasing concentration of ADH. Significance was assessed using one-way ANOVA with Tukey's multiple comparisons test, giving p values, ** $p < 0.01$ of Young's modulus, †† $p < 0.01$ of the gelation time.

The formation of DAH was confirmed by FTIR and ^{1}H-NMR spectroscopies, as shown in Figure 2A,B. In the FTIR spectrum, the disappearance of the C=C signals indicated the consumption of furan and maleimide groups. A new peak emerges at 1448 cm^{-1}, which corresponds to the C=C bond in the Diels–Alder adduct. The peaks at 6.43, 6.56, and 7.57 ppm in the ^{1}H-NMR spectrum correspond to the protons of the furan groups, confirming the successful grafting of furan onto HA. The substitution degree of HA-Furan was determined to be 69.5% ± 1.5%. The volume ratios of HA-Furan and Mal-PEG-Mal significantly affected compression properties. When DAH3 was obtained with a volume ratio of HA-Furan to Mal-PEG-Mal of 0.7, as shown in Figure 2C,D, the maximum yield strength was 174.23 kPa under 59% compression strain, and Young's modulus also reached the highest value (92.62 kPa). Moreover, the gelation time of DAH also decreased to 52 ± 8 min under the volume ratio of HA-Furan to Mal-PEG-Mal of 0.7 from 78 ± 12 min under the volume ratio of HA-Furan to Mal-PEG-Mal of 0.3. Young's modulus decreased, whereas the gelation time increased significantly as the volume ratio increased from 0.7 to 11. Compared with maleimide groups, excess furan groups are safe for biological application [16]. Therefore, DAH3, DAH4, DAH5, DAH6, and DAH7 were selected for further investigation of DNH. All DAH samples were transparent, which is advantageous for observing the wound healing process, as shown in Figure 2E.

Following the optimization of HYH and DAH gelation, the results for DNH prepared with different HA-Furan and Mal-PEG-Mal ratios are shown in Figure 3. The results showed that the mechanical properties of DNH could be adjusted by changing the volume ratio of HA-Furan and Mal-PEG-Mal. DNH1 had the highest ultimate yield strength (371.25 kPa) under 71% compressive strain (Figure 3A). This is higher than that when acrylamide/N,N′-methylene bisacrylamide (242.4 ± 2.43 kPa) is used as wound healing hydrogel [21]. This indicates that DNH1 possesses sufficient mechanical properties for wound healing. The Young's modulus of DNH1 (82.33 ± 26.67 kPa, Figure 3B) was between the HYH4 (28.87 ± 1.92) and DAH3 (92.62 ± 3.88 kPa). The DNH gelation time (25 s) did not change significantly with DAH variation. Although the time was longer than that of HYH4 (10.00 ± 1.73 s), it was still acceptable for the in situ gelation for wound healing and less than that reported by Fang (30 s) [22]. DNH was transparent, as shown in Figure 3C, which facilitated wound observation during healing. Based on its gelation properties, DNH1 was selected for further investigation of wound healing.

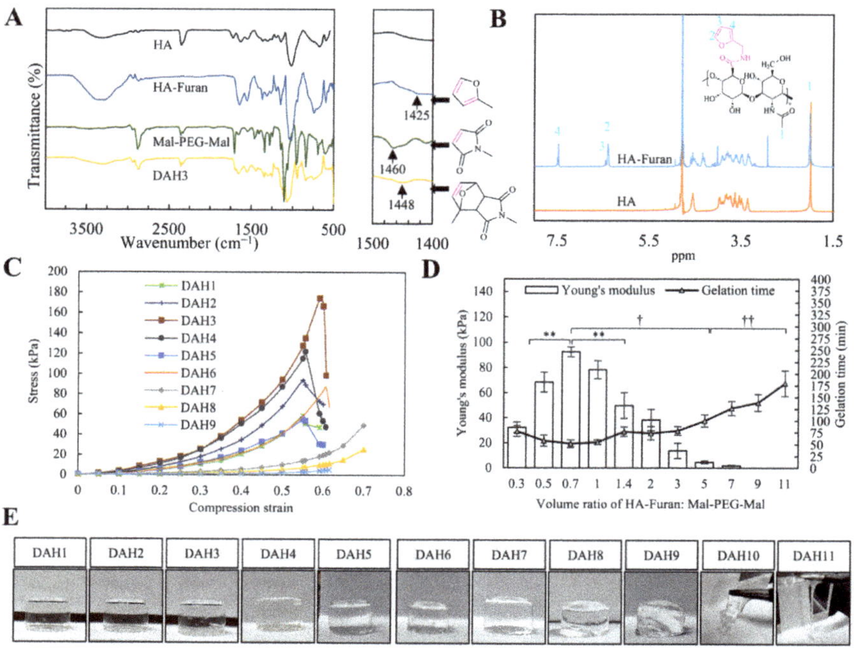

Figure 2. Properties of DAH. (**A**) ATR-FTIR spectra of HA, HA-Furan, Mal-PEG-Mal, and DAH3. (**B**) ^{1}H-NMR of HA and HA-Furan. (**C**) Compressive stress–strain curves for DAH. (**D**) Young's modulus and gelation time of DAH (n = 3, the value indicates mean ± S.D.). (**E**) Appearance of DAH with different volume ratios between HA-Furan and Mal-PEG-Mal. Significance was assessed using one-way ANOVA with Tukey's multiple comparisons test, giving p values, ** $p < 0.01$ of Young's modulus, † $p < 0.05$ of the gelation time, †† $p < 0.01$ of the gelation time.

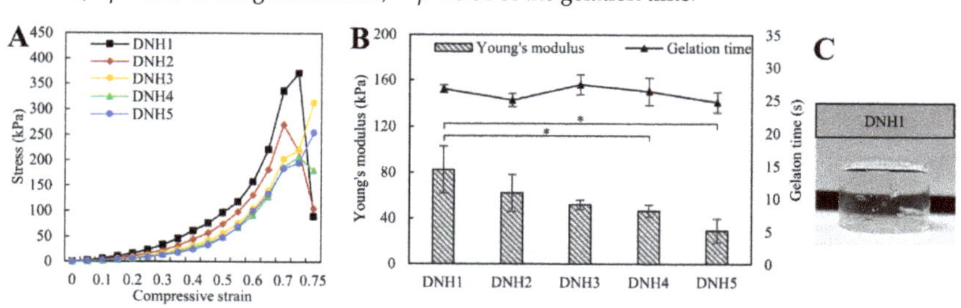

Figure 3. Properties of DNH. (**A**) Compressive stress–strain curves. (**B**) Young's modulus and gelation time (n = 3, the value indicates mean ± S.D.). (**C**) Appearance of DNH1. Significance was assessed using one-way ANOVA with Tukey's multiple comparisons test, giving p values, * $p < 0.05$ of the Young's modulus.

2.2. Swelling Rate, Resilience, and Morphology of Hydrogel

Healthy skin has a slightly acidic pH, whereas wounds are mainly alkaline [23,24]. Thus, hydrogel swelling properties were investigated at different pH values. As shown in Figure 4,

HYH4, DAH3, and DNH1 had different swelling properties. HYH4 absorbed phosphate-buffered saline (PBS) at a maximum of 5.42 times its weight at pH 11. However, it was completely degraded after 6 h at pH 3.0 and 5.0. The swelling rate of HYH under acidic conditions was higher than that under alkaline conditions, which resulted from the hydrolysis of hydrazone [25]. In addition, the degradation rate was higher than the swelling rate, leading to faster degradation [26]. DAH3 had a low swelling ratio under acidic conditions and fast swelling under alkaline conditions, which can be attributed to the opening of the furan rings under alkaline conditions [27]. When DNH1 was tested, the hydrogel maintained a swelling ratio close to 2 from pH 5.0 to 11.0 within the first 6 h, which demonstrated that DNH absorbed an equivalent volume solution over a wide pH range (pH 5.0–11.0). This range fully covers the skin pH range (pH 5.0–9.0). Therefore, DNH1 easily satisfies the requirements for absorbing exudates while maintaining its structural integrity. In summary, compared with HYH and DAH, DNH1 exhibited optimal swelling properties, ensuring efficient absorption of the wound exudate while maintaining its original shape after swelling.

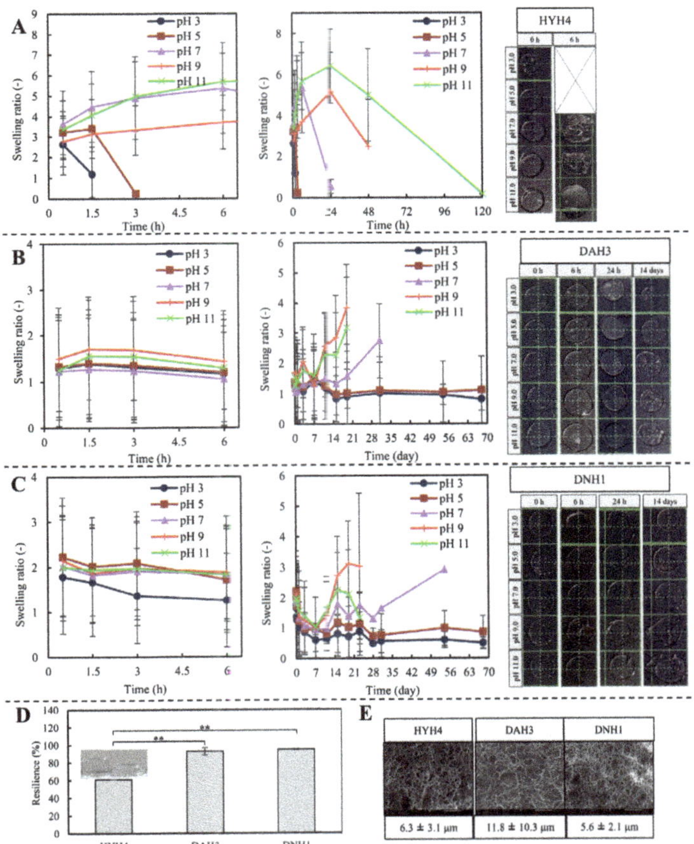

Figure 4. Swelling rate and related appearance of the swollen hydrogel at different pH values ($n = 3$, the value indicates mean ± S.D.). (**A**) HYH4. (**B**) DAH3. (**C**) DNH1. (**D**) Resilience of the hydrogel ($n = 3$, the value indicates mean ± S.D.). (**E**) SEM images of cross-section hydrogel and the corresponding mean diameter ($n = 10$, the value indicates mean ± SD.). Significance was assessed using one-way ANOVA with Tukey's multiple comparisons test, giving p values, ** $p < 0.01$.

The elasticity of the hydrogel can protect the wound from rupture owing to compression or movement of the body. To further investigate the elasticity of the hydrogels, they were compressed six times under 60% compression strain to test their ability to recover their original form. As shown in Figure 4D, HYH4 deformed severely but did not break, indicating that HYH4 had weak elasticity, which can be attributed to its dynamic hydrazone bonds. DAH3 and DNH1 exhibited excellent elasticity. This can be attributed to the flexible PEG linkages in HYH and DNH [28]. Pore size analysis revealed that HYH and DNH had smaller pore sizes than DAH (Figure 4D). The smaller pore size of DNH1 facilitated the absorption of wound exudates while maintaining the moisture content of the hydrogel, which is beneficial for wound healing.

2.3. Cytocompatibility of DNH1

The cellular-level biocompatibility of DNH1 was evaluated using NIH/3T3 fibroblast cells before wound healing in vivo. As shown in Figure 5A, cells treated with DNH1 exhibited normal adhesion and spreading, with a morphology similar to that of the control group, indicating no apparent cytotoxic effects. Cell viability assays confirmed that DNH-treated cells maintained over 95% viability, even at concentrations up to 20% (v/v). There were no significant differences in cell viability between the treated and control groups, suggesting that DNH1 does not adversely affect cell proliferation. These results demonstrate that DNH1 is biocompatible and suitable for wound healing applications, providing a safe microenvironment for fibroblast growth.

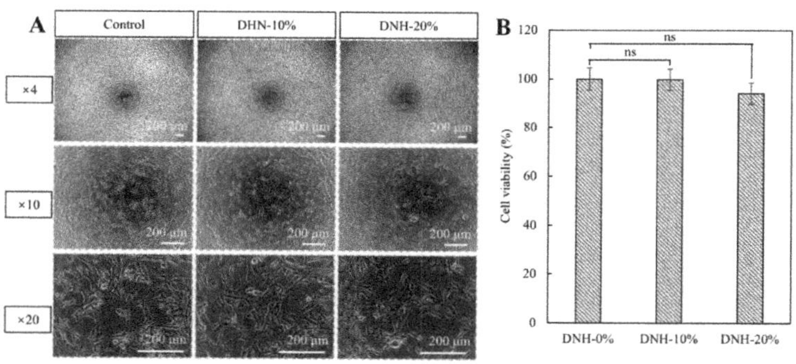

Figure 5. (**A**) The morphology of NIH/3T3 cells. (**B**) The cell viability ratio of NIH/3T3 cells. Significance was assessed using one-way ANOVA with Tukey's multiple comparisons tests (n = 3, the value indicates mean ± S.D.); ns represents no significance.

2.4. Wound Healing with DNH

The healing capability of the hydrogel was also evaluated. Figure 6A displays the gross wound appearance on days 0, 3, 6, 10, and 14, indicating complete wound closure with DNH1 treatment for 2 weeks. A comparison using gauze is shown in Figure 6B. It can be seen that the wound treated with gauze remained visible on day 14. However, DNH1-treated wounds were nearly healed. This demonstrated faster wound healing with DNH. Further analysis (Figure 6C) shows no significant differences within the first 3 days. Between days 3 and 8, the wound closure rate with DNH was significantly higher than that with gauze treatment. Importantly, on day 8, the wound closure rate reached 91% with the DNH1 treatment, whereas the wound closure rate with gauze was only 78%. On day 10, wound closure rates in both groups were comparable (92%). These results indicated that DNH1 treatment shortened the wound healing time.

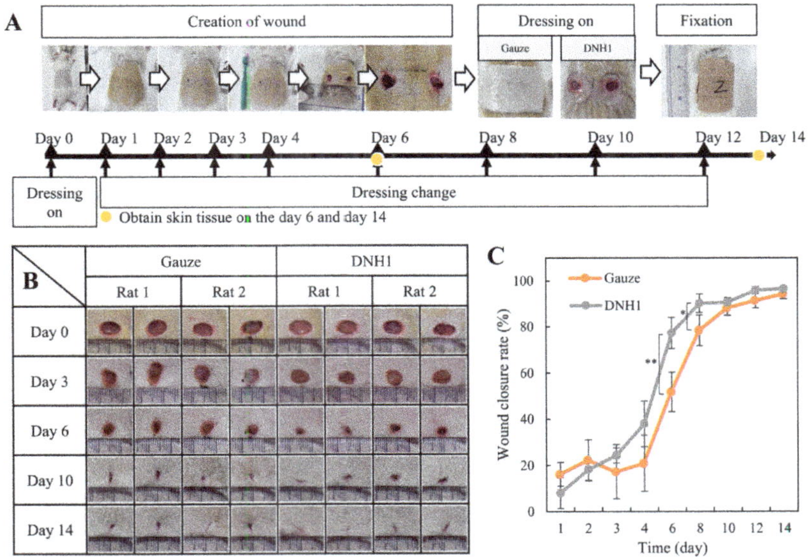

Figure 6. Wound healing procedure. (**A**) Detailed stage of wound healing. (**B**) Appearance of the wound treated with gauze and DNH1 on days 0, 3, 6, 10, and 14. (**C**) Quantification of wound closure ($n = 3$, the value indicates mean ± S.D.). Significance was assessed using one-way ANOVA with Tukey's multiple comparisons test, giving p values, * $p < 0.05$, ** $p < 0.01$.

Tissue sections from the wound center were stained and analyzed on days 6 and 14, as shown in Figure 7. On day 6, the wounds treated with gauze exhibited a larger granulation area with significant immune cell infiltration and lacked the formation of skin appendages such as hair follicles and pores (Figure 7A,B). In contrast, the wounds treated with DNH1 displayed characteristics of the wound maturation stage, with a small number of hair follicles and pores. However, no significant differences in granulation epidermis and dermis thickness were observed between the gauze and DNH1 groups on day 6 (Table 1).

Table 1. Quantification of histological wound samples obtained on days 6 and 14 ($n = 2$, the value indicates mean ± S.D.).

	Day 6		Day 14	
	Gauze	**DNH1**	**Gauze**	**DNH1**
Epidermis thickness (μm)	77.33 ± 4.33	64.59 ± 32.12	47.98 ± 3.27	45.6 ± 3.83
Dermis thickness (μm)	1198.99 ± 636.69	1126.83 ± 23.31	993.70 ± 90.29	1323.16 ± 654.28
Vessel number (-)	116.5 ± 37.48	36.00 ± 21.21	75.50 ± 3.54	38.50 ± 2.12
CD68 (-)	172 ± 8.49	106.5 ± 71.42	28.5 ± 10.61	24.50 ± 26.16
CD206 (-)	234 ± 50.91	135 ± 94.75	19.50 ± 3.54	25.5 ± 13.44

By day 14, the wounds in all groups had generally healed based on hematoxylin and eosin (H&E) staining (Figure 7A). The granulation epidermis thickness had reached a healthy skin epidermis range (41.97 ± 12.70 μm), indicating complete re-epithelialization. Importantly, the wounds treated with DNH1 exhibited smaller granulation areas and more mature hair follicles and pores than those treated with gauze. This H&E staining result was consistent with the wound closure rate results shown in Figure 6C, demonstrating that DNH1 not only accelerated wound closure but also improved the overall quality of healing by promoting skin appendage regeneration.

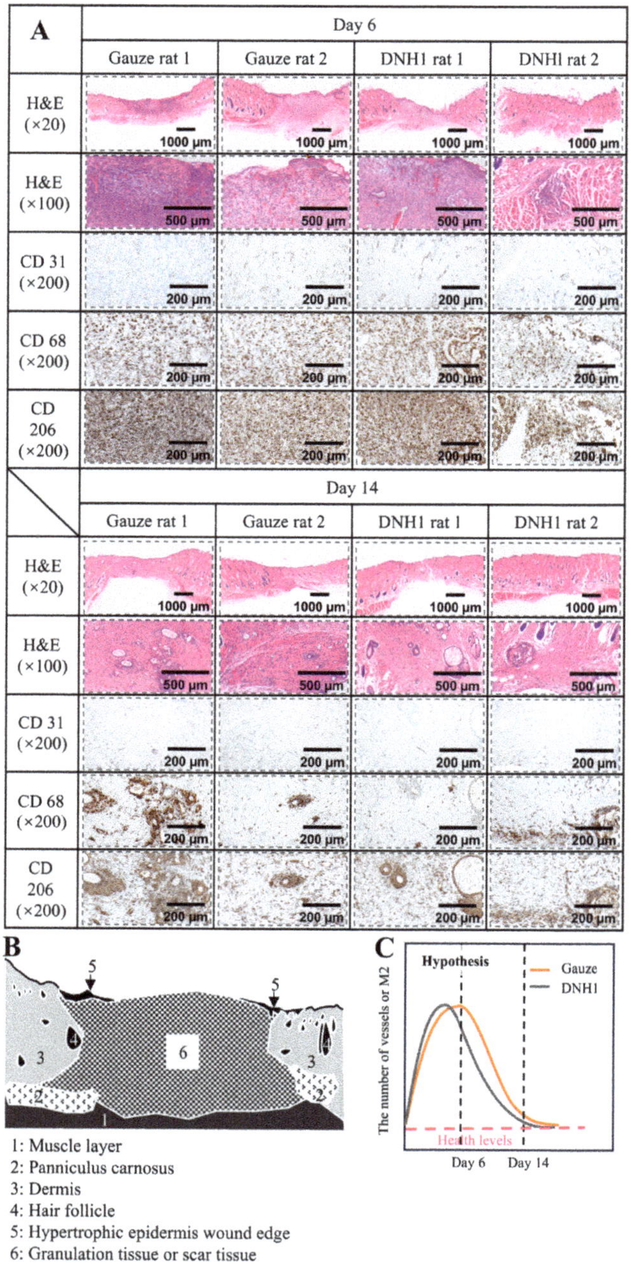

Figure 7. Histological evaluations of wounds. (**A**) Histological evaluations at 6 and 14 days. Scale bars, 1000 μm for H&E (×20), 500 μm for H&E (×100), and 200 μm for CD 31, CD68, and CD 206 (×200). (**B**) Wound structure. (**C**) Hypothesis of wound healing.

The number of vessels and macrophages on days 6 and 14 were recorded. M1 macrophages are angiogenesis initiators and secrete inflammatory cytokines [29], whereas M2 macrophages support angiogenesis to turn into mature vessels and secrete anti-inflammatory cytokines [29]. The number of CD68- (total macrophages) and CD206-positive

(M2 macrophages) vessels were analyzed under the same magnification (Figure 7A). The numbers of vessels positive for CD68 and CD 206 on day 6 were higher than those on day 14 (Table 1). Wounds treated with DNH1 exhibited lower vessel numbers than the gauze group on days 6 and 14, whereas there was no difference between CD68 and CD206 on days 6 and 14. This tendency is similar to that previously reported [30,31]. Combined with the results of the wound closure rate shown in Figure 6C, it can be concluded that the number of CD68- and CD86-positive vessels in wounds treated with DNH1 was lower than those in wounds treated with gauze. Based on this observation, we concluded that DNH1 accelerated macrophage polarization from M1 to M2 before day 6, as illustrated in Figure 7C. Given that M2 macrophages promote the formation of mature blood vessels and secrete anti-inflammatory cytokines, their early polarization may contribute to improved healing outcomes. In summary, DNH1 not only shortened the wound healing time but also enhanced the quality of tissue regeneration.

3. Conclusions

In this study, we developed a DNH combined with hydrazone and Diels–Alder crosslinking to address the rapid degradation and weak mechanical properties of HA hydrogels. DNH was systematically tested and compared to HYH and DAH. HYH provided rapid crosslinking, whereas DAH contributed to the mechanical properties and extended the degradation time of the hydrogel. Among them, DNH1 exhibited optimal swelling properties, facilitating the efficient absorption of wound exudates while maintaining hydrogel integrity after swelling. To evaluate the therapeutic potential of DNH1, it was applied to a full-thickness rat skin wound model. The results demonstrated that DNH1 not only accelerated wound healing but also improved tissue regeneration. These findings highlight DNH as a promising candidate for wound dressings and offer an effective solution for enhanced wound repair.

4. Materials and Methods

4.1. Materials and Instrument

HA (20–50 kDa, HA0-01025) was purchased from Kishida Chemical (Osaka, Japan). Sodium periodate ($NaIO_4$, 199-02401), ethylene glycol (054-00983), 2-morpholinethanesulfonic acid (MES, 343-01626), deuterium oxide (D_2O, 99.8%, 049-34242), PBS, and isoflurane (099-06571) were purchased from Fujifilm Wako Chemicals Japan Corporation (Fukuoka, Japan). 4-(4,6-Dimethoxy-1,3,5-triazin-2-yl)-4-methylmorpholinium (DMTMM, D2919) and adipic dihydrazide (ADH, A0170) were purchased from TCI Tokyo Chemical Industry (Tokyo, Japan). Furfurylamine (F20009), hydroxylamine hydrochloride (255580), and fetal bovine serum (FBS, 173012) were purchased from Sigma-Aldrich (Tokyo, Japan). Maleimide-PEG-maleimide (Mal-PEG-Mal, HO022022-2K) was purchased from Funakoshi (Tokyo, Japan). NIH/3T3 clone 5611 (JCRB0615) was purchased from JCRB cell bank (Osaka, Japan). Dulbecco's modified eagle's medium (DMEM, SH30002.03) was purchased from Pall Life Sciences (Tokyo, Japan). Cell counting kit-8 (AJ006) was purchased from Dojindo Laboratories (Kumamoto, Japan). IHC-CD 31 (ab28364), CD-68 (ab125212), and CD 206 (ab234000) antibodies were purchased from Abcam (Cambridge, UK).

Water purification system (Direct-Q 3UV, Merck, Darmstadt, Germany); freezer dryer (WYELA FDU-1200, Tokyo Rikakikai, Tokyo, Japan); benchtop scanning electron microscope (JCM-7000, JEOL Ltd., Tokyo, Japan); tensile and compression test machine (LTTU-50 NB, Minebeamitsumi, Tokyo, Japan); ATR-FTIR spectrometer (4700, Jasco, Tokyo, Japan); ^{1}H-NMR (JNM-ECZ400, JEOL Ltd., Tokyo, Japan); 8 mm skin biopsy puncture (Kai Industries, Tokyo, Japan); small animal anesthetizer (TK-36, Bio machinery, Seki-shi, Japan); inverted light microscope (CKX53, Olympus, Tokyo, Japan).

4.2. Synthesis of DNH

HYH and DAH were mixed in a one-pot method (Scheme 1). OHA, ADH, HA-Furan, and Mal-PEG-Mal were treated with PBS as the solvent. The optimal conditions were obtained based on the synthesis of HYH and DAH.

HYH + DAH → dual network hydrogel (DNH)

Scheme 1. Synthesis of DNH.

HYH was prepared by mixing OHA and ADH (Scheme 2).

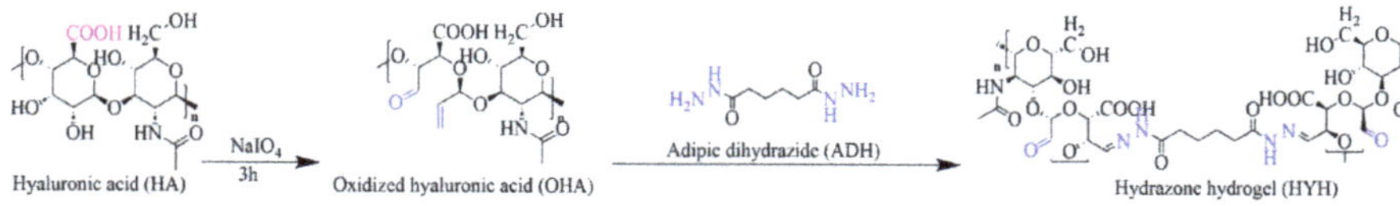

Scheme 2. Synthesis of HYH.

OHA was obtained using HA as a reactant by adding $NaIO_4$ in the dark, following a previous report [32]. Briefly, HA (2.0 g) in 100 mL distilled water was kept at 4 °C overnight. $NaIO_4$ (10.0 mL) in water (80 mg/mL) was added to HA solution and mixed under 25 °C for 3 h. A volume of 1396.4 μL of ethylene glycol was added to stop the reaction for 1 h under 25 °C. The solution was dialyzed with a dialysis bag (cutoff 1 kDa) against distilled water for 3 days to remove $NaIO_4$ and ethylene glycol, during which the distilled water was changed three times each day. The dialysate was distributed into 6-well plates and pre-frozen at −80 °C overnight. OHA was obtained after lyophilization.

HYH was obtained by mixing OHA (192.0 μL) and ADH (48.0 μL) at 25 °C [33]. A total of 100 mg OHA in 1 mL PBS (pH 7.40) was used as a reactant. ADH in PBS at different concentrations (2, 4, 6, 8, 10, 20, 40, and 60 mg/mL) was evaluated. HYHs were named HYH1, HYH2, HYH3, HYH4, HYH5, HYH6, HYH7, and HYH8 following the related ADH concentrations.

DAH was synthesized by mixing HA-Furan with Mal-PEG-Mal in Scheme 3.

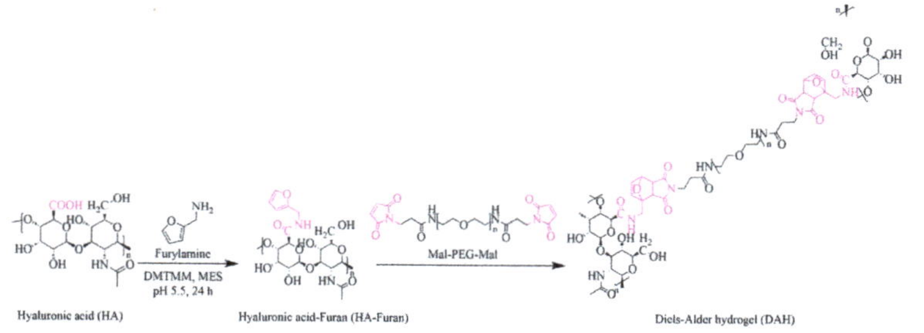

Scheme 3. Synthesis of DAH.

HA-Furan was prepared according to a previously reported method [34]. HA (2.0 g) dissolved in 100 mL MES buffer (100 mM, pH 5.50) was kept at 4 °C overnight. A total of 2.8 g DMTMM was added, followed by a dropwise addition of 486.0 μL furfurylamine after 10 min. The mixture was stirred for 24 h at 25 °C. The product was dialyzed against water for three days (cutoff: 1 kDa). HA-Furan was obtained by freeze-drying.

DAH was prepared by mixing HA-Furan and Mal-PEG-Mal in different volume ratios. Newly prepared lyophilized HA-Furan (100 mg/mL) and Mal-PEG-Mal (100 mg/mL) in PBS (pH 7.40) were mixed and kept overnight at 4 °C. DAH was formed at 37 °C overnight by vertexing [35]. Different volume ratios of HA-Furan and Mal-PEG-Mal (0.3,

0.5, 0.7, 1, 1.4, 2, 3, 5, 7, 9, and 11) were investigated and the results are listed in Table S3. Newly formed DAHs was named as DAH1, DAH2, DAH3, DAH4, DAH5, DAH6, DAH7, DAH8, DAH9, DAH10, and DAH11 following the different ratios between HA-Furan and Mal-PEG-Mal.

The DNH was obtained by mixing HA-Furan (100 mg/mL) and Mal-PEG-Mal (100 mg/mL), OHA (100 mg/mL, 64.0 μL), and ADH (8 mg/mL, 16.0 μL) with PBS as solvent. The sequence was added according to a previous report at 37 °C [17]. Different volume ratios of OHA and HA-Furan were evaluated (Table S1). The newly formed DNHs were named DNH1, DNH2, DNH3, DNH4, and DNH5.

4.3. *Characterization of Hydrogel*

4.3.1. Determination of Oxidation Degree (OD) of OHA

The OD of OAH was determined by titration and is listed in Table S4 to determine the optimal ratio of HA to $NaIO_4$. OHA (≈50 mg) was dissolved in 25 mL hydroxylamine hydrochloride solution (0.25 mmol/L, in H_2O) and incubated for 5 h. The samples were titrated with NaOH (0.2 mol/L) until the pH increased to 9.97.

The reaction conducted during titration is shown below.

$$\text{HA-(CHO)n} + \text{n } H_2NOH \cdot HCl \rightarrow \text{HA-(CH=NOH)n} + \text{n HCl}$$
$$HCl + NaOH \rightarrow NaCl + H_2O$$

OD was calculated based on Equation (1):

$$OD\ (\%) = Mw \times C_{NaOH} \times (V_{OHA} - V_{blank})/(2 \times m_{OHA}) \times 100\% \quad (1)$$

where Mw = 400 g/mol, the monomeric molecular weight of OHA; C_{NaOH} = 0.2 mol/L; V_{OHA} is the NaOH volume (mL) consumed with OAH; V_{blank} is the NaOH volume (mL) consumed without OAH; and m_{OHA} is the mass of OHA in mg.

4.3.2. Determination of OHA and HA-Furan Using FT-IR and ^{1}H-NMR

ATR-FTIR was used to determine the aldehyde content of OHA, furan groups of HA-Furan, and the crosslinking bonds of HYH and DAH. The analyses were obtained from 400 to 4000 cm^{-1} with a resolution of 4 cm^{-1}. Eight scans were performed for each sample.

^{1}H-NMR was used to determine the OHA formation, the furan group of HA-Furan, and the modification degree of HA-Furan. The 0.6 mL of HA, OHA, and HA-Furan (10 mg/mL) were detected using D_2O as the solvent. Chemical shifts were recorded.

4.3.3. Characterization of Mechanical Properties and Gelation Time

The mechanical properties of the hydrogels were tested using a tensile- and compression-testing machine at a load of 50 N. The sample size was Ø = 7 mm and H = 5 mm. Stress–strain curve was plotted at a compression speed of 2 mm/s with the compression strain increasing from 0 to 70%. The maximum value indicates the maximum compressive stress. Young's modulus was calculated based on the linear part of the stress–strain curve between 10% and 20%. The HYH results are presented in Table S2.

The gelation time was visually recorded. The time until no fluid remained in the inverted tube was recorded. The HYH results are presented in Table S2.

4.3.4. Swelling Properties of Hydrogels Under Different pH

The samples for the swelling ratio of HYH, DAH, and DNH were prepared in a silicon mold (Ø = 8 mm, H= 0.5 mm), and the results were tested at varying pH. Samples were accurately weighted (*Wi*). The samples were immersed in PBS buffer with different pH values (3.0, 5.0, 7.0, 9.0, and 11.0, adjusted using HCl (1 M) and NaOH (1 M)), and the PBS was changed weekly. The treated samples were weighed (*Wt*) after the buffer solution was removed and wiped with paper towels. The swelling ratio *(SW)* was calculated using Equation (2):

$$SW = Wt/Wi \tag{2}$$

4.3.5. Characterization of Resilience

The resilience of HYH4, DAH3, and DNH1 cells was evaluated using tensile and compression testing. The resilience was recorded by measuring the height change in the samples after six compression cycles, with each cycle involving a compression strain of 60% and a compression speed of 2 mm/s.

4.3.6. Characterization of Morphology of Hydrogels

The pore size and morphology of the hydrogels were determined using SEM. The hydrogels were placed in liquid nitrogen for 5 min and vertically removed. The samples were then freeze-dried and coated with gold for further analysis. Images of each sample were captured at $\times$500.

4.4. Cytocompatibility Testing of DNH1 In Vitro

The cytocompatibility of the DNH1 was evaluated using a leachate-based assay with mouse embryonic fibroblast NIH/3T3 cells. Hydrogel components were sterilized using 0.22 μm filters before hydrogel preparation. Hydrogels (Φ =16 mm, h = 0.5 mm) were soaked in a DMEM with 10% FBS to prepare leachates at two concentrations: 10% (100 μL hydrogel in 900 μL medium) and 20% (200 μL hydrogel in 800 μL medium). Prior to extraction, the medium was refreshed three times at 3 h intervals. The final leachates were collected after 24 h of incubation at 37 °C in 5% carbon dioxide. A control group with no hydrogel present (DMEM with 10% FBS only) was used as the control and designated as DNH-0%. NIH/3T3 cells were seeded at 6400 cells per well in 96-well plates and incubated with the leachates for 24 h. After removing the leachates, Cell Counting Kit-8 working solution (10% v/v in culture medium) was added and incubated for 2 h. Absorbance at 450 nanometers was measured using a microplate reader. Untreated cells served as the control. Cell viability was calculated according to Equation (3):

$$Cell\ viability\ (\%) = \frac{As - Ab}{Ac - Ab} \times 100\% \tag{3}$$

where *As*, *Ab*, and *Ac* refer to sample, blank, and control absorbance, respectively. Cell morphology was observed using an inverted microscope.

4.5. Wound Healing Using DNH1

Eight male rats (6-week-old, Slc: S.D, SLC, Shizuoka, Japan) were acclimated to the laboratory conditions for 3 days. The average weight of all rats was 260 $\pm$ 10 g. All animals received humane care in accordance with the guidelines of the Animal Investigation Ethics Committee of Kyushu University. Animals (n = 2) were randomly divided into two groups according to treatment: gauze as the control group and DNH1 as the experimental group. Rats were housed in pairs (two rats per cage) in a temperature-controlled environment (25 $\pm$ 1 °C) with a 12 h light/dark cycle. The rats were provided with clean tap water and standard laboratory chow ad libitum. Anesthesia was induced using inhalational isoflurane during all surgical procedures to minimize pain and distress.

On day 0, the back of each rat was shaved. Double-circular full-thickness skin wounds were made on the back of each rat using a disposable 8 mm skin biopsy puncture. The wound circle was 9.5 cm from the tail and 1.5 cm from the mid-axis of the back of each rat.

The skin was disinfected with a 50% (v/v) iodine solution in alcohol. Gauze or DNH1 was placed directly on the wound. The wound was additionally covered with a 5 cm × 10 cm piece of 3M Tegaderm to keep the dressing in place. The rat's body was wrapped in an elastic bandage. The dressing was changed daily for the first 4 days. The dressing was changed every 2 days. The wounds were cleaned with a sterile 0.9% sodium chloride solution before each vehicle was changed. The DNH1 sheet was set to a diameter of 20 mm and 1.5 mm in thickness for the first 6 days. The DNH1 sheets were reduced to a diameter of 10 mm before sacrifice.

The wound closure rate was determined using wound photo analysis. The wound area of each rat was analyzed using ImageJ software (version 1.54d). The wound closure rate was expressed as a percentage of the original area on day 0 using Equation (4) [36]:

$$\text{Wound closure rate (\%)} = [1 - \text{wound area (day t)}/\text{wound area (day 0)}] \times 100\% \quad (4)$$

where t represents the time point of wound healing.

The rats were sacrificed on days 6 and 14 after wound healing. Wound sites on the healthy skin were excised and fixed in 10% formalin. Tissues were treated with a graded alcohol series and xylene and then embedded in paraffin. Central wound sections (4 mm) were mounted on glass slides and stained with H&E or with IHC for CD 31, CD-68, and CD 206. Images of each sample were acquired using an optical microscope with the ×20, ×100, and ×200 objectives.

4.6. Statistical Analysis

All experimental data were expressed as mean ± standard division (S.D.). Statistical significance was determined using one-way ANOVA, followed by Tukey's Honest significant difference (HSD) test for multiple comparisons.

Supplementary Materials: The following supporting information can be downloaded at: https://www.mdpi.com/article/10.3390/gels11040266/s1, Table S1: Volume ratios of HA-Furan, Mal-PEG-Mal, OHA, and ADH during DNH preparation; Table S2: HYH properties (n = 3, the value indicates mean ± S.D.; Table S3: Volume ratios of HA-Furan and Mal-PEG-Mal during DAH preparation; Table S4: Effect of the molar ratio between HA and $NaIO_4$ on gelation time (n = 3, the value indicates mean ± S.D.).

Author Contributions: Conceptualization, data curation, formal analysis, methodology, writing—original draft preparation, L.W.; investigation, Y.Z. (Yu Zhou); investigation, Y.Z. (Yi Zhang); investigation, J.H.; writing—review and editing, Y.I.; histological staining and analysis, S.A.; supervision and project administration, H.I. All authors have read and agreed to the published version of the manuscript.

Funding: Part of this study was supported by Grants-in-Aid for Scientific Research (KAKENHI, grant numbers [JP21H01732 and JP24K01274]) from the Japan Society for the Promotion of Science (JSPS).

Institutional Review Board Statement: The study was conducted in accordance with the guidelines of the Declaration of Helsinki and approved by the Institutional Review Board of Kyushu University (protocol codes A23-478-0 and 21 December 2023 of approval).

Informed Consent Statement: Not applicable.

Data Availability Statement: Data are contained within the article and Supplementary Materials.

Acknowledgments: We express our sincere gratitude to Yusuke Sakai for his invaluable assistance with the SEM analysis. His expertise and guidance significantly contributed to the successful completion of this research project. We express our sincere gratitude to Lihua Wu of Kuantan Sunny Scientific Collaboration Sdn for her review and editing of this manuscript. We thank the members of the Biomaterial & Medical Engineering Laboratories (Ijima Lab) for their continued support and feedback. This study was financially supported by the Chinese Scholarship Council (Grant No. 202206610001).

Conflicts of Interest: The authors declare no conflicts of interest.

Abbreviations

The following abbreviations are used in this manuscript:

DNH	Double network hydrogel
HA	Hyaluronic acid
ECM	Extracellular matrix
HYH	Hydrazone hydrogel
OHA	Oxidized hyaluronic acid
ADH	Adipic dihydrazide
DAH	Diels–Alder hydrogel
HA-Furan	Hyaluronic acid with furan group
Mal-PEG-Mal	Maleimide-PEG-maleimide

References

1. Farahani, M.; Shafiee, A. Wound Healing: From Passive to Smart Dressings. *Adv. Healthc. Mater.* **2021**, *10*, 2100477. [CrossRef]
2. Cai, C.; Zhu, H.; Chen, Y.; Guo, Y.; Yang, Z.; Li, H.; Liu, H. Mechanoactive Nanocomposite Hydrogel to Accelerate Wound Repair in Movable Parts. *ACS Nano* **2022**, *16*, 20044–20056. [CrossRef]
3. Bakadia, B.M.; Qaed Ahmed, A.A.; Lamboni, L.; Shi, Z.; Mutu Mukole, B.; Zheng, R.; Pierre Mbang, M.; Zhang, B.; Gauthier, M.; Yang, G. Engineering Homologous Platelet-Rich Plasma, Platelet-Rich Plasma-Derived Exosomes, and Mesenchymal Stem Cell-Derived Exosomes-Based Dual-Crosslinked Hydrogels as Bioactive Diabetic Wound Dressings. *Bioact. Mater.* **2023**, *28*, 74–94. [CrossRef]
4. Liu, H.; Wang, C.; Li, C.; Qin, Y.; Wang, Z.; Yang, F.; Li, Z.; Wang, J. A Functional Chitosan-Based Hydrogel as a Wound Dressing and Drug Delivery System in the Treatment of Wound Healing. *RSC Adv.* **2018**, *8*, 7533–7549. [CrossRef]
5. Graça, M.F.P.; Miguel, S.P.; Cabral, C.S.D.; Correia, I.J. Hyaluronic Acid—Based Wound Dressings: A Review. *Carbohydr. Polym.* **2020**, *241*, 116364. [CrossRef]
6. Sha, Q.; Wang, Y.; Zhu, Z.; Wang, H.; Qiu, H.; Niu, W.; Li, X.; Qian, J. A Hyaluronic Acid/Silk Fibroin/Poly-Dopamine-Coated Biomimetic Hydrogel Scaffold with Incorporated Neurotrophin-3 for Spinal Cord Injury Repair. *Acta Biomater.* **2023**, *167*, 219–233. [CrossRef]
7. Mirjalili, F.; Mahmoodi, M. Controlled Release of Protein from Gelatin/Chitosan Hydrogel Containing Platelet-Rich Fibrin Encapsulated in Chitosan Nanoparticles for Accelerated Wound Healing in an Animal Model. *Int. J. Biol. Macromol.* **2023**, *225*, 588–604. [CrossRef]
8. Wang, J.; Liu, S.; Huang, J.; Ren, K.; Zhu, Y.; Yang, S. Alginate: Microbial Production, Functionalization, and Biomedical Applications. *Int. J. Biol. Macromol.* **2023**, *242*, 125048. [CrossRef]
9. Kong, F.; Mehwish, N.; Lee, B.H. Emerging Albumin Hydrogels as Personalized Biomaterials. *Acta Biomater.* **2023**, *157*, 67–90. [CrossRef]
10. Nelson, D.W.; Gilbert, R.J. Extracellular Matrix-Mimetic Hydrogels for Treating Neural Tissue Injury: A Focus on Fibrin, Hyaluronic Acid, and Elastin-Like Polypeptide Hydrogels. *Adv. Healthc. Mater.* **2021**, *10*, 2101329. [CrossRef]
11. Muir, V.G.; Burdick, J.A. Chemically Modified Biopolymers for the Formation of Biomedical Hydrogels. *Chem. Rev.* **2021**, *121*, 10908–10949. [CrossRef]
12. Jha, A.K.; Hule, R.A.; Jiao, T.; Teller, S.S.; Clifton, R.J.; Duncan, R.L.; Pochan, D.J.; Jia, X. Structural Analysis and Mechanical Characterization of Hyaluronic Acid-Based Doubly Cross-Linked Networks. *Macromolecules* **2009**, *42*, 537–546. [CrossRef]
13. Fan, P.; Dong, Q.; Yang, J.; Chen, Y.; Yang, H.; Gu, S.; Xu, W.; Zhou, Y. Flexible Dual-Functionalized Hyaluronic Acid Hydrogel Adhesives Formed in Situ for Rapid Hemostasis. *Carbohydr. Polym.* **2023**, *313*, 120854. [CrossRef]
14. Yang, R.; Liu, X.; Ren, Y.; Xue, W.; Liu, S.; Wang, P.; Zhao, M.; Xu, H.; Chi, B. Injectable Adaptive Self-Healing Hyaluronic Acid/Poly (γ-Glutamic Acid) Hydrogel for Cutaneous Wound Healing. *Acta Biomater.* **2021**, *127*, 102–115. [CrossRef]
15. Jia, Y.; Zhang, X.; Yang, W.; Lin, C.; Tao, B.; Deng, Z.; Gao, P.; Yang, Y.; Cai, K. A pH-Responsive Hyaluronic Acid Hydrogel for Regulating the Inflammation and Remodeling of the ECM in Diabetic Wounds. *J. Mater. Chem. B* **2022**, *10*, 2875–2888. [CrossRef]
16. Nonoyama, T.; Gong, J.P. Tough Double Network Hydrogel and Its Biomedical Applications. *Annu. Rev. Chem. Biomol. Eng.* **2021**, *12*, 393–410. [CrossRef]

17. Mihajlovic, M.; Rikkers, M.; Mihajlovic, M.; Viola, M.; Schuiringa, G.; Ilochonwu, B.C.; Masereeuw, R.; Vonk, L.; Malda, J.; Ito, K.; et al. Viscoelastic Chondroitin Sulfate and Hyaluronic Acid Double-Network Hydrogels with Reversible Cross-Links. *Biomacromolecules* **2022**, *23*, 1350–1365. [CrossRef]
18. Wang, Y.; Chen, Y.; Zheng, J.; Liu, L.; Zhang, Q. Three-Dimensional Printing Self-Healing Dynamic/Photocrosslinking Gelatin-Hyaluronic Acid Double-Network Hydrogel for Tissue Engineering. *ACS Omega* **2022**, *7*, 12076–12088. [CrossRef]
19. Nair, S.; Remya, N.S.; Remya, S.; Nair, P.D. A Biodegradable in Situ Injectable Hydrogel Based on Chitosan and Oxidized Hyaluronic Acid for Tissue Engineering Applications. *Carbohydr. Polym.* **2011**, *85*, 838–844. [CrossRef]
20. Yu, F.; Cao, X.; Du, J.; Wang, G.; Chen, X. Multifunctional Hydrogel with Good Structure Integrity, Self-Healing, and Tissue-Adhesive Property Formed by Combining Diels–Alder Click Reaction and Acylhydrazone Bond. *ACS Appl. Mater. Interfaces* **2015**, *7*, 24023–24031. [CrossRef]
21. Xie, X.; Lei, H.; Fan, D. Antibacterial Hydrogel with pH-Responsive Microcarriers of Slow-Release VEGF for Bacterial Infected Wounds Repair. *J. Mater. Sci. Technol.* **2023**, *144*, 198–212. [CrossRef]
22. Chen, G.; Yu, Y.; Wu, X.; Wang, G.; Ren, J.; Zhao, Y. Bioinspired Multifunctional Hybrid Hydrogel Promotes Wound Healing. *Adv. Funct. Mater.* **2018**, *28*, 1801386. [CrossRef]
23. Schneider, L.A.; Korber, A.; Grabbe, S.; Dissemond, J. Influence of pH on Wound-Healing: A New Perspective for Wound-Therapy? *Arch. Dermatol. Res.* **2007**, *298*, 413–420. [CrossRef]
24. Bennison, L.R.; Miller, C.N.; Summers, R.J.; Minnis, A.M.B.; Sussman, G.; McGuiness, W. The pH of Wounds during Healing and Infection: A Descriptive Literature Review. *Wound Pract. Res. J. Aust. Wound Manag. Assoc.* **2017**, *25*, 63–69.
25. Sonawane, S.J.; Kalhapure, R.S.; Govender, T. Hydrazone Linkages in pH Responsive Drug Delivery Systems. *Eur. J. Pharm. Sci.* **2017**, *99*, 45–65. [CrossRef]
26. Tawagi, E.; Ung, T.; Cheng, H.-L.M.; Santerre, J.P. Arrhenius-Model-Based Degradable Oligourethane Hydrogels for Controlled Growth Factor Release. *Acta Biomater.* **2023**, *166*, 167–186. [CrossRef]
27. Kirchhof, S.; Strasser, A.; Wittmann, H.-J.; Messmann, V.; Hammer, N.; Goepferich, A.M.; Brandl, F.P. New Insights into the Cross-Linking and Degradation Mechanism of Diels–Alder Hydrogels. *J. Mater. Chem. B* **2014**, *3*, 449–457. [CrossRef]
28. Shi, J.; Yu, L.; Ding, J. PEG-Based Thermosensitive and Biodegradable Hydrogels. *Acta Biomater.* **2021**, *128*, 42–59. [CrossRef]
29. Spiller, K.L.; Anfang, R.; Spiller, K.J.; Ng, J.; Nakazawa, K.R.; Daulton, J.W.; Vunjak-Novakovic, G. The Role of Macrophage Phenotype in Vascularization of Tissue Engineering Scaffolds. *Biomaterials* **2014**, *35*, 4477–4488. [CrossRef]
30. Kuninaka, Y.; Ishida, Y.; Ishigami, A.; Nosaka, M.; Matsuki, J.; Yasuda, H.; Kofuna, A.; Kimura, A.; Furukawa, F.; Kondo, T. Macrophage Polarity and Wound Age Determination. *Sci. Rep.* **2022**, *12*, 20327. [CrossRef]
31. Wang, Y.; Zhou, C.; Li, Z.; Li, G.; Zou, Y.; Li, X.; Gu, P.; Liu, J.; Bai, L.; Yan, H.; et al. Injectable Immunoregulatory Hydrogels Sequentially Drive Phenotypic Polarization of Macrophages for Infected Wound Healing. *Bioact. Mater.* **2024**, *41*, 193–206. [CrossRef]
32. Nguyen, N.T.-P.; Nguyen, L.V.-H.; Tran, N.M.-P.; Nguyen, D.T.; Nguyen, T.N.-T.; Tran, H.A.; Dang, N.N.-T.; Vo, T.V.; Nguyen, T.-H. The Effect of Oxidation Degree and Volume Ratio of Components on Properties and Applications of in Situ Cross-Linking Hydrogels Based on Chitosan and Hyaluronic Acid. *Mater. Sci. Eng. C* **2019**, *103*, 109670. [CrossRef]
33. França, C.G.; Plaza, T.; Naveas, N.; Andrade Santana, M.H.; Manso-Silván, M.; Recio, G.; Hernandez-Montelongo, J. Nanoporous Silicon Microparticles Embedded into Oxidized Hyaluronic Acid/Adipic Acid Dihydrazide Hydrogel for Enhanced Controlled Drug Delivery. *Microporous Mesoporous Mater.* **2021**, *310*, 110634. [CrossRef]
34. Nimmo, C.M.; Owen, S.C.; Shoichet, M.S. Diels–Alder Click Cross-Linked Hyaluronic Acid Hydrogels for Tissue Engineering. *Biomacromolecules* **2011**, *12*, 824–830. [CrossRef]
35. Ilochonwu, B.C.; Mihajlovic, M.; Maas-Bakker, R.F.; Rousou, C.; Tang, M.; Chen, M.; Hennink, W.E.; Vermonden, T. Hyaluronic Acid-PEG-Based Diels–Alder In Situ Forming Hydrogels for Sustained Intraocular Delivery of Bevacizumab. *Biomacromolecules* **2022**, *23*, 2914–2929. [CrossRef]
36. Li, M.; Liang, Y.; He, J.; Zhang, H.; Guo, B. Two-Pronged Strategy of Biomechanically Active and Biochemically Multifunctional Hydrogel Wound Dressing to Accelerate Wound Closure and Wound Healing. *Chem. Mater.* **2020**, *32*, 9937–9953. [CrossRef]

gels

MDPI

Article

Assessment of Flurbiprofen Suspension and Composite Gel Pre- and Post Skin Perforation: Effectiveness in Managing Inflammatory Responses in Ear Tags and Periocular Piercings

Sheimah El Bejjaji [1,†], Gladys Ramos-Yacasi [2,†], Valeri Domínguez-Villegas [3], Délia Chaves Moreira Dos Santos [4], Antonio Braza [1,*], Lilian Sosa [5,6], Maria José Rodríguez-Lagunas [7], Ana Cristina Calpena [1,*], Mireia Zelaya [8] and Alexander Parra [1]

[1] Department of Pharmacy, Pharmaceutical Technology and Physical Chemistry, Faculty of Pharmacy and Food Sciences, University of Barcelona, 08028 Barcelona, Spain; shelbejb7@alumnes.ub.edu (S.E.B.); drecoar@gmail.com (A.P.)
[2] Facultad de Ciencias Farmacéuticas, Bioquímicas y Biotecnológicas, Universidad Católica de Santa María (UCSM), Arequipa 04001, Peru; glramos011@hotmail.es
[3] Facultad de Ciencias Químicas e Ingeniería, Universidad Autónoma del Estado de Morelos (UAEM), Cuernavaca, Morelos 62209, Mexico; valeri.dominguez@uaem.mx
[4] Department of Pharmacy and Nutrition, Federal University of Espírito Santo (UFES), Goiabeiras, Vitória 29075-910, Brazil; deliachavesmoreira@gmail.com
[5] Centro Experimental en Biociencia (CENBIO), Facultad de Ciencias Químicas y Farmacia, Universidad Nacional Autónoma de Honduras (UNAH), Tegucigalpa 11101, Honduras; liliansosa2012@gmail.com
[6] Instituto de Investigaciones en Microbiología (IIM), Facultad de Ciencias, Universidad Nacional Autónoma de Honduras (UNAH), Tegucigalpa 11101, Honduras
[7] Department of Biochemistry and Physiology, Faculty of Pharmacy and Food Science, University of Barcelona (UB), 08028 Barcelona, Spain; mjrodriguez@ub.edu
[8] Laboratory of Plant and Animal Histology, School of Biology, Faculty of Sciences, National Autonomous University of Honduras (UNAH), Tegucigalpa 11101, Honduras; mireya.zelaya@unah.edu.hn
* Correspondence: braza@ub.edu (A.B.); anacalpena@ub.edu (A.C.C.)
† These authors contributed equally to this work.

Abstract: (1) Background: Controlled skin perforations, such as ear tags, piercings, and microdermal implants, induce inflammation and stress in individuals undergoing these procedures. This localized trauma requires care to optimize healing, reduce inflammation, and prevent infections. (2) Methods: Two formulations were developed: an FB-suspension and an FB-gel. Their in vivo efficacy was evaluated, along with drug retention in porcine and human skin after 30 min of administration, chemical stability at different temperatures, cytotoxicity, histological changes induced via transdermal application, and irritative potential, assessed using the HET-CAM assay. (3) Results: Both formulations reduced inflammation when applied 30 min before perforation compared to the positive control. The FB-suspension demonstrated no cytotoxicity and exhibited greater efficacy than the free flurbiprofen solution, highlighting the advantages of using nanoparticle-mediated drug delivery. Moreover, the FB-gel maintained chemical stability for up to 3 months across a temperature range of 4 to 40 °C. Histologically, no significant changes in skin composition were observed. (4) Conclusions: The FB-suspension is viable for both pre- and post-perforation application, as it is a sterile formulation. In contrast, the FB-gel is a convenient and easy application, making it a practical alternative for use in both clinical and veterinary settings.

Keywords: flurbiprofen; suspension; hydrogel; polyethylene glycol 3350; human skin; porcine skin; NSAID; transdermal drug delivery; controlled skin perforation; inflammation management

1. Introduction

Identification ear tags serve as perforated attachments that enable individual animal identification, supporting traceability and compliance with sanitary and commercial regulations throughout an animal's life cycle from birth to slaughter or market [1,2]. Many countries mandate the use of these tags to meet regulatory requirements for animal welfare and to effectively monitor livestock movements and commercial transactions.

The ear tagging procedure for pigs requires a specialized applicator that pierces the ear and fastens the tag securely. The process begins with cleaning the intended application site to reduce the infection risk. The operator then positions the loaded applicator in the central portion of the ear, carefully avoiding major vascular structures, before applying firm pressure to perforate the tissue and secure the identification device. Ensuring proper attachment and readability is critical, as these tags provide essential information required for herd management and traceability protocols.

Researchers have investigated the impact of ear tagging on tissue integrity, emphasizing that factors such as the animal's age at tagging, environmental housing conditions, and precise tag placement play crucial roles in the risk of lesion development. Studies have underscored the importance of conducting ear examinations approximately two weeks after tagging to detect and address potential tissue complications before they escalate into more severe conditions, ultimately improving overall animal welfare. Furthermore, research findings have indicated that the tagging process triggers measurable stress responses and discomfort, highlighting the need for improved application techniques and exploring alternative identification methods.

In a related work, Barz and colleagues [3] studied how combined meloxicam and iron administration affects piglets undergoing castration. Their results indicate that meloxicam administration meaningfully reduces pain indicators and stress responses following the procedure, with positive implications for animal welfare standards. Further investigations remain essential to fully characterize the pathophysiological mechanisms underlying ear necrosis due to ear tagging and to develop targeted preventative interventions [4–6].

On the other hand, skin perforations in humans vary significantly. Unlike identification ear tags, whose perforation location is standardized, piercings can be applied to various areas of the body, leading to different healing processes depending on the pierced area. In this case, piercings serve esthetic and cultural purposes. The piercing process involves creating a controlled wound in the skin or cartilage using a sterile needle or catheter, followed by inserting a biocompatible piece of jewelry. Another method of inserting jewelry into human skin is via microdermal implants or dermal punches. The main difference from piercings is that piercings have both an entry and an exit point, whereas microdermal implants are anchored beneath the epidermis [7,8].

This study analyzes and compares the efficacy of two formulations characterized by El Bejjaji et al. [9] and Ramos et al. [10,11], designed for both preventing and treating inflammation associated with skin perforation procedures, such as ear tagging, piercings, or microdermal implants. To optimize their effectiveness and ensure user comfort, the formulations were administered 30 min before the procedure. This carefully planned timeframe was an essential component of the experimental design, allowing for thoroughly evaluating their anti-inflammatory properties in both preventative and therapeutic contexts.

These formulations contain flurbiprofen (FB), a nonsteroidal water-insoluble anti-inflammatory drug (NSAID) widely used in clinical practice for treating pain and inflammation, as the active ingredient. FB is encapsulated in freeze-dried polymeric nanoparticles of poly(ε-caprolactone) (PεCL), incorporating polyethylene glycol 3350 (PEG) as a cryoprotectant and being sterilized using gamma (γ) irradiation.

The first formulation corresponds to the suspension described by Ramos et al. and is sterile [9] and Ramos et al. [10,11]. The second formulation is obtained by incorporating the suspension into a hydrogel (Sepigel®), following the procedure described by El Bejjaji et al. [9]. A graphical representation of the study's formulations is shown in Figure 1.

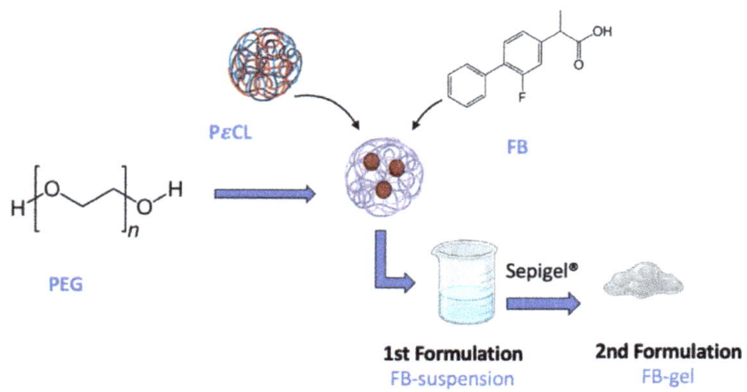

Figure 1. Preparing the formulations under study.

The selected active ingredient, FB, functions by inhibiting cyclooxygenase (COX) enzymes 1 and 2. COX-1 maintains constant levels in most tissues and facilitates the physiological production of prostaglandins. Its metabolites from arachidonic acid regulate critical functions, including gastrointestinal protection, vascular homeostasis, renal hemodynamics, and platelet activity [12]. In contrast, COX-2 appears primarily at inflammatory sites, in neoplasms, or under specific physiological conditions. This enzyme primarily synthesizes prostaglandins involved in inflammatory, pathological, and stress responses. Pro-inflammatory factors can induce its expression, potentially increasing the baseline concentration by 10 to 80 times [13].

FB demonstrates stronger inhibitory effects on the COX-1 enzyme, which accounts for its adverse effects, particularly in the gastrointestinal system. With a relatively brief half-life of approximately 4 h, FB requires frequent administration [14,15]. These limitations highlight the need to develop pharmaceutical formulations that bypass the gastrointestinal tract, such as dermal and transdermal delivery systems.

One key advantage of transdermal administration is its ability to bypass first-pass metabolism and factors associated with the gastrointestinal tract, such as pH and the gastric emptying time. Furthermore, it enables sustained and controlled drug release, reduces side effects related to systemic toxicity by minimizing fluctuations in blood concentration, facilitates direct access to the therapeutic target, and enhances treatment adherence, contributing to an overall reduction in therapy costs. Incorporating the drug through nanoparticles aims to improve skin penetration and reduce skin irritation. These nanotechnology-based carriers have shown better performance in dermal delivery compared to conventional systems, facilitating more efficient drug vehiculation in the nanoencapsulated form than in the free form [16,17]. Incorporating FB into nanoencapsulated systems has been explored by other researchers, such as Kim et al. [18], who, in contrast to this study, integrated the drug into lipid nanoparticles. This nanoencapsulation approach provides significant benefits for drug release [Figure 2] [19].

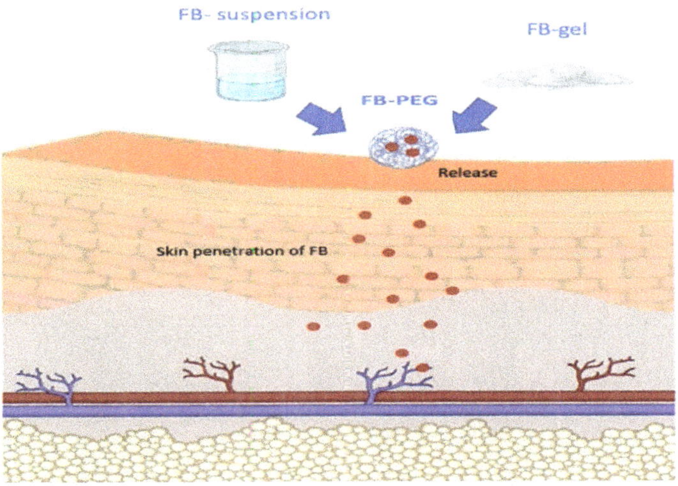

Figure 2. Flurbiprofen release process (red spheres) from nanoparticles.

Although nanoparticles offer significant benefits, we also explore incorporating formulations into hydrogels, as hydrogel-based drug delivery systems pose challenges in achieving optimal stability, biodegradability, and targeted efficacy while maintaining biocompatibility. Additionally, hydrogels provide user-friendly administration and convenient storage. Incorporating drugs into hydrogels further enhances the stability and bioavailability of compounds through controlled release mechanisms. Ultimately, this review highlights the potential of hydrogels in biomedical research and underscores the need for continuous innovation to overcome existing challenges and expand their clinical applications [20].

The in vivo efficacy of the formulations was evaluated by performing controlled perforations in mouse ears, simulating the application of an ear tag. Additionally, cytotoxicity and ocular tolerance studies were conducted to assess potential formulation-induced damage and irritability in cases of accidental eye contact, which is particularly relevant for eyebrow piercings. Histological analyses were also performed, and the chemical stability of the formulations was assessed over a three-month period at three different temperatures: 4 °C, 25 °C, and 40 °C.

Both formulations demonstrated enhanced transcutaneous permeability in porcine skin and human skin [8]. Consequently, their anti-inflammatory efficacies were evaluated in both species, along with determining drug retention in porcine skin and human skin.

2. Results and Discussion

2.1. *Effectiveness Studies on Inflammation Prevention*

To verify that the formulations under study have a preventive effect on reducing inflammation resulting from a perforation process, an in vivo study was conducted, with four groups of mice ($n = 6$) being evaluated in parallel (Figure 3).

The results shown in Table 1 concern the ear thickness of the rats. Regarding the studied formulations, it can be observed that group 3, corresponding to animals that received FB-suspension, and group 4, corresponding to animals that received FB-gels, exhibited a preventative effect against inflammation, as significant differences were observed among all groups. The graphical representation corresponds to Figure 4a.

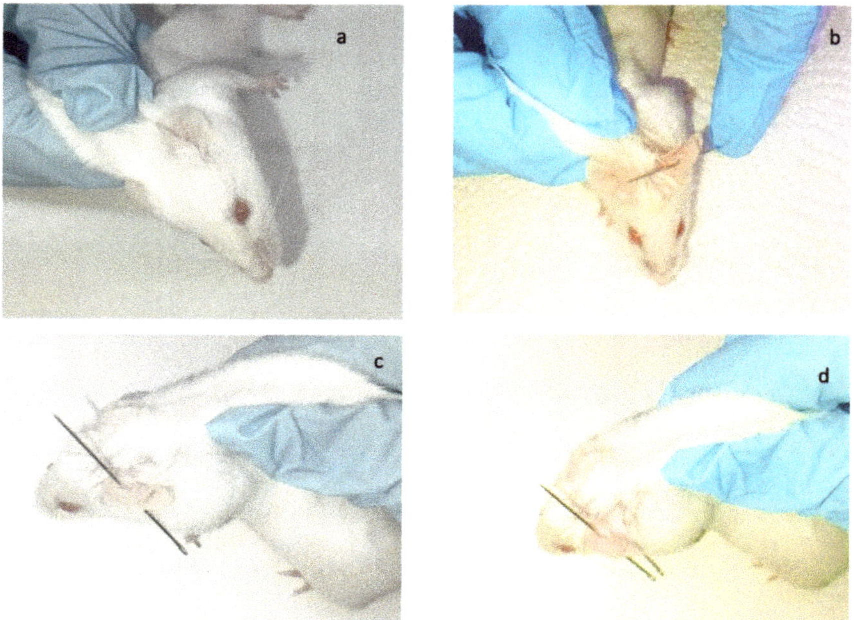

Figure 3. (**a**) Group 1: negative control, without perforation or preventive anti-inflammatory treatment; (**b**) group 2: positive control, which received only perforation; (**c**) group 3: treated with FB-suspension; (**d**) group 4: treated with FB-gel.

Table 1. Measurements of ear thickness.

	Thickness (mm)	*p* Value
Group 1	0.06 ± 0.05	Not applicable
Group 2	0.24 ± 0.02	$p < 0.001$
Group 3	0.09 ± 0.03 [a]	$p < 0.001$
Group 4	0.14 ± 0.01 [a,b]	$p < 0.001$

[a] Statistically significant difference compared to group 2; [b] statistically significant difference compared to group 3.

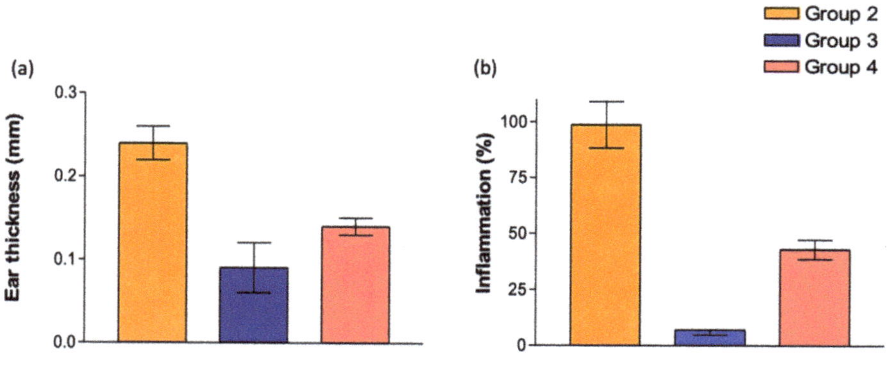

Figure 4. Graphical representations of (**a**) ear thickness and (**b**) percentage of inflammation.

After euthanasia was performed on the animals, the ears were weighed, and the percentage of ear inflammation was calculated, considering the weight of the positive control to represent 100% inflammation. On the other hand, since the negative control group did not undergo any induced inflammatory process or receive treatment, the values from this group were deemed not applicable for subsequent statistical analysis.

As indicated in Table 2, significant differences were observed among all groups. Groups 3 and 4, treated with the studied formulations, exhibited inflammation levels below 50% compared to the positive control, demonstrating the efficacy of the preventative action. Furthermore, in both cases, the FB-suspension showed the highest anti-inflammatory effect.

Table 2. Inflammation percentage.

	Inflammation (%)	p Value
Group 1	Not applicable	Not applicable
Group 2	98.74 ± 10.2 [a]	$p < 0.001$
Group 3	6.98 ± 2.24 [a]	$p < 0.001$
Group 4	43.02 ± 4.25 [a,b]	$p < 0.01$

[a] Statistically significant difference compared to group 2; [b] statistically significant difference compared to group 3.

The formulations followed the same activity pattern in both analyzed experiments. Based on ear thickness measurements, statistically significant differences were observed across all groups. These results are also reflected in Figure 4b.

El Moussaoui et al. [21] obtained results similar to ours. They studied inflammation inhibition and found results comparable to those of our study, showing that drugs formulated with nanoparticles in suspension exhibit better anti-inflammatory effects than those formulated in gels.

In the mouse ear tissue studies, we observed the positive control (Figure 5A), where the skin is damaged, with a broken epidermis identified. Inflammation in the skin (I) and some inflammatory infiltrates were also observed. This ear corresponds to the mouse that experienced perforation without any treatment. Figure 5B shows the negative control, where the epidermis (E) and dermis (D) are intact and without inflammation. In Figure 5C, we can observe the ear of the mouse to which the gel loaded with flurbiprofen nanoparticles was applied before the piercing. Here, we can see that the formulation improved the inflammation generated by the applied piercing, but not entirely. In the lower zone, we observe an area with inflammatory infiltrates and skin thickening, indicating an inflammatory process (the analyzed skin pieces are sourced from around the piercing site). Finally, in Figure 5D, we see the ear of the mouse where the suspension of flurbiprofen nanoparticles was applied. The skin is structurally well observed, without any inflammation. It is very similar to the negative control, indicating that this formulation prevented inflammation more effectively than the gel studied (this analyzed skin is a piece taken from around the perforation).

Pain and inflammation are among the most common disorders in contemporary clinical medicine, often occurring in processes where piercings are applied [22]. To prevent these inflammatory processes when performing skin piercings, two formulas have been used: a suspension and a gel loaded with flurbiprofen nanoparticles. These formulas were applied in mouse models, specifically in the ears, after piercing with a sterile pin to demonstrate their anti-inflammatory capacities.

In the histological studies of the skin in mouse ears, it was observed that the suspension of flurbiprofen nanoparticles generated more marked and effective prevention of inflammation processes compared to applying the gel, which prevented inflammation, albeit not completely (Figure 5C,D). This is consistent with the permeation studies of both formulations, where more of the flurbiprofen nanoparticle suspension was retained in the skin compared to the gel, which would explain this more marked anti-inflammatory effect in the ears of the mice. It is known that inflammation is an essential protective response of the body, which develops against tissue damage caused by physical trauma, chemical irritants, etc., and to eliminate harmful metabolites and repair tissues. The widespread oral use of NSAIDs such as flurbiprofen has been undertaken, but this can cause some undesirable side effects [23]; thus, nowadays, dermal and transdermal formulations are preferred, as carried out in this study [24].

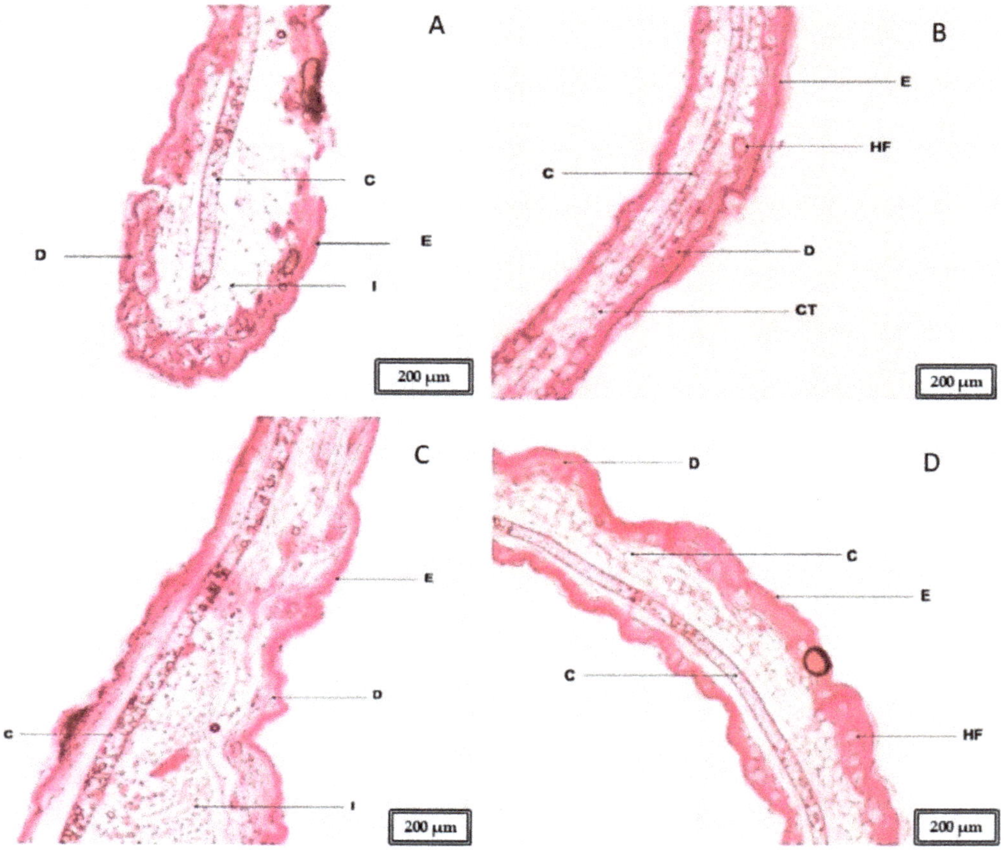

Figure 5. (**A**) Positive control of mouse ear; (**B**) negative control of mouse ear; (**C**) mouse ear where the FB-gel was applied; (**D**) mouse ear where the FB-suspension was applied; E: epidermis; D: dermis; I: inflammation; C: cartilage; CT: connective tissue; HF: hair follicle. Scale bar: 200 µm.

2.2. *Cumulative Retained Drug Amount Studies in Porcine and Human Skin (Qr)*

The retained amount of FB (Qr) in human and porcine skin from both formulations is determined using Franz cells after administering the formulations topically at 15 and 30 min and subsequently analyzed via HPLC [25]. The objective is to determine the amount of drug retained in the skin of the target species. This assessment is based on applying the product 30 min before perforation or piercing, following the same methodology used to evaluate inflammation prevention.

As shown in Figure 6, Figure 6a,b represent pig skin, while Figure 6c,d correspond to human skin.

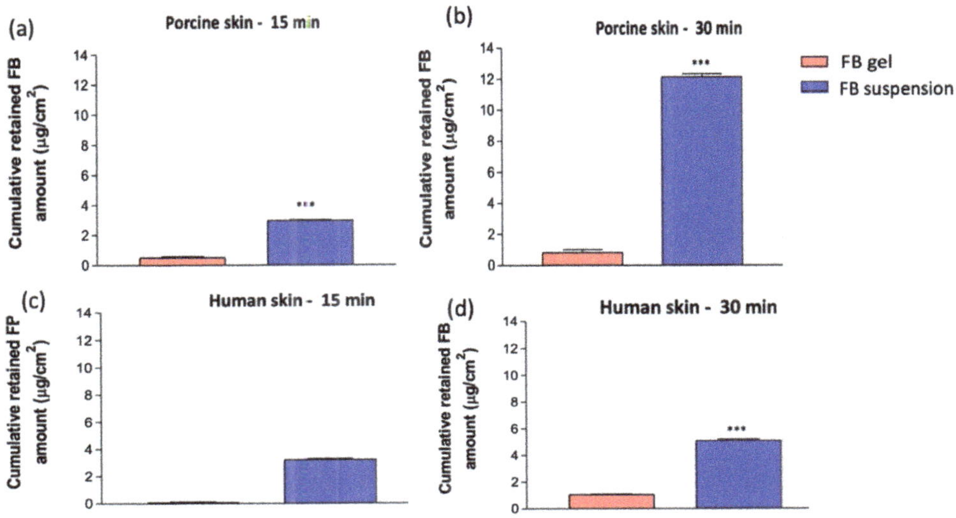

Figure 6. The amount of FB retained in porcine skin (**a**,**b**) and in human skin (**c**,**d**) at 15 and 30 min; ***: $p < 0.0001$.

Regarding the sampling time, a directly proportional relationship is observed between the duration and the amount retained in both species. At 15 and 30 min, both formulations remain in the susceptible tissues to exert their therapeutic effect. However, from a comparative perspective, the tissue evidently acts as a reservoir, progressively absorbing the active ingredient. This results in a higher amount being retained at 30 min.

Regarding the type of formulation, statistically significant differences are observed, showing that FB incorporated into nanoparticles in suspension is retained much more effectively compared to its delivery via hydrogel. The drug's ability to be retained by the skin is higher when the FB-suspension is administered in both species. It is well recognized that a drug's diffusion through a biological membrane is affected not only by its physicochemical traits but also by how the formulation interacts with the skin. Ultimately, when comparing species, porcine skin demonstrates a greater capacity for drug retention after 30 min of the experiment.

Regarding the gel, the amount of drug retained is of the same order, with no significant differences between 15 and 30 min. These results align with those obtained in the previous section, where the inhibition percentage of the FB-suspension is higher than that of the FB-gel. Although the retained amount is lower, this does not exclude the therapeutic efficacy of FB-gel, as seen in the previous section, where both formulations show an efficacy greater than 50% in inflammation inhibition.

2.3. Stability Essay

The stability results of the suspensions were published by Ramos et al. [10,11]. Chemical stability studies on the gels were conducted by quantifying the drug content at different temperatures using HPLC analysis to determine the optimal storage temperature range for the formulation. A sample of FB-gel, prepared at room temperature (25 °C), was stored for three months at three different temperatures: 4 °C (refrigerated), 25 °C (ambient), and 40 °C (incubator). At the end of this period, the FB content was assessed using HPLC analysis. The results indicate no decrease in concentration over time, with no statistically significant differences observed in the obtained values (Figure 7).

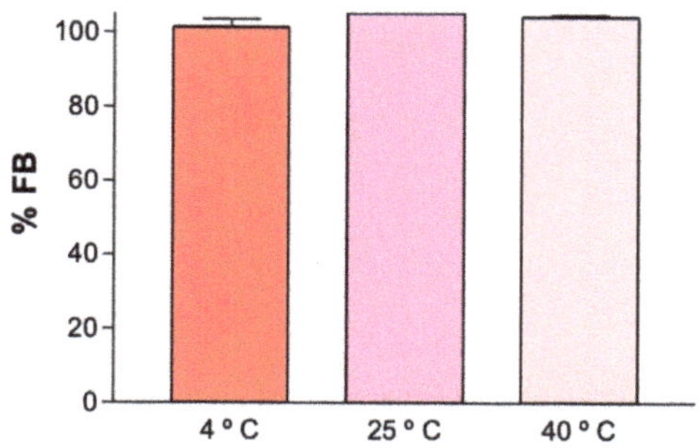

Figure 7. The percentage of FB in the FB-gel formulation stored at different temperatures after three months. No significant differences are observed.

This stability may be attributed to the presence of the cryoprotectant PEG in the formulation. Studies conducted by Ramos et al. utilized PEG as a cryoprotectant for developing FB nanoparticles intended for ocular administration. It was concluded that formulations containing FB-PEG exhibited superior permeation in human and porcine skin compared to other cryoprotectants [9].

PEG is widely used in biomedical applications, including drug formulations, vaccine stabilization, and regenerative medicine. It is an effective cryoprotectant that prevents ice crystal formation, stabilizes cell membranes and proteins, and reduces osmotic stress during freezing. Its biocompatibility and ability to maintain enzymatic activity make it ideal for the cryopreservation of cells, proteins, and nanoparticles [26,27]. Gupta et al. reported that incorporating PEG reduced aggregation and facilitated the redispersion of the final product after freezing and dehydration processes [28]. Similarly, other studies have evaluated the effect of PEG on liposomes. For example, Kim et al. [29] demonstrated that PEG-coated liposomes exhibit fewer changes in particle size and reduced aggregate formation during freeze drying, ultimately resulting in enhanced system stability.

It is also important to consider that hydrogels function as highly effective cryoprotectants by regulating temperature fluctuations, inhibiting ice crystal formation, and mitigating thermal stress during freezing and thawing processes. Their polymeric network facilitates water retention, thereby minimizing dehydration at low temperatures, while their viscoelastic properties absorb mechanical stress induced through thermal expansion and contraction [30,31]. Furthermore, hydrogels offer extended thermal stability, creating a protective microenvironment that safeguards cells and biomolecules from extreme temperature variations [32,33].

The FB-gel exhibits stability within a temperature range of 4 °C to 40 °C. However, for user convenience, storage at room temperature (25 °C) is recommended. If presented as a magistral formulation, no specific temperature conditions would be required. As previously mentioned, the excipients accompanying FB are likely responsible for providing thermal protection for three months, consistent with the study period.

2.4. *In Vitro Cytotoxicity Assesment in HaCaT Cells*

Assessing cell viability is essential for ensuring the safety of the developed formulation and to prevent potential cytotoxic effects that could disrupt skin barriers and compromise the body's primary protective layer. As shown in Figure 8, the FB-solution does not affect cell viability after 24 h of exposure at any of the tested concentrations. The two formulations shown in Figure 8 exhibit cell viability above 80%, indicating that free FB is not considered cytotoxic. However, the statistically significant differences observed between free FB and FB incorporated into nanoparticles suggest that nanoparticles can reduce cytotoxicity.

Therefore, this approach represents a safe and effective drug delivery system. Nanoparticles play a crucial role in reducing drug toxicity by enhancing targeted delivery, controlling release, and improving bioavailability. Encapsulating drugs in nanoparticles minimizes their exposure to healthy tissues, reducing systemic toxicity and side effects. Polymer-based nanoparticles, such as poly(ε-caprolactone), provide sustained drug delivery, decreasing toxicity peaks [34].

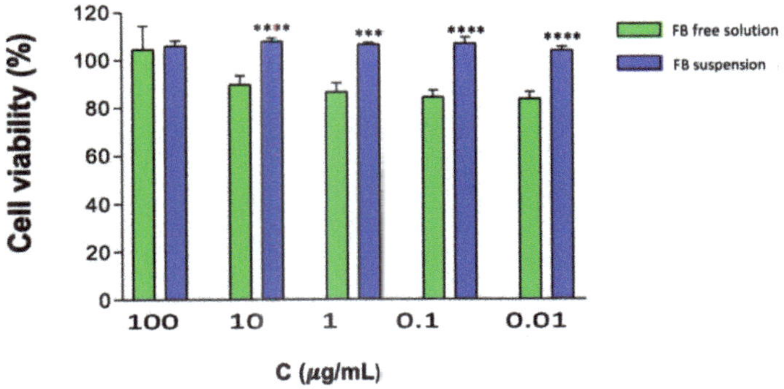

Figure 8. The cell viability of primary human keratinocytes (HaCaT) after 24 h of exposure. The data show significant differences compared to the suspension: *** $p < 0.0001$; **** $p < 0.00001$.

FB-NP-mediated drug release may prevent cell death induced by the free drug solution. Since FB is encapsulated with PCL, it reduces the direct contact surface of the active ingredient with HaCat cells, which may explain the statistically significant differences between the two formulations.

In accordance with previous in vitro studies evaluating FB cytotoxicity [35,36], the FB-suspension did not induce cell death rates exceeding 10%, suggesting that it may be safe for dermal administration.

Regarding the gel formulation (Figure 9), a lower cell viability was observed, which improves as the formulation is diluted. However, the gel was not considered toxic, as its cytotoxicity was previously evaluated by Berenguer et al. [37]. They assessed the cytotoxicity of Sepigel® hydrogel on HaCat cells, observing that the inhibitory concentration in the analyzed cell lines was higher than 75 μg/mL, concluding that this excipient is not cytotoxic.

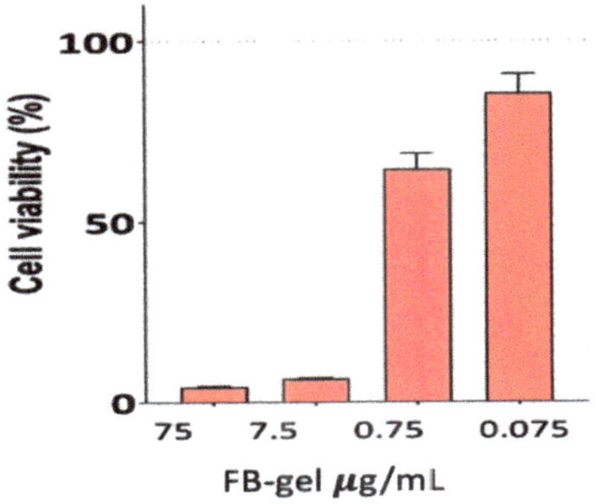

Figure 9. The cell viability of primary human keratinocytes (HaCaT) after 24 h of exposure to the composite gel.

2.5. Histological Analysis of Human and Porcine Skin Treated with Different Formulations

Histological images (Figure 10) provide a comparative analysis of human (A, B, C) and porcine (D, E, F) skin following treatment with different formulations. Control samples (A, D), treated only with saline solution (SF), exhibit a well-preserved epidermal and dermal structure, serving as a baseline for comparison. In the gel-treated samples (B, E), the stratum corneum in the human skin (B) appears structurally similar to the control (A), whereas porcine skin I shows signs of epidermal dehydration. Conversely, in the suspension-treated samples (C, F), human skin (C) exhibits more pronounced dehydration compared to its control, while porcine skin (F) maintains better hydration relative to the gel-treated group.

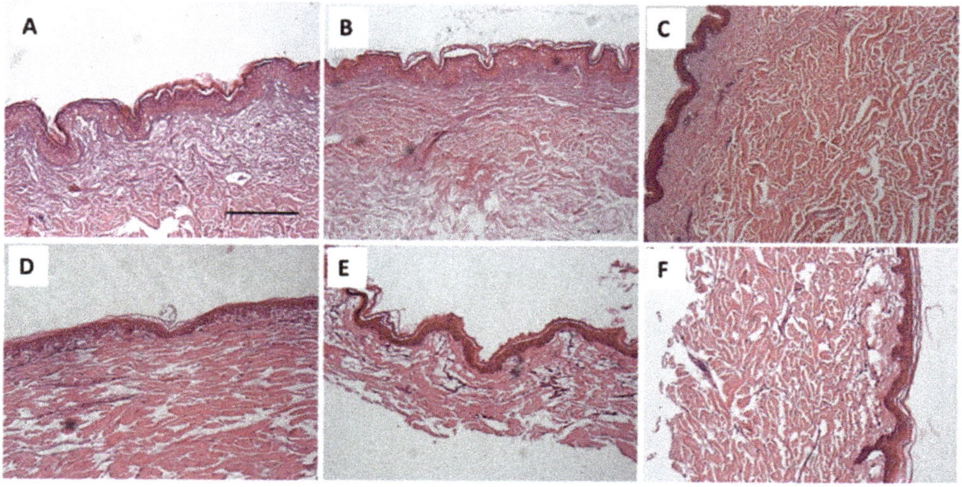

Figure 10. Histological images of human (**A–C**) and porcine (**D–F**) skin after treatment. Controls (**A,D**) show intact structure. FB-gel treated samples (**B,E**) maintain human skin integrity but dehydrate porcine skin. FB-suspension (**C,F**) causes more dehydration in human skin but better preserves porcine skin. Scale bar = 200 μm.

These findings suggest that the gel formulation preserves human skin integrity more effectively, whereas the suspension performs better in porcine skin. However, as the treatments were applied prophylactically over a short period, these histological differences are unlikely to significantly impact overall efficacy.

Notably, our previous study [9] evaluated the effect of these formulations on skin barrier function using trans-epidermal water loss (TEWL) measurements in vivo. The results demonstrated that any temporary disruptions in barrier integrity were reversible within a few hours, suggesting that despite the histological differences observed in ex vivo conditions, the formulations do not significantly affect the long-term barrier function of the skin [30,38].

Human skin and porcine skin share several similarities that make them useful in comparative studies, such as the presence of essential lipids—ceramides, cholesterol, and fatty acids—that play key roles in skin barrier function, as well as a generally similar structure composed of the epidermis, dermis, and hypodermis. However, they also present important differences in lipid composition. For example, the specific composition of ceramides varies between the two species, which may influence skin permeability and hydration. Likewise, the proportion and type of fatty acids differ, which can affect the texture of the skin and its response to environmental factors. Pigs have a higher density of sebaceous glands compared to humans, increasing the production of sebum and oilier skin. Moreover, although both types of skin perform a barrier function, there are structural and functional differences that make human skin generally more permeable than porcine skin. Finally, water content also varies, with human skin typically containing a higher amount of water.

Liu et al. [39] point out that the age of pigs is an important factor to consider when comparing species. The skin of young pigs appears to be more similar to that of adult humans than the skin of adult pigs.

On the other hand, as shown in Figure 10, corresponding to the histological analysis section, image E corresponds to porcine skin treated with the FB-gel. We believe that the observed dehydration is due to the gel drawing water from the skin in vitro, where there is no dynamic exchange in the process. Additionally, although the main purpose of PEG-3350 in the formulation is to act as a cryoprotectant agent, its high osmotic capacity should be considered, as this explains its ability to effectively attract and retain water. Since it is a non-absorbable compound, it tends to accumulate in the outermost tissues [40].

2.6. *In Vitro Tolerance Study: Hen's Egg Test on the Chorioallantoic Membrane (HET-CAM)*

The results for the FB-suspension have already been published by Ramos et al. [11]. Their findings indicate that the FB-suspension does not induce hemorrhage, lysis, or coagulation, suggesting its suitability for periocular application.

As we developed a topical gel formulation for periocular administration, it was essential to evaluate its ocular tolerance. Although our formulations are not intended for direct ocular application, assessing their tolerance was necessary in case of accidental contact with the cornea during periocular piercing. In this context, we conducted an in vitro HET-CAM assay to evaluate the ocular tolerance of the FB-gel formulation. The images corresponding to the experimental process are shown in Figure 11.

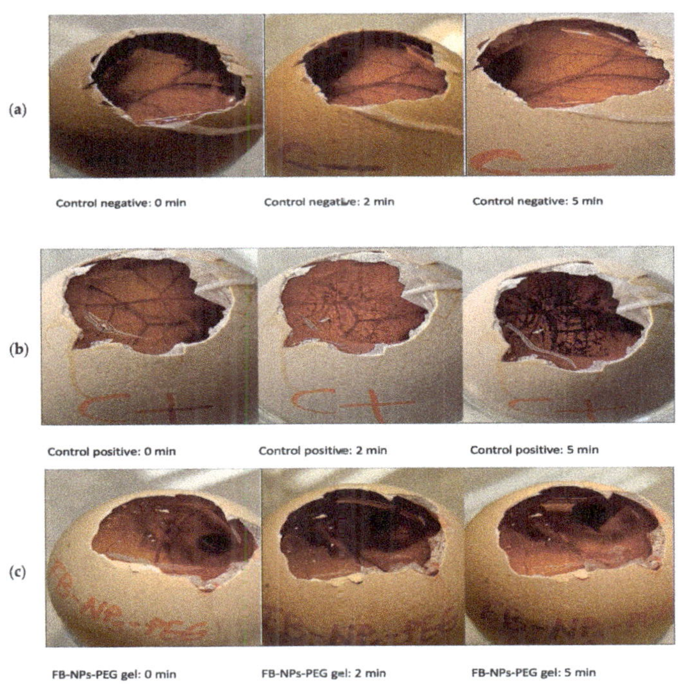

Figure 11. An assessment of the irritant potential of the formulations using the HET-CAM method: (**a**) negative control (saline solution); (**b**) positive control (sodium hydroxide solution 0.1 N; (**c**) FB-gel (study formulation).

The irritation score for the positive control was 20.49 ± 1.95. In contrast, no signs of ocular irritation, such as coagulation, vascular lysis, or hemorrhage, were observed within 5 min when testing the various nanoparticles (including the FB-PEG formulation) and the negative control (0.9% NaCl). The irritancy score remained below 0.1 (Table 3), confirming the safety of the nanoparticle formulations for ocular use, as no reactions were recorded during the test period [41].

Table 3. Irritation score (IS) of the composite gel tested using the HET-CAM technique.

Formulation	Irritation Score (IS)
FB-gel	0.03

In El Bejjaji et al.'s study [9], the NP gel formulation without FB incorporation was evaluated through trans-epidermal water loss (TEWL) in an in vivo experiment with healthy human volunteers to determine whether the excipients in the formulation irritate when administered topically. The TEWL experiment serves as an indirect biomarker of skin irritation, as alterations in the skin barrier are often associated with inflammation, dryness, sensitivity, and increased permeability to irritant substances. It is a key tool for the safety and efficacy assessment of topical products [30,38]. The study concluded that applying the composite gel did not disrupt the stratum corneum and was well tolerated by the skin.

3. Conclusions

Our investigation demonstrates the efficacy of flurbiprofen suspension and composite gel formulations in attenuating the inflammation associated with cutaneous perforations, including ear tagging procedures in livestock and periocular piercings or microdermal implants in clinical settings. The FB-suspension exhibited superior anti-inflammatory activity, attributed to enhanced permeability and retention mechanisms. At the same time, the gel formulation offered distinct advantages in terms of application convenience and prolonged storage stability. Both preparations effectively suppressed inflammatory responses, with comprehensive stability and safety assessments confirming their suitability for dermal and periocular applications.

Numerous investigations underscore the need for alternative approaches to minimize distress in animals subjected to ear tagging procedures, as substantial evidence indicates that this practice induces both acute and persistent discomfort. Research consistently demonstrates that ear tagging elicits marked stress responses, emphasizing the necessity for developing less invasive methodologies [3,42]. Analogously, the clinical literature documents vasovagal reactions and anxiety-related manifestations associated with needle phobia in patients undergoing ear piercing. The prophylactic application of these formulations may substantially reduce stress, improve recipient comfort, and facilitate a more conducive procedural environment for practitioners [43].

These results substantiate the therapeutic potential of nanotechnology-based drug delivery systems for managing inflammatory sequelae associated with cutaneous perforations. Future research should focus on refining these formulations for broader dermatological applications while exploring innovative drug delivery mechanisms that enhance bioavailability, improve patient compliance, and, ultimately, lead to better therapeutic outcomes.

In vivo models have demonstrated the pre- and post-procedural anti-inflammatory efficacy of these formulations in mice, and drug retention has been confirmed 30 min post-application in porcine and human skin models.

Accordingly, subsequent research endeavors should emphasize preclinical evaluations using porcine models, given their greater physiological similarity to human integument. Conducting additional preclinical studies in pigs would enable evaluating in vivo kinetics and the impact of the formulation on wound healing. Additionally, its impact on reducing the risk of infection associated with ear tag use can be analyzed, as can its potential contribution to lowering stress levels in animals during this process. These studies would provide valuable information with translational relevance and could help to optimize application protocols.

For prophylactic inflammation management, directly applying liquid formulations to the perforation site prior to intervention warrants further exploration, potentially facilitating incorporation into spray delivery systems [44]. Post-procedural care might benefit from sterile single-dose ampoule packaging to minimize infection risk and optimize healing outcomes.

Alternatively, the gel formulation presents as an optimal candidate, offering practical administration, cost advantages compared to spray systems, and superior stability profiles.

Sepigel® has gained widespread acceptance as a gelling agent in cosmetic preparations, demonstrating remarkable efficacy across diverse pH environments. Beyond its primary thickening function, it confers additional stabilizing and texturizing benefits. Its moderate viscosity characteristics and distinctive opalescent properties contribute to a refreshing, non-occlusive sensory profile, rendering it particularly suitable for dermo-cosmetic applications.

These formulations represent valuable adjunctive therapies for pre- and post-procedurally managing cutaneous perforations, complementing established hygiene protocols designed to prevent infection and enhance wound resolution. While controlled perforations are performed under aseptic conditions, pain and discomfort remain inherent physiological responses to localized tissue trauma. This investigation aims to develop an accessible over-the-counter pharmaceutical preparation addressing these specific requirements in both veterinary and clinical practice.

4. Materials and Methods

4.1. Materials and Reagents

Flurbiprofen, poly(ε-caprolactone) (Mw $\approx$ 14,000 g/mol, Mn $\approx$ 10,000 g/mol, dispersity = 1.4), PEG-3350, and acetone were obtained from Sigma-Aldrich Co. (St. Louis, MO, USA). Poloxamer 188 (P188; Lutrol® F68) was sourced from BASF (Barcelona, Spain), while Sepigel® 305 (polyacrylamide, C13-14 isoparaffin, laureth-7) was procured from Acofarma (Barcelona, Spain). Phosphate-buffered saline (PBS) tablets were purchased from Sigma-Aldrich Chemie (Steinheim, Germany) and prepared per manufacturer's instructions. Double-distilled water was filtered using a Millipore® system (EMD Millipore, Billerica, MA, USA). High-performance liquid chromatography (HPLC) reagents were obtained from Fisher Scientific (Leicestershire, UK).

4.2. Preparation of the Gels Loading Flurbiprofen Nanoparticles

Flurbiprofen nanoparticles were developed using a solvent displacement technique, as described by Fessi et al. [45] and modified by Ramos et al. [10,11]. In total, 15 mg of FB and 49.5 mg of PεCL dissolved in 30 mL of acetone were inserted dropwise into 60 mL of an aqueous P188 solution at a pH of 3.5 under moderate magnetic stirring. The suspension of FB-PεCL was concentrated to 15 mL using a rotary evaporator (R-144; Buchi, Flawil, Switzerland), removing acetone. This process allowed us to obtain an average particle size smaller than 200 nm and a polydispersity index value of 0.088 $\pm$ 0.011, common in a monodisperse colloidal suspension, at 25 °C using a Zetasizer Nano ZS (Malvern Instrument, Malvern, UK). Following this line, PEG (160 mg/mL) was added to 15 mL of formulation to protect nanoparticles from stressful processes such as lyophilization.

The nanosuspension (NPs-PEG) was lyophilized, irradiated at 25 KGy, and reconstituted in water. After that, average particle size and the polydispersity index were determined again; they were 187.5 ± 1.5 nm and 0.076 ± 0.013, respectively. In accordance with TEM analysis results, these nanoparticles' sizes were assayed by El Bejjaji et al. [9].

For the gel formulation, 0.55 g of Sepigel® 305 was added to 5 mL of the nanoparticle suspension under constant agitation, forming a homogenous semisolid nanocomposite gel with a final flurbiprofen concentration of 0.75 mg/mL.

The suspension (NPs-PEG), after being lyophilized and irradiated, was reconstituted in water to prepare two composite gels. To complete the gelation process, 0.55 g of Sepigel® 305 was added to 5 mL of each suspension and mixed thoroughly under agitation to ensure uniform consistency. This resulted in the formation of thin, yet distinct, semisolid composite gel structures [9]. The drug concentration obtained in the final formulation was equivalent to 0.75 mg/mL of flurbiprofen.

4.3. Biological Materials

Human and porcine skin were used to determine drug retention in the skin (Qr). This study was approved by the Bioethics Committee of the Barcelona SCIAS Hospital (Protocol N°002; 17 January 2020). The flank skin of Yorkshire-Landrace pigs was obtained from the animal facility at the Bellvitge Campus of the University of Barcelona (Barcelona, Spain) immediately after the animals were sacrificed for various purposes. The studies were conducted following a protocol approved by the Committee of Animal Experimentation of the Regional Autonomous Government of Catalonia (Spain) and the Animal Experimentation Ethics Committee of the University of Barcelona (Barcelona, Spain) under reference number 7428. Skin samples were sectioned using a GA 630 dermatome (Aesculap, Tuttlingen, Germany) at different thicknesses depending on the skin type—400 and 700 μm for human and porcine skin, respectively, after being frozen at −20 °C.

For in vivo studies, adult male CD-1 mice (20–23 g) were obtained from the Experimental Center for Bioscience at the Faculty of Chemical Sciences and Pharmacy, the National Autonomous University of Honduras (CENBIO-UNAH). The animals were housed in plastic cages with soft bedding, provided with a controlled diet, and given tap water ad libitum. Environmental conditions were maintained at 24 ± 1 °C, with relative humidity between 50 and 60%. Additionally, light conditions followed a 12 h light/12 h dark cycle within each 24 h period. This experiment received approval from the Ethics Committee CENBIO-UNAH (Protocol code CICUAL 002-2025, approved on 14 March 2025).

4.4. Drug Analysis and Quantification

The determination of FB was conducted using high-performance liquid chromatography (HPLC). The conditions are explained in Table 4, and the process is represented in Figure 12.

Table 4. The chromatographic conditions for the determination of flurbiprofen.

Parameter	Conditions
Chromatographic column	Luna ® 5 μm C18 (2) 100 Å (150 × 4.6 mm)
Mobile phase	Acetonitrile: PBS (pH 2.5) (65:25) (*v*/*v*)
Flux	1.5 mL/min
Injection volume for Qr	10 μL
Run time	3.2 min
Wavelength	245 nm
Stability range	400–0.391 μg/mL
Qr range	100–6.25 μg/mL

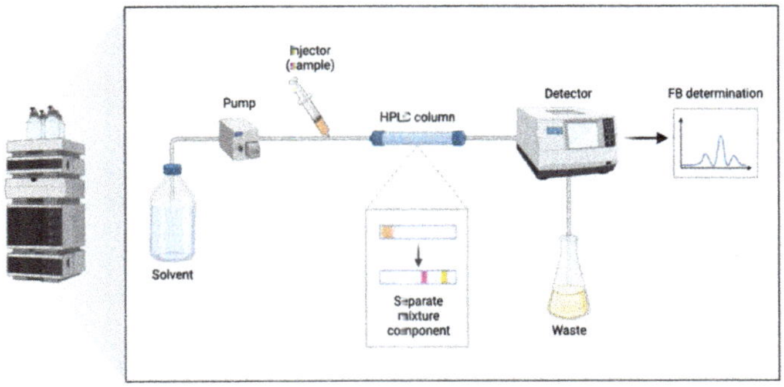

Figure 12. Drug analysis and quantification of flurbiprofen in skin samples using RP-HPLC.

4.5. Methods

4.5.1. Inflammation Prevention Assessment

To evaluate the in vivo efficacy of the formulations, four groups of male CD-1 mice (n = 6) were formed and kept under standard animal facility conditions according to regulations. The experimental conditions received by each study group are described in Table 5, and the visual representation of the method is detailed in Figure 13.

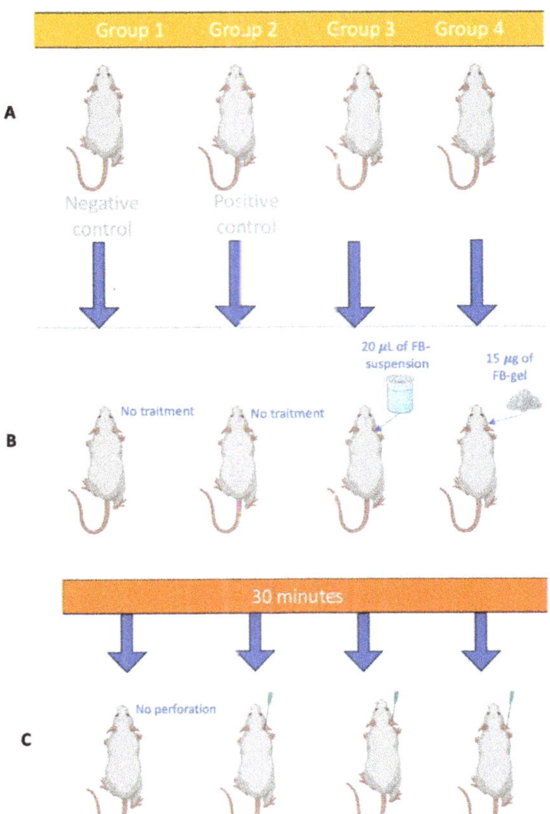

Figure 13. The representation of the in vivo procedure: (**A**) Distribution of the groups; (**B**) the preventative administration of the study formulations to the study groups; (**C**) ear perforation.

Table 5. The classification of the groups.

	Preventive Treatment	Perforation	
Group 1	No	No	Negative control
Group 2	No	Yes	Positive control
Group 3	FB-suspension	Yes	Study group
Group 4	FB-gel	Yes	Study group

Groups 3 and 4 received preventative anti-inflammatory treatment topically on the left ear, and after 30 min, perforation was performed on the study ears of groups 2, 3, and 4. Group 1, the negative control, received neither anti-inflammatory treatment nor perforation.

The representation of the procedure is described in Figure 13.

Once the preventative action and the trauma were enacted, the degree of effectiveness was evaluated using two methods. On one hand, the thickness of the ears was measured with a Mitutoyo® 547-561S Thickness Gage Steel digital clipper (Kawasaki, Japan) 45 min after perforation.

Then, after 4 h, the animals were slaughtered, and 7 mm circular sections of the left ear were cut to determine anti-inflammatory activity (Figure 14). The percentage of inhibition was calculated using the following formula:

$$\text{Inflammation}(\%) = \frac{(Weight\ of\ study)}{Weight\ control\ positive} \times 100 \quad (1)$$

where the weight of the positive control is considered to represent the maximum inflammation (100%), and the weight of the negative control is considered to represent the absence of inflammation (0%), as it did not undergo any process that altered its morphology. As a reference, a 7 mm circular section in the negative control group weighed 0.02 ± 0.01 mg.

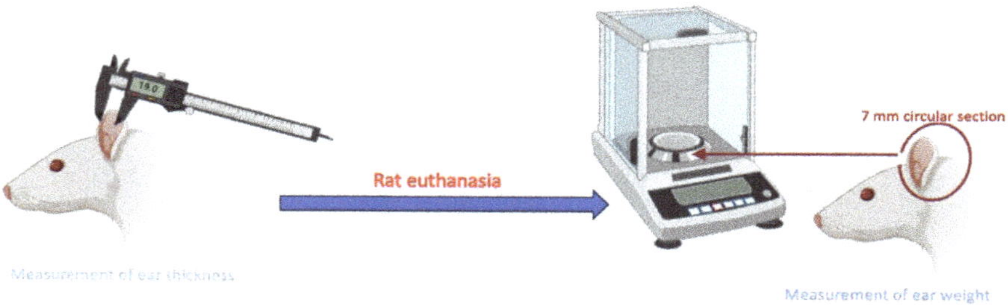

Figure 14. Efficacy analysis process.

4.5.2. Retained Drug Determination

The experiments were conducted in independent vertical Franz diffusion cells with a diffusional surface area of 0.64 cm^2. Skin tissues were positioned between the two compartments of a Franz cell, with the dermal side in contact with the receptor medium and the epidermal side in contact with the donor chamber. The skin was covered with laboratory film (parafilm, Chicago) to prevent evaporation during this study. Phosphate-buffered saline (PBS) solution at a pH of 7.4 was used as the receptor medium. The permeation study was conducted for 30 min at 32 ± 0.5 °C under continuous stirring in accordance with sink conditions. For the donor compartment, 100 µL of the solution formulation and 125 mg of the gel was applied once the temperature of the skin surface equilibrated to 32 ± 0.5 °C [9]. A saturated solution of FB in PBS was also assayed.

The porcine and human skin tissues were carefully removed from the Franz cell to determine the retained drug concentration. The skin surface was washed three times with gauze soaked in a 0.05% solution of sodium lauryl sulfate and distilled water. Excess skin surrounding the diffusion area was trimmed, and the exposed surface was gently blotted with filter paper to ensure that it was dry before weighing. The retained FB was extracted using an acetonitrile/water solution (50:50, *v:v*) via sonication for 15 min in an ultrasound bath. The resulting solutions were analyzed via RP-HPLC to determine the amount of FB retained in the skin, expressed as Qr (μg/cm^2). The results were normalized based on the tissue weight and the diffusion area (0.64 cm^2) and then multiplied by the drug recovery factor. The retained amount of drug in the tissue (Qr, μg/cm^2) was determined using the following equation:

$$Q_R = \frac{(Ex)R}{A \times 100} \quad (2)$$

where *Ex* (μg) is the quantity of drug extracted; *A* (cm^2) is the effective surface area accessible for diffusion; and R is the drug recovery percentage [46,47]. The experimental conditions are outlined in Table 6 and Figure 15.

Table 6. The experimental conditions for the ex vivo permeation test (*Qret*).

Parameter	Conditions
Receptor fluid	PBS (pH 7.4)
Cell volume	6 mL
Diffusion area	0.64 cm^2
Membrane	Pig and human skin
Replicates	5 replicates
Temperature	32 ± 0.5 °C
Stirring	500 r.p.m.
Dose	1 mg/mL
Sample volume (solution)	100 μL
Sample weight (gel)	125 mg
Sampling times	0 (pre-sampling), 15 min, 30 min

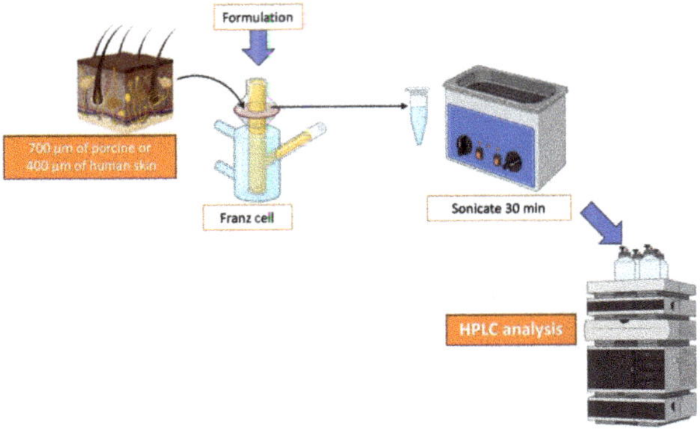

Figure 15. The experimental setup and conditions for the in vitro skin permeation study using Franz diffusion cells.

4.5.3. Chemical Stability in Different Temperatures

The stability of the suspension was reported by Ramos et al., who concluded that there were no statistically significant differences in stability over the period in which the sample was monitored [10,11].

For the composite gel, stability studies were conducted to determine whether exposure to different temperatures affected the active ingredient. A sample of the composite gel was used as the starting material. For the analysis, 0.1 g of the gel was dissolved in 1 mL of acetonitrile. The samples were stored in Eppendorf tubes at three different temperatures (40 °C, 25 °C, and 4 °C) and analyzed every month for three months. The analysis was performed under validated conditions, as specified in Table 4, with a sample size of n = 5 for each time point.

4.5.4. Cytotoxicity Study in HaCaT Cells

Cell viability in response to the flurbiprofen composite gel and solution was assessed using a 3-(4,5-dimethylthiazol-2-yl)-2,5-diphenyltetrazolium bromide (MTT) assay. The cytotoxicity potential of the free drug was also analyzed.

The immortalized keratinocyte cell line HaCaT was seeded at 2×10^5 cells/mL in 96-well plates (Corning) and incubated at 37 °C with 5% CO_2 for 24 h to allow for cell adhesion. Experiments were conducted once the cell confluence reached 80–90%. HaCaT cells were cultured in Dulbecco's Modified Eagle's Medium (DMEM) with high glucose content, supplemented with 25 mM HEPES, 1% non-essential amino acids, 100 U/mL of penicillin, 100 mg/mL of streptomycin, and 10% heat-inactivated fetal bovine serum (FBS).

Various dilutions of the formulations were tested (1/10, 1/100, 1/1000, and 1/10,000).

After 24 h of incubation, the HaCaT cells were washed with 1% sterile PBS and treated with an MTT solution (5 mg/mL) for 2 h at 37 °C. Following this, the medium was carefully aspirated, and 0.1 mL of 99% pure dimethyl sulfoxide (DMSO) was added to lyse the cells and dissolve the purple MTT crystals. The resulting cell lysate was transferred to a fresh 96-well plate, and the absorbance was measured at 540/630 nm excitation/emission wavelengths using an Automatic Microplate Reader (Modulus Microplate Multimode Reader, Turner Biosystems, Sunnyvale, CA, USA).

A negative control consisting of untreated cells was included for comparison. Absorbance values were directly proportional to cell viability, and the percentage of cell viability was calculated using the following equation:

$$\text{Cell viability} = \frac{\text{ABS treated cells}}{\text{ABS control cells}} \times 100 \quad (3)$$

The experiment was conducted in parallel since the two formulations had different concentrations. The suspension and free drug were tested at a concentration of 1 mg/mL (Figure 8), whereas the composite gel was formulated at 0.75 mg/mL (Figure 9).

4.5.5. Histological Analysis of Mouse Ear Tissue Following Perforation

For fixing the samples, several pieces cut from the treated mouse ears, positive control, and negative control were immersed for 24 h in a fixative mixture called Orth-ER liquid, which is a combination of potassium dichromate (5 g), glacial acetic acid (5 mL), commercial formalin (5 mL) (MERK, Darmstadt, Germany), and distilled water (90 mL). This solution must remain protected from light and heat. Next, the excess fixative was washed off with constant running water for 4 h. To dehydrate without causing damage to the tissues, the samples were immersed in a gradual series of ethanol solutions at different concentrations (50%, 60%, 70%, 80%, 90%, 95%, and 99%) for an average of 6 h at each concentration (DIMELAB, Tegucigalpa, Honduras).

The tissue had to be de-alcoholized, for which a lightening substance was used, making the tissue transparent and soluble in kerosene, where the sample to be cut was placed. In this case, Xylol (DIMELAB, Tegucigalpa, Honduras) was used for 6 h. Introducing the tissues in the kerosene had to be gradual, so the tissue was introduced in liquid kerosene of three parts xylol and one part kerosene (Química Industrial, Tegucigalpa, Honduras), then, in equal parts, one part solvent and three parts kerosene, ending with 100% kerosene; these changes were made at a temperature of 56 to 60 °C, and each change took, on average, from

3 to 6 h. The definitive inclusion was performed using metallic lead bars to position the tissue in the plane to be cut (transverse cut). Once the kerosene with the tissue was solidified at room temperature, the wooden block was placed as a support to perform the cuts. A Minot-type microtome (AO Scientific Instruments) was used to make the cuts, cutting the tissues in a thickness of 10 μm and performing kerosene stretching in water baths with thermal regulation. The cut sections were placed on a slide previously treated with an adhesive solution (previously prepared Haupt glue). For tissue staining, the hematoxylin/eosin combination staining battery technique (MERK, Darmstadt, Germany) was used. Finally, the coverslip was cleaned and glued with Entellan resin (MERK, Darmstadt, Germany), placed in an oven at a temperature of 28 to 35 °C, and left to cool for at least 24 h. It was then viewed and analyzed under an Olympus CX31 microscope equipped with a camera. The results are shown in Figure 5.

4.5.6. Histological Analysis of Porcine and Human Skin

Skin samples were processed using standard histological techniques. The samples were fixed in 10% neutral-buffered formalin at room temperature for 24 h, followed by dehydration through a graded ethanol series, clearing in xylene, and embedding in paraffin. Thin sections of 5 μm thickness were obtained using a rotary microtome and stained with hematoxylin and eosin (H&E) for microscopic evaluation.

The stained sections were examined under an Olympus BX41 light microscope equipped with an Olympus XC50 digital camera (Olympus Co., Tokyo, Japan) to assess tissue morphology and structural integrity. Images were captured and analyzed to identify any histopathological alterations. Quantitative evaluations were performed using ImageJ software, version 1.54k (NIH, Bethesda, MD, USA) in a blinded manner to ensure objectivity. The results correspond to Figure 10.

4.5.7. In Vitro Tolerance Study: Hen's Egg Test on the Chorioallantoic Membrane (HET-CAM)

The potential ocular irritation of the flurbiprofen composite gel was assessed using the hen's egg test on the chorioallantoic membrane (HET-CAM). This in vitro test evaluated toxicity by observing its effects on the chorioallantoic membrane (CAM) of 10-day-old embryonated hen's eggs, sourced from the G.A.L.L.S.A. farm, Tarragona, Spain. The reactions were monitored at two time points, 2 min and 5 min, within a 5 min observation window, focusing on the onset of hemorrhage (bleeding), coagulation (disintegration of blood vessels), and vessel lysis (protein denaturation both within and outside of blood vessels) [48].

Each response was analyzed separately, and the irritation score (IS) was calculated by combining the individual effects to classify the irritancy level of the substance. The irritation score was determined using the following equation:

$$IS = \frac{301 - H}{300} \times 5 + \frac{301 - L}{300} \times 7 + \frac{301 - C}{300} \times 9 \quad (4)$$

where H stands for hemorrhage; L stands for vessel lysis; C stands for coagulation; and time (s) stands for the number of seconds after which each reaction was observed. A total of 300 μL of the test substance was applied to the CAM, and the membrane was observed for 2 and 5 min to assess the severity of each reaction in accordance with the INVITTOX protocol [49,50]. NaOH (0.1 N) was used as the positive control, while a 0.9% NaCl solution served as the negative control. The interpretation of the results from Equation (4) can be derived from Table 7.

Table 7. Classification parameters for irritation severity.

Irritation Measurement	Severity Categorization
0–0.9	Non-Irritant
1–4.9	Slight Irritant
5–8.9	Moderate Irritant
9–21	Severe Irritant

4.6. Statistical Analysis

The results are reported as mean ± standard deviation. Statistical differences were determined using one-way analysis of variance (ANOVA) in GraphPad Prism® software v. 5.0 (GraphPad Software Inc., San Diego, CA, USA).

For in vivo efficacy and stability studies, the Tukey post hoc test was applied, while Student's *t*-test was used to assess cytotoxicity, as shown in Figure 8. A p-value < 0.05 was considered statistically significant.

Author Contributions: Conceptualization A.C.C., S.E.B. and A.P.; methodology, A.C.C., S.E.B. and G.R.-Y.; validation, D.C.M.D.S., L.S. and A.B.; formal analysis, S.E.B., A.C.C., D.C.M.D.S., A.B., L.S. and V.D.-V.; investigation, V.D.-V., D.C.M.D.S., A.B., L.S., M.J.R.-L. and M.Z.; resources, A.C.C.; data curation, A.C.C., S.E.B. and G.R.-Y.; writing—original draft preparation, S.E.B. and G.R.-Y.; writing—review and editing, S.E.B., G.R.-Y., D.C.M.D.S. and A.C.C.; visualization, A.C.C. and A.P.; supervision, A.C.C. and A.P.; project administration, A.C.C. All authors have read and agreed to the published version of the manuscript.

Funding: This research received no external funding.

Institutional Review Board Statement: This study was conducted in accordance with the Declaration of Helsinki and approved by the University of Barcelona (protocol code IRB00003099 ON 30 January 2019). The animal study protocol was approved by the Experimental Center for Bioscience at the Faculty of Chemical Sciences and Pharmacy, the National Autonomous University of Honduras (CENBIO-UNAH) (protocol code 002-2025, 14 March 2025).

Informed Consent Statement: Not applicable.

Data Availability Statement: The data presented in this study are available upon request from the corresponding author. As the data are part of a doctoral thesis, they will not be available until the thesis has been published.

Acknowledgments: The authors thank Josefa Badia and Sergio Martínez for their support with the cytotoxic studies and Salima El Moussaoui and Negar Beirampour for their contributions and suggestions. For the creation of Figure 1, Figure 2, Figure 12, Figure 13, Figure 14, and Figure 15, the authors acknowledge BioRender.com (accessed on 21 February 2025). We thank ChatGPT-4o for its assistance with grammatical and idiomatic corrections throughout the manuscript (accessed on 10 March 2025).

Conflicts of Interest: The authors declare no conflicts of interest.

References

1. Caja, G.; Hernández-Jover, M.; Conill, C.; Garín, D.; Alabern, X.; Farriol, B.; Ghirardi, J. Use of ear tags and injectable transponders for the identification and traceability of pigs from birth to the end of the slaughter line. *J. Anim. Sci.* **2005**, *83*, 2215–2224. [CrossRef] [PubMed]
2. Harmon, M.L.; Downey, B.C.; Drwencke, A.M.; Tucker, C.B. Development and Application of a Novel Approach to Scoring Ear Tag Wounds in Dairy Calves. *J. Dairy Sci.* **2023**, *106*, 5043–5053. [CrossRef] [PubMed]
3. Barz, A.; Ritzmann, M.; Breitinger, I.; Langhoff, R.; Zols, S.; Palzer, A.; Heinritzi, K. Examination of different options for combined administration of an NSAID (meloxicam) and iron for piglets being castrated. *Tierärztliche Praxis. Ausg. G Grosstiere/Nutztiere* **2010**, *38*, 23–30.

4. Hayer, J.J.; Nysar, D.; Schmitz, A.; Leubner, C.D.; Heinemann, C.; Steinhoff-Wagner, J. Wound lesions caused by ear tagging in unweaned calves: Assessing the prevalence of wound lesions and identifying risk factors. *Animmal* **2022**, *16*, 100454. [CrossRef]
5. Malik, M.; Chiers, K.; Theuns, S.; Vereecke, N.; Chantziaras, I.; Croubels, S.; Maes, D. Porcine ear necrosis: Characterization of lesions and associated pathogens. *Vet. Res.* **2023**, *54*, 85. [CrossRef]
6. Numberger, J.; Ritzmann, M.; Übel, N.; Eddicks, M.; Reese, S.; Zöls, S. Ear tagging in piglets: The cortisol response with and without analgesia in comparison with castration and tail docking. *Animal* **2016**, *10*, 1864–1870. [CrossRef]
7. Van Hoover, C.; Rademayer, C.A.; Farley, C.L. Body Piercing: Motivations and Implications for Health. *J. Midwifery Womens Health* **2017**, *62*, 521–530. [CrossRef]
8. Sheldon, R.R.; Loughren, M.J.; Marenco, C.W.; Winters, J.R.; Bingham, J.R.; Martin, M.J.; Eckert, M.J.; Burney, R.O. Microdermal Implants Show No Effect on Surrounding Tissue During Surgery With Electrocautery. *J. Surg. Res.* **2019**, *241*, 72–77. [CrossRef]
9. El Bejjaji, S.; Ramos-Yacasi, G.; Suñer-Carbó, J.; Mallandrich, M.; Goršek, L.; Quilchez, C.; Calpena, A.C. Nanocomposite Gels Loaded with Flurbiprofen: Characterization and Skin Permeability Assessment in Different Skin Species. *Gels* **2024**, *10*, 362. [CrossRef]
10. Ramos-Yacasi, G.R.; Calpena-Campmany, A.C.; Egea-Gras, M.A.; Espina-García, M.; García-López, M.L. Freeze Drying Optimization of Polymeric Nanoparticles for Ocular Flurbiprofen Delivery: Effect of Protectant Agents and Critical Process Parameters on Long-Term Stability. *Drug Dev. Ind. Pharm.* **2017**, *43*, 637–651. [CrossRef]
11. Ramos Yacasi, G.R.; García Lopéz, M.L.; Espina García, M.; Parra Coca, A.; Calpena Campmany, A.C. The Influence of Freeze Drying and Gamma-Irradiation in Pre-Clinical Studies of Flurbiprofen Polymeric Nanoparticles for Ocular Delivery Using D-(+)-Trehalose and Polyethylene Glycol. *Int. J. Nanomed.* **2016**, *11*, 4093–4106. [CrossRef]
12. Oscanoa-Espinoza, T.; Lizaraso-Soto, F. Antiinflamatorios No Esteroides: Seguridad Gastrointestinal, Cardiovascular y Renal. *Rev. Gastroenterol.* **2015**, *35*, 63–71.
13. Salido, A.; Abásolo, M.; Bañares, L. Revisión de los antiinflamatorios inhibidores selectivos de la ciclooxigenasa-2. *Inf. Ter. Sist. Nac. Salud* **2001**, *25*, 46–52.
14. Dorbandt, D.M.; Labelle, A.L.; Mitchell, M.A.; Hamor, R.E. The Effects of Topical Diclofenac, Topical Flurbiprofen, and Humidity on Corneal Sensitivity in Normal Dogs. *Vet. Ophthalmol.* **2017**, *20*, 160–170. [CrossRef] [PubMed]
15. Roberts, B.M.; Geddis, A.V.; Matheny, R.W., Jr. The dose-response effects of flurbiprofen, indomethacin, ibuprofen, and naproxen on primary skeletal muscle cells. *J. Int. Soc. Sports Nutr.* **2024**, *21*, 2302046. [CrossRef]
16. Siafaka, P.I.; Özcan Bülbül, E.; Okur, M.E.; Karantas, I.D.; Üstündağ Okur, N. The Application of Nanogels as Efficient Drug Delivery Platforms for Dermal/Transdermal Delivery. *Gels* **2023**, *9*, 753. [CrossRef]
17. Berthet, M.; Gauthier, Y.; Lacroix, C.; Verrier, B.; Monge, C. Nanoparticle-Based Dressing: The Future of Wound Treatment? *Trends Biotechnol.* **2017**, *35*, 770–784. [CrossRef]
18. Kim, R.M.; Jang, D.-J.; Kim, Y.C.; Yoon, J.-H.; Min, K.A.; Maeng, H.-J.; Cho, K.H. Flurbiprofen-Loaded Solid SNEDDS Preconcentrate for the Enhanced Solubility, In-Vitro Dissolution and Bioavailability in Rats. *Pharmaceutics* **2018**, *10*, 247. [CrossRef]
19. Alves, G.L.; Teixeira, F.V.; da Rocha, P.B.R.; Krawczyk-Santos, A.P.; Andrade, L.M.; Cunha-Filho, M.; Marreto, R.N.; Taveira, S.F. Preformulation and characterization of raloxifene-loaded lipid nanoparticles for transdermal administration. *Drug Deliv. Transl. Res.* **2022**, *12*, 526–537. [CrossRef]
20. Liu, B.; Chen, K. Advances in Hydrogel-Based Drug Delivery Systems. *Gels* **2024**, *10*, 262. [CrossRef]
21. El Moussaoui, S.; Abo-Horan, I.; Halbaut, L.; Alonso, C.; Coderch, L.; Garduño-Ramírez, M.L.; Clares, B.; Soriano, J.L.; Calpena, A.C.; Fernández-Campos, F.; et al. Polymeric Nanoparticles and Chitosan Gel Loading Ketorolac Tromethamine to Alleviate Pain Associated with Condyloma Acuminata during the Pre- and Post-Ablation. *Pharmaceutics* **2021**, *13*, 1784. [CrossRef] [PubMed]
22. Oktay, A.N.; Tamer, S.I.; Han, S.; Uludag, O.; Celebi, N. Preparation and In Vitro/In Vivo Evaluation of Flurbiprofen Nanosuspension-Based Gel for Dermal Application. *Eur. J. Pharm. Sci.* **2020**, *155*, 105548. [PubMed]
23. Pireddu, R.; Caddeo, C.; Valenti, D.; Marongiu, F.; Scano, A.; Ennas, G.; Sinico, C. Diclofenac acid nanocrystals as an effective strategy to reduce in vivo skin inflammation by improving dermal drug bioavailability. *Colloids Surf. B Biointerfaces* **2016**, *143*, 64–70. [CrossRef] [PubMed]
24. Kawadkar, J.; Pathak, A.; Kishore, R.; Chauhan, M.K. Formulation, characterization and in vitro–in vivo evaluation of flurbiprofen-loaded nanostructured lipid carriers for transdermal delivery. *Drug Dev. Ind. Pharm.* **2012**, *39*, 569–578. [CrossRef]
25. Parra, A.; Clares, B.; Rosselló, A.; Garduño-Ramírez, M.L.; Abrego, G.; García, M.L.; Calpena, A.C. Ex vivo permeation of carprofen from nanoparticles: A comprehensive study through human, porcine and bovine skin as anti-inflammatory agent. *Int. J. Pharm.* **2016**, *501*, 10–17. [CrossRef]
26. Zhou, C.; Sun, M.; Wang, D.; Yang, M.; Loh, J.L.C.; Xu, Y.; Zhang, R. In vitro antibacterial and anti-inflammatory properties of imidazolium poly (ionic liquids) microspheres loaded in GelMA-PEG hydrogels. *Gels* **2024**, *10*, 278. [CrossRef]
27. Thangavelu, M.; Kim, P.-Y.; Cho, H.; Song, J.-E.; Park, S.; Bucciarelli, A.; Khang, G. A gellan gum, polyethylene glycol, hydroxyapatite composite scaffold with the addition of ginseng-derived compound K with possible applications in bone regeneration. *Gels* **2024**, *10*, 257. [CrossRef]

28. Gupta, A.; Smith, B.; Kumar, C.; Lee, D.; Johnson, E. Evaluation of PEGylation as a cryoprotectant in PLGA nanoparticles. *J. Control. Release* **2012**, *158*, 123–130. [CrossRef]
29. Kim, S.; Park, J.; Lee, Y.; Choi, H.; Cho, S. Enhancement of Liposomal Stability by PEGylation: Effects on Cryoprotection and Redispersibility. *Int. J. Pharm.* **2014**, *475*, 120–127. [CrossRef]
30. Chirikhina, E.; Chirikhin, A.; Xiao, P.; Dewsbury-Ennis, S.; Bianconi, F. In vivo assessment of water content, trans-epidermal water loss and thickness in human facial skin. *Appl. Sci.* **2020**, *10*, 6139. [CrossRef]
31. Lin, C.-C. Recent advances in crosslinking chemistry of biomimetic poly (ethylene glycol) hydrogels. *RSC Adv.* **2015**, *5*, 39844–39853. [CrossRef] [PubMed]
32. Wang, X.; Xu, H. Incorporation of DMSO and dextran-40 into a gelatin/alginate hydrogel for controlled assembled cell cryopreservation. *Cryobiology* **2010**, *61*, 345–351. [CrossRef] [PubMed]
33. Zhang, C.; Zhou, Y.; Zhang, L.; Wu, L.; Chen, Y.; Xie, D.; Chen, W. Hydrogel cryopreservation system: An effective method for cell storage. *Int. J. Mol. Sci.* **2018**, *19*, 3330. [CrossRef] [PubMed]
34. Beirampour, N.; Bustos-Salgado, P.; Garrós, N.; Mohammadi-Meyabadi, R.; Domènech, Ò.; Suñer-Carbó, J.; Rodríguez-Lagunas, M.J.; Kapravelou, G.; Montes, M.J.; Calpena, A.; et al. Formulation of Polymeric Nanoparticles Loading Baricitinib as a Topical Approach in Ocular Application. *Pharmaceutics* **2024**, *16*, 1092. [CrossRef]
35. Tanino, T.; Funakami, Y.; Nagai, N.; Kato, Y. Cyclosporin A-sensitive cytotoxicity of flurbiprofen non-stereoselectively mediated by cytochrome P450 metabolism in three-dimensional cultured rat hepatocytes. *J. Pharm. Pharmacol.* **2015**, *67*, 1406–1415. [CrossRef]
36. Wang, L.; Bao, S.H.; Pan, P.P.; Xia, M.M.; Chen, M.C.; Liang, B.Q.; Dai, D.P.; Cai, J.P.; Hu, G.X. Effect of CYP2C9 genetic polymorphism on the metabolism of flurbiprofen in vitro. *Drug Dev. Ind. Pharm.* **2015**, *41*, 1363–1367. [CrossRef]
37. Berenguer, D.; Alcover, M.M.; Sessa, M.; Halbaut, L.; Guillén, C.; Boix-Montañés, A.; Fisa, R.; Calpena-Campmany, A.C.; Riera, C.; Sosa, L. Topical Amphotericin B Semisolid Dosage Form for Cutaneous Leishmaniasis: Physicochemical Characterization, Ex Vivo Skin Permeation and Biological Activity. *Pharmaceutics* **2020**, *12*, 149. [CrossRef]
38. Rauscher, M.; Rauscher, A.; Hu, L.Y.; Schlitt, H.J.; Krauß, S.; Illg, C.; Reis Wolfertstetter, P.; Hofmann, A.; Knorr, C.; Denzinger, M. Influence of Accumulation of Humidity under Wound Dressings and Effects on Transepidermal Water Loss (TEWL) and Skin Hydration. *Appl. Sci.* **2024**, *14*, 7739. [CrossRef]
39. Liu, Y.; Chen, J.Y.; Shang, H.T.; Liu, C.E.; Wang, Y.; Niu, R.; Wei, H. Light microscopic, electron microscopic, and immunohistochemical comparison of Bama minipig (Sus scrofa domestica) and human skin. *Comp. Med.* **2010**, *60*, 142–148.
40. Schiller, L.R.; Emmett, M.; Santa Ana, C.A.; Fordtran, J.S. Osmotic effects of polyethylene glycol. *Gastroenterology* **1988**, *94*, 933–941. [CrossRef]
41. Avram, Ș.; Bora, L.; Vlaia, L.L.; Muț, A.M.; Olteanu, G.-E.; Olariu, I.; Magyari-Pavel, I.Z.; Minda, D.; Diaconeasa, Z.; Sfirloaga, P.; et al. Cutaneous Polymeric-Micelles-Based Hydrogel Containing *Origanum vulgare* L. Essential Oil: In Vitro Release and Permeation, Angiogenesis, and Safety Profile In Ovo. *Pharmaceuticals* **2023**, *16*, 940. [CrossRef] [PubMed]
42. Sutherland, M.A.; Davis, B.L.; Brooks, T.A.; Coetzee, J.F. The Physiological and Behavioral Response of Pigs Castrated with and without Anesthesia or Analgesia. *J. Anim. Sci.* **2012**, *90*, 2211–2221. [CrossRef] [PubMed]
43. Xie, Z.A.; Zhang, K.L.; Han, F.; Tang, M.Y.; Chen, J.W.; Liu, G.P. The Incidence of Vasovagal Reactions during Earlobe Piercing. *Front. Med. (Lausanne)* **2023**, *10*, 1103071. [CrossRef] [PubMed]
44. Angsusing, J.; Samee, W.; Tadtong, S.; Mangmool, S.; Okonogi, S.; Toolmal, N.; Chittasupho, C. Development, Optimization, and Stability Study of a Yataprasen Film-Forming Spray for Musculoskeletal Pain Management. *Gels* **2025**, *11*, 64. [CrossRef]
45. Fessi, H.; Puisieux, F.; Devissaguet, J.P.; Ammoury, N.; Benita, S. Nanocapsule formation by interfacial polymer deposition following solvent displacement. *Int. J. Pharm.* **1989**, *55*, R1–R4. [CrossRef]
46. Bouwstra, J.A.; Honeywell-Nguye, P.L. Skin structure and mode of action of vesicles. *Adv. Drug Deliv. Rev.* **2002**, *54*, 141–155. [CrossRef]
47. Ferrero, C.; Massuelle, D.; Doelker, E. Towards elucidation of the drug release mechanism from compressed hydrophilic matrices made of cellulose ethers. II. Evaluation of a possible swelling-controlled drug release mechanism using dimensionless analysis. *J. Control Release.* **2010**, *141*, 223–233. [CrossRef]
48. Lorenzo-Veiga, B.; Diaz-Rodriguez, P.; Alvarez-Lorenzo, C.; Loftsson, T.; Sigurdsson, H.H. In Vitro and Ex Vivo Evaluation of Nepafenac-Based Cyclodextrin Microparticles for Treatment of Eye Inflammation. *Nanomaterials* **2020**, *10*, 709. [CrossRef]
49. Interagency Coordinating Committee on the Validation of Alternative Methods (ICCVAM). *Test Method Protocol: Hen's Egg Test-Chorioallantoic Membrane (HET-CAM) Test Method*; Interagency Coordinating Committee on the Validation of Alternative Methods (ICCVAM): Research Triangle Park, NC, USA, 2010.
50. Spielmann, H. HET-CAM Test. *Methods Mol. Biol.* **1995**, *43*, 199–204.

MDPI AG
Grosspeteranlage 5
4052 Basel
Switzerland
Tel.: +41 61 683 77 34

Gels Editorial Office
E-mail: gels@mdpi.com
www.mdpi.com/journal/gels

www.ingramcontent.com/pod-product-compliance
Lightning Source LLC
Chambersburg PA
CBHW061459210926
51850CB00008B/1766